MINING EQUIPMENT AND SYSTEMS

Mining Equipment and Systems

Theory and Practice of Exploitation and Reliability

Jacek M. Czaplicki
Mining Mechanization Institute, Silesian University of Technology, Gliwice, Poland

CRC Press
Taylor & Francis Group
Boca Raton London New York Leiden

CRC Press is an imprint of the
Taylor & Francis Group, an **informa** business

A BALKEMA BOOK

First issued in paperback 2017

CRC Press/Balkema is an imprint of the Taylor & Francis Group, an informa business

© 2010 Taylor & Francis Group, London, UK

Typeset by Vikatan Publishing Solutions (P) Ltd., Chennai, India

Published by: CRC Press/Balkema
    P.O. Box 447, 2300 AK Leiden, The Netherlands
    e-mail: Pub.NL@taylorandfrancis.com
    www.crcpress.com – www.taylorandfrancis.co.uk – www.balkema.nl

Library of Congress Cataloging-in-Publication Data

Czaplicki, Jacek M.
 Mining equipment and systems : theory and practice of exploitation and reliability / Jacek M. Czaplicki.
   p. cm.
 Includes bibliographical references and index.
 ISBN 978-0-415-87731-2 (hardcover : alk. paper) -- ISBN 978-0-203-85280-4 (e-book) 1. Mining engineering. 2. Mining machinery. I. Title.

 TN146.C93 2010
 622.028'4--dc22

                  2009044089

ISBN 13: 978-1-138-11403-6 (pbk)
ISBN 13: 978-0-415-87731-2 (hbk)

Contents

Preface and acknowledgements

In Poland the mining industry is several centuries old. It boomed in the nineteenth and twentieth centuries and still has good prospects for the future. As a result, mining engineers in this country have undertaken a lot of good research. Unfortunately this work is sometimes unknown in the world literature. For this reason I feel that it is useful to publish the results of some of this research in English.

It is almost fifty years since the first papers on the application of reliability theory to mining problems were published in the United States. Developing rapidly in the late 1950s and 1960s, reliability theory quickly found a wide application in mining engineering. However, in a relatively short time, Central European researchers engaged in reliability investigations found that very often these examinations should encompass a wider scope of considerations, wider than the area connected strictly with reliability problems. In such a way the theory of exploitation came into being.

My experience in this area began in 1973 when I worked on my M.Sc. thesis on the reliability of the head ropes of mine hoists. At the beginning of my scientific career I was engaged exclusively with reliability problems. Time passed and my interests enlarged beyond the analysis of the reliability of mine machines and their systems and found myself in the area of the theory of exploitation. After several years of experience in this field I feel that it is time to share the knowledge gained in a wider forum.

This book presents my view on reliability and exploitation problems in mining—both practical and theoretical. The composition of this book is somewhat different from others in this regard. Most texts are usually ordered according to the theoretical scheme exclusively. In this book theoretical problems are formulated in connection with practical problems and appropriate literature is cited so that readers may increase their knowledge about a given aspect. Furthermore, the data presented here is gathered from several mines located in different countries. Some readers may treat this book as a presented set of mathematical tools useful in the analysis of problems of the exploitation and reliability of mining equipment and its systems.

Incidentally, where literature from the field of reliability in mining is concerned, it appears that there is a lack of an internationally recognised textbook in this field. I hope and is intended to help to fill this gap.

In my two books written in English, I presented an outline of the theory of modelling, analysis and calculation of mine equipment systems apart from methods for the calculation of rail transport systems or mine readiness systems.

This book is dedicated—first of all—to students of mining faculties and schools of mining with a specialisation in mine mechanisation all over the world. They have always been close to my heart, no matter in which part of the world I have lectured. Some chapters of the book may be interesting for those who specialise in earthmoving enterprises. Students of mathematics searching for practical applications of models of mathematical statistics, reliability theory, renewal theory and queue theory might find some chapters interesting as well. I presume that mining engineers will also find some parts helpful. Additionally, I hope that my dear academic colleagues working at mining universities will also find this book interesting, and useful in their educational work.

I would like to express my very warm thanks to Ms Michele Simmons for language editing and Janjaap Blom, Senior Publisher CRC Press/Balkema - Taylor and Francis Group and his open and extremely professional team for their reliable and efficient cooperation.

Jacek M. Czaplicki
Mining Mechanization Institute
Silesian University of Technology, Gliwice, Poland

About the author

Jacek M. Czaplicki has been an academic lecturer for more than thirty years and is continuously associated with his home University. He did however leave his school for a couple of years' lecturing in African universities.

He worked for three years at the Kwara State College of Technology, Ilorin, Nigeria on a UNESCO project. A few years later he was appointed to Zambia Consolidated Copper Mines Ltd and worked as a lecturer at the School of Mines of the University of Zambia as part of a World Bank project.

Jacek Czaplicki received a Master of Science in Mine Mechanization from the Silesian University of Technology, Gliwice, Poland. He also obtained a Doctorate degree in Technical Sciences. Later he submitted a thesis and passed all requirements, obtaining a D.Sc. degree in Mining and Geological Engineering with a specialization in Mine Machinery at the same home University. Actually, he is a Professor of Mining Engineering.

He has published more than a hundred and thirty papers and about ten books in Poland and abroad. His specialization comprises mine transport, reliability and computation of mine machinery and their systems and reliability of hoist head ropes. He is an internationally recognized specialist in mine mechanization.

List of major notations

A – steady-state availability

B – accessibility coefficient

E – expected value, mathematical hope

G – general case, general distribution

H_0, H_1 – statistical hypotheses

$H(t)$ – renewal function

k – number of repair stands

L – rope length, the expected daily mass of rock hoisted out from the production level

L_t – number of theoretically possible states

m – number of trucks directed to accomplish transportation task

M – (Markov), exponential distribution

n – number of shovels in system

N – sample size

P, p – probability

$P_{kd}^{(p)}$ – probability that d power shovels are in work state

$P_{wb}^{(p)}$ – probability that b trucks are in work state

q – number of winds executed by hoist/rope

r – reserve size, number of spare trucks

$r^{(a)}$ – autocorrelation coefficient

$r^{(S)}$ – Spearman rank correlation coefficient

$r_\alpha(n)$ – critical value of Spearman rank correlation coefficient

R_{XY} – empirical Pearson correlation coefficient, estimator

$R_0(\varphi)$ – reliability (survival) function of rope

S – empirical standard deviation

S_n – area of rope cross-section, metallic area of rope

t – time

$t_\alpha(N)$ – quintile of order α of Student's t-distribution with N degrees of freedom

T_c – mean time of truck work cycle

$T_c^{(h)}$ – mean time of hoist work cycle

T_d – hoist disposal time per day

T_t – mean time of truck travel (haulage + dump + return)

T_n, T_p – mean time: repair, work

xi

u – residual

u_α – quintile of order α of the standardised normal distribution

v, V – rank

y – shaft bin volume

Z – random variable, total daily mass delivered to shaft in given underground level

Z' – mean adjusted loading time by shovel

α – level of significance

Δ – time loss function for truck

θ – empirical total number of cracks of wires of hoist head rope

Θ – theoretical total number of cracks of wires of hoist head rope

κ – repair rate (fault coefficient)

λ – intensity of object failures, parameter of exponential density function

$\lambda(t)$ – hazard function

μ – intensity of object repair

ρ – correlation coefficient in population, flow intensity rate in service system

$\rho^{(P)}_{X,Y}$ – the Pearson correlation coefficient

σ – standard deviation of random variable

τ – proportional coefficient indicating how many times is longer the mean time of truck loading by front-end loader compared to the adjusted mean loading time by shovel

ϑ – random variable, total number of cracks of wires in a moment of rope withdrawal

$\varphi_N(X)$ – the standardized density function of random variable $N(0, 1)$

$\Phi_N(X)$ – the standardized distribution function of random variable $N(0, 1)$

$B(a, b)$ – beta distribution with parameters a, b

$F(2N, 2N)$ – F-Snedecor probability distribution with $(2N, 2N)$ degrees of freedom

$Ga(k, \varsigma)$ – gamma distribution with parameters k, ς

$N(m, \sigma)$ – normal distribution with parameters m, σ

$t(N)$ – Student's distribution with N degrees of freedom

$\chi^2(N)$ – chi-square distribution with N degrees of freedom

$\mathfrak{D}$ – decisions made by system controller (truck dispatcher)

$\mathfrak{E}$ – exploitation repertoire

$\mathfrak{H}$ – set of power shovels

$\mathfrak{M}$ – method of utilization

$\mathfrak{R}$ – set of natural numbers

$\mathfrak{P}(\mathfrak{S}, \mathfrak{D})$ – exploitation process of shovel-truck system

$\mathfrak{O}$ – maintenance system

$\mathfrak{O}_1$ – sufficient maintenance system

$\mathfrak{O}_2$ – insufficient maintenance system

$\mathfrak{R}$ – set of real numbers

$\mathfrak{r}$ – set of repair stands

$\mathfrak{s}$ – state of object

$\mathfrak{S}$ – system

$\mathfrak{W}$ – set of trucks

Random variables are usually marked in bold; this does not apply to gothic letters.

Each example considered commences with a sign ■ and finishes with a sign ◄

CHAPTER 1

Introduction: terotechnology and the theory of exploitation

Technology has given engineers a very wide range of technical objects and items with vast abilities to fulfil variety of requirements. Usually, these devices are set up in systems to accomplish set tasks, such as the extraction or transportation of mass material, dumping, mechanical, chemical or thermal processing, and so on. The equipment utilised in these systems usually has complicated construction and must be operated according to the manufacturer's recommendations and precise advice if an appropriate level of performance is to be achieved. Users of technical objects should pay attention to the realisation of object exploitation—understood here as the process of changes of the object's properties running in time. These changes are also interesting for designers and manufacturers of these objects. They are a point of interest for object users—after spending money to purchase technical equipment, users expect it to fulfil its duties and functions in the most convenient way.

In the above paragraph, the term *exploitation* is used. This term may be a bit strange for native English speakers. Therefore, it would be advantageous to commence a discussion about the subject of this book by giving some explanations of the mutual relationship between *terotechnology* and the *theory of exploitation*. This relationship is not as well known in the English-speaking part of engineering world as it is in Central Europe where the theory of exploitation was born.

The theory of the exploitation of technical objects and items is presently well developed in Central and Eastern Europe, and especially in Poland. The scope of the theory—the definitions of basic terms, principles, methods of investigation, rules and methods of description—is well defined, but communication with the English-speaking world in this area is difficult. Direct translation of the term *teoria eksploatacji* as 'exploitation theory' almost always fails to work in practice. And, conversely, the term 'terotechnology' is not very popular in the engineering world in this part of Europe. A similar situation arises with the term *proces eksploatacji*, which is directly translated as 'exploitation process'. There are also some problems with the English term 'operation'. This is a peculiar situation that still creates problems. Some explanations appear to be inevitable.

Let us start with a few notes on the history of the development of the theory of exploitation in Poland.

The first papers that examined the exploitation/operation problems of machines were published in the 1960s (cf. Ziemba 1973). The term 'exploitation theory' was initially associated with machinery, and later expanded to cover a more general application—the exploitation of technical objects. At this time rapid progress was observed in the field of tribology, and the theory of reliability was coming into being. At the end of the 1960s the Mechanical Engineering Committee of the Polish Academy of Sciences created a Section of Exploitation of Machines with three topic groups: tribology, reliability and exploitation. A fourth topic group, technical diagnostics, came later (1980). A relatively new arrival is 'safety', and this topic group—together with the other four—has existed for almost ten years. At the end of the 1960s and beginning of the 1970s, the quarterly *Exploitation Problems of Machines* was established by the Polish Academy of Sciences. Subsequently, some further technical periodicals that focus on exploitation have been founded.

Generally, the rapid growth of the theory of reliability noted in the 1960s and 1970s accelerated the development of the theory of exploitation. Researchers involved in penetrative reliability considerations quickly came to the conclusion that many problems associated with the usage of an object/item (element or system according to the theory of reliability) in real conditions needed to be considered within a wider scope. It was necessary to take into account some properties of the object not connected with reliability in a direct way, but rather connected with the process of the object's application in particular operating conditions. Researchers were interested in what happened with many object properties during the utilisation of the object and during its maintenance. A large part of reliability theory, therefore, belongs within the scope of the theory of exploitation; however, the theory of exploitation has a much wider scope than that of reliability theory.

Pure theoretical investigations connected with the theory of object utilisation and maintenance (that is the exploitation of an object) were carried out, resulting in some significant—at that time—books, such as Koźniewska 1965, Bojarski 1967, Konieczny et al. 1969, Konieczny 1971, and Olearczuk 1972. In the second half of the 1960s, the first National Symposia on Exploitation of Technical Objects were held. In 1973 the study of exploitation for a doctoral degree was offered (Konieczny 1973) for the first time. The theory of exploitation proved to be very useful in practice. Evidence for this statement can be found by considering the problems of the operation of mine machinery systems for instance (see, for example, Sajkiewicz 1974, 1975, Czaplicki and Lutyński 1982, 1987).

The term exploitation comes from the French word *exploitation*, which means, first of all, *usage*. In connection with engineering problems, **exploitation** is based on the statement that it is *usage of something in a rational way*.

In the 1970s much was published on exploitation in Central and Eastern Europe, and after several years of intensive development the theory was well established. Researchers involved in these kinds of investigations came to the conclusion that *the exploitation of a technical object is a set of intentional actions running in time of a technical, economic and organisational nature directed at this object, as well as mutual relationships existing between them (people cooperating and the object) from the moment of the object's first usage until its withdrawal and disposal* (Polish Standard PN-82/N-04001). Similar definitions can be found in the standards of neighbouring countries.

For more than twenty years, the theory of exploitation has been offered as a separate lecture subject in connection with the utilisation of various classes of technical objects at several technical universities, such as Warsaw University of Technology and Wrocław University of Technology. Many books and textbooks have been published (for example, Adamkiewicz 1982, Downarowicz 1997, Kaźmierczak 2000, Będkowski and Dąbrowski 2006). The theory of exploitation has proved both essential and useful.

In the English-speaking world the term 'exploitation' in this sense is non-existent. The word in the English language has many negative meanings. According to thesaurus.com synonyms include corruption, wrong, dishonesty, crime, barbarism, misuse, cheat and even 'unwanted sexual advance', and Wikipedia gives an association with Marxist theory. Generally, the term has many unpleasant associations in the English language.[1] For this reason, it is extremely difficult to get acceptance of the term *exploitation*—in the precise engineering sense—in English-speaking parts of the world.

A generally accepted interchangeable term for exploitation theory is terotechnology. Both terms mean roughly the same thing, but they are not identical in their meaning. There are differences in how a particular problem should be addressed by exploitation theory and terotechnology, as well as in the scope of consideration of both theories and their main points of interest.

The history of terotechnology's early development was different in the United Kingdom, where the discipline was born, and in Central Europe. However, the period of this early development is roughly the same—the end of the 1960s. In the United Kingdom the problem of the maintenance of objects has its own history, and it partly separated from reliability when national conferences were held on this topic in the 1960s. In 1967 the British Council of Maintenance Associations was founded. After several years, in which vast numbers of works of both an empirical and a theoretical nature were published, the Committee for Terotechnology was formed. It took two years to agree on a definition of *terotechnology*.

---

[1] If a given word or phrase has a good connotation, English-speakers have no objections to using it. In mining science we have a type of mine called a 'glory hole'.

This is a combination of management, financial, engineering and other practices applied to physical assets in pursuit of economic life cycle costs. The practice of terotechnology is concerned with the specification and design for reliability and maintainability of plant, machinery, equipment, building and structure, with their installation, commissioning, maintenance, modification and replacement, and with feedback of information, performance and cost (Hewgill and Parkes 1979).

In the future terotechnology will be an essential element of good husbandry, of quality and of the ability to understand that an artefact commits resources both in its making and in its subsequent use ... the outcome of such an approach may result in a product which has a high initial cost and long reliable life, or which is cheap with a short life and anticipated replacement or breakdown ... Terotechnology has a simple objective—that of minimising the whole life cost of ownership—but its practice can be complex, involving interdependencies and relationships of a diversity of resources—people, money, material, ideas and techniques (Darnell 1979).

It was presumed that the papers by Finniston (1973), Rappini (1973) and Wiegel (1974) could be taken as reports of pioneering work in the area of terotechnology. All the authors were associated with integrated iron and steelworks. Therefore, terotechnology is a recent practice, with its formal definition developed in 1972, and its beginnings were as a practical method for efficient and timely maintenance in an integrated iron and steelworks. One of the pioneers was Evans, whose paper was presented at a conference in Durham, UK in 1974. The first International Congress on Terotechnology was held in London in May 1979.

In 1981 Tamaki formulated the term total productive maintenance (TPM). It has been stated that this is the application of terotechnology by the Japanese in their own way. In its pursuit of economy in life cycle costs, TPM has an identical purpose to terotechnology, but it is different in its approach. TPM is based on the assumption that causes of equipment failure and poor quality are interdisciplinary and that it is necessary to have a plant-oriented management organisation, and it stresses the need for the total participation of the workforce. Following this line of reasoning, Takahashi (1981) identified specific motives for advocating the subsequent adoption of TPM in Japan. The concept of TPM looks vital and it seems as though it has a future (Smith and Mobley 2007).

To get some idea about terotechnology today some quotations should be taken into consideration. In 1993 BS3811, *Glossary of terms used in Terotechnology*, was published. This British standard gives the following definition:

a combination of management, financial, engineering, building and other practices applied to physical assets in pursuit of economic life cycle cost.

Bhandury and Basu's book 2003 on terotechnology—probably the first example of such a comprehensive elaboration in English—stated succinctly:

Terotechnology—a concept, nay, a philosophy.

According to the online MSN dictionary (2007) terotechnology is:

a branch of technology that uses managerial and financial expertise as well as engineering skills when installing and running machinery.

Investopedia (2007) states:

A word derived from the Greek root word 'tero', or 'I care', that is now used with the term 'technology' to refer to the study of the costs associated with an asset throughout its life cycle—from acquisition to disposal. The goals of this approach are to reduce the different costs incurred at the various stages of the asset's life and to derive methods that will help extend the asset's life span. Terotechnology uses tools such as net present value, internal rate of return and discounted cash flow in an attempt to minimise the costs associated with the asset in the future. These costs can include engineering, maintenance, and wages payable to operate the equipment, operating costs and even disposal costs. Also known as 'life-cycle costing'.

For example, let's say an oil company is attempting to map out the costs of an offshore oil platform. It would use terotechnology to map out the exact costs associated with assembly, transportation, maintenance and dismantling of the platform, and finally a calculation of salvage value. This study is not an exact science: there are many different variables that need to be estimated and approximated. However, a company that does not use this kind of study may be worse off than one that approaches an asset's life cycle in a more ad hoc manner.

At en.wikipedia.org (2008) we can find the following statements:

Terotechnology is the economic management of assets. It is a combination of management, financial, engineering and other practices applied to physical assets such as plant, machinery, equipment, buildings and structures in pursuit of economic life cycle costs. It is concerned with the reliability and maintainability of physical assets and also takes into account the processes of installation, commissioning, operation, maintenance, modification and replacement. Decisions are influenced by feedback on design, performance and costs information throughout the life cycle of a project. It can be applied equally to products, as the product of one organisation often becomes the asset of another.

The term terotechnology can also be found in an online business dictionary (www.businessdictionary.com, accessed 2008), where the following statement is given:

Multidisciplinary approach to obtaining maximum economic benefit from physical assets. Developed in the UK in the early 1970s, it involves systematic application of engineering, financial, and management expertise in the assessment of the lifecycle impact of an acquisition (buildings, equipment, machines, plants, structures) on the revenues and expenses of the acquiring organisation. Practice of terotechnology is a continuous cycle that begins with the design and selection of the required item, follows through with its installation, commissioning, operation, and maintenance—until the item's removal and disposal—and then restarts with its replacement. From the Greek word 'terein' to guard or to care for. Terotechnology is in the accounting & auditing industries, manufacturing & technology, and [in] purchasing & procurement.

Terotechnology has become a separate academic subject. The School of Oil and Gas Engineering at the University of Western Australia, for example, offers a terotechnology course (course code ASST8577), though the recommended course reading comprises books on maintenance, replacement and reliability exclusively.

Monash University, Australia has a course on terotechnology and life-cycle costs (code GEG7014) in the Postgraduate Faculty of Engineering (course duration: 150 hours). The first sentence in the course prospectus reads 'Introduction to asset management and terotechnology', suggesting that these two matters are different.

Växjö University, Sweden offers a Masters programme in terotechnology in which students should have a profile in production systems, that is be production managers, creative quality developers or maintenance designers. In the course leaflet one can find the statement: 'Terotechnology is about how to optimise the manufacturing part of an organisation.'

To conclude this discussion on the definition of terotechnology two remarks should be mentioned.

In 2004 and 2005 Belak from Croatia published two papers on the problem of terotechnology. In his 2004 work he analysed the position of terotechnology in the overall approach to the planning, design, manufacturing, installation, service, maintenance, decommissioning and recycling of a technical system. He indicated that terotechnology is also related to economics, life-cycle cost and maintenance. According to this author terotechnology should be treated as an *optimisation process*, one optimising the ratio between the total effective or potential production of the system and the cumulative cost in its lifetime. As a static phenomenon the ratio can be optimised without regard to the market influence and business strategy and policy. The *optimisation* process must include the reliability, adaptability and availability of the system.

In the 2005 paper Belak—together with his co-author Čičin-Šain—reviewed the definitions of terotechnology. They indicated differences in the definitions that are the source of misunderstandings. According to the authors, all definitions have the aspect of costs examined throughout the life cycle of business operation systems in common; however, the concept of terotechnology also includes the earnings produced by the activity of the business operation system, a fact that most terotechnology definitions disregard.

A second remark should be made in association with the paper written by Ibrahim and Brack 2004. They presented a new concept and implementation of inter-continental flexible training of terotechnology and life-cycle costs (suggesting that these two things are different), based on the experience obtained in Australia and the USA.

In connection with the relationship between terotechnology and the theory of exploitation one more disagreement is often noticed. In Central and Eastern Europe one hears the statement 'during the exploitation'. English-speaking researchers tend to say 'during the operation', indicating that this is a more appropriate statement. Is this so? There is no doubt that in common English such a statement is easily understood and communicative. However, when living among English-speakers it is easy to perceive that the word 'operation' is used frequently and has a great number of synonyms. One impression arises repeatedly—this word is *over-used*. In common usage, everyday language, there is no objection. Nevertheless, if strict, precise scientific language is to be employed, this word looks like a *picklock*. Thesaurus.com (2008) gives 48 meanings of the word 'operation', from service, progression, force, happening, movement, affair, deal, enterprise, achievement, deed, stroke to space, size and surgery. This is an extremely wide range.

BS3811 gives the following definition of the term *operation: the combination of all technical and administrative actions intended to enable an item to perform a required function, recognising necessary adaptation to changes in external condition*. Looking at both definitions for 'operation' and 'exploitation' a clear impression looks inevitable—both terms mean approximately the same thing. Thus, neglecting the fact that this word has perhaps too many denotations and that in science precise terms are needed, in this book the terms will be used interchangeably. For native English-speakers it will be less troublesome.

Let us briefly summarise.

Terotechnology is:

a concept, a combination of practices, a kind of technology, the economic management of assets, a multidisciplinary approach, a philosophy, an optimisation process.

Terotechnology has these objectives:

to minimise the whole cost of ownership over the object's life,
to obtain maximum economic benefit from physical assets,
to reduce the different costs incurred at the various stages of the asset's life and to derive methods that will help extend the asset's life span,
to optimise the ratio between the total effective or potential production of the system and the cumulative cost in the lifetime of the system.

The theory of exploitation is the strictly defined *field of science* determining the fundamentals of the exploitation/operation of objects in a rational way

The theory of exploitation has two objectives: theoretical and practical. The theoretical goal comprehends the mechanisms governing the course of the changes of an object's properties during its utilisation and maintenance. The practical aim is to maximise the economic benefit from object exploitation during the object's life, provided that the safety requirements are fulfilled.

Terotechnology has existed for almost forty years now, and yet there is no statement that it is a field of science! To date it has been the focus of many different definitions and different approaches. The main points of interest are located in slightly different areas. Some statements do not agree. The only precise statement that can be formulated is that 'we roughly know what it is all about'. And that is almost all. There have been no spectacular theoretical achievements. Many quite good elaborations have a partly scientific, partly technical character, but they do not solve problems in the way they should according to the high requirements formulated by terotechnology *sensu stricto*. What causes such a situation?

Analysing the various definitions we find that the main point of interest is connected with the whole life-cycle cost of the object. And this 'object' in terotechnology is a system—a complicated structure. For a person who is familiar with the theory of exploitation, it is obvious that statements such as 'achievement of maximum profit from an object's exploitation during the object's life cycle', 'achievement of minimum costs spent on the object during its whole life', or 'to find the optimum ... during the object's life cycle' are too ambitious! From a theoretical point of view we can construct many mathematical patterns describing relationships between different variables that have an influence on the 'object costs'. But is it necessary to take into consideration a considerable number of variables, many of them random, some of them with unidentified distributions, and some of them with a great dispersion. Mutual stochastic relationships between the variables are usually unknown. We frequently have functions of many of these variables. If we add that a great many of these considerations should be located in the theory of prediction (as a rule, terotechnology considerations are made when the object still exists), and that many components of the relevant patterns are estimated with varying degrees of accuracy and that many elements of cost vary over time, it will be obvious that from a practical point of view the usefulness of such divagations is doubtful, especially if we have the entire object life in mind.

The theory of exploitation has also existed for almost forty years. It is a field (discipline) of science with its own well-defined nomenclature, a determined field of interest, verified mathematical methods, and laws and principles. In many areas the theory of exploitation has proven to be very useful both from theoretical and practical points of view. It is worth noting that in exploitation theory we are not interested in the maximum of a certain function describing a certain object property during the entire life of an object; we are interested in 'maximising', that is 'going towards the maximum'. This is more practical and easier to attain.[2] The theory of exploitation has positive effects in this way.

Before we commence our considerations, one additional item needs an explanation. This is the relation between the terms 'a piece of equipment' and a 'machine'.

This book deals with the properties of technical objects applied in mining and the mathematical models that describe their operation. All these objects are pieces of equipment or—using a different term—devices. An important category of devices are machines. Therefore all machines are pieces of equipment, but there are some pieces of equipment that are not machines. A machine usually consists of a certain number of mechanisms located in a common body and their task is the transmission of some motion and forces to perform some type of work. For this reason draglines, winders, wheel loaders and shearers are machines, but powered supports, belt conveyors and hoisting installations are types of devices but not machines.

---

[2] A note on a problem with the mutual relationship between terotechnology and the theory of exploitation is indicated in Czaplicki's paper 2008c.

CHAPTER 2

Exploitation theory

2.1 FUNDAMENTALS

A general goal of human activity is to make changes in the surrounding reality in such a way that it will be more convenient for human beings. This change should be done optimally—if possible—and one of its purposes is to satisfy human needs. In order to attain this goal, humans develop processes by making use of their knowledge and using appropriately selected and constructed technical objects. A general requirement in designing these objects is to have some knowledge of the target process—the process that will be generated using the constructed object. But this is not enough. For a good and proper design of the object (and, further, in order to attain the determined purpose), it is necessary to have some knowledge of the changes in the properties of the object that will be observed during the realisation of this target process. This knowledge is necessary to achieve the goal. If such a process is not controlled, it will lead to the object's disorganisation according to the second law of thermodynamics. In effect, it is very likely that the goal will not be attained or will not be completely achieved. Thus, getting to know the regularities that govern the course of a process is a vital problem from the point of view of human activity.

The field of science that examines the process of changes in an object's properties during its life is termed the theory of exploitation in many countries (see, for example, Piasecki 1973, Kaźmierczak 2000).

The main goal of this theory is the determination of such activities that allow the object to be kept in a state that assures the most convenient course of the target process. Therefore, a main point of interest in the theory is the rational arrangement of the system of the object's operation and optimal control of the process of exploitation.

The theory of exploitation deals with, or tries to find answers to, the following questions:

– What does the exploitation process consist of? How is it possible to identify its components?
– What kind of influence does the object's environment have on the course of its exploitation process?
– What are the technical and economic possibilities for changing the course of the process towards one that is more convenient for its realisation?
– What type of changes will be the most profitable?
– What kind of modifications in the construction of the object or system structure should be carried out to improve its performance?
– How can the production of the object be improved based on the information about the object's operation?
– How should the course of the process be arranged to ensure an appropriate level of operational safety?

Changes of properties in an object depend on:

- the features of the object obtained during its design and production
- operational conditions—the environment of the object during its operation
- methods of the object's utilisation and the maintenance applied.

Generally, each technical object has three phases in its life.[1] These are the:

- design stage
- production stage
- exploitation/operation stage.

The starting point for *generating* an object is a *need,* a *necessity.*[2] Somewhere in the sphere of human activity a call for a particular item occurs—a call for a new article that will be able to fulfil certain requirements. In other words, there is *a requirement* to create an object that accomplishes these needs. Frequently, the answer to this call is a decision to make an object. At this moment, the first phase of the object's existence begins.

During this first stage the object does not exist physically. It comes into being through a series of concepts, ideas, drawings, drafts and calculations. At the end of this period complete technical documentation is ready, usually in digital form. The important point is that the designers have given the object its virtual properties.

When this documentation reaches manufacturers' hands and production starts, the second phase of the object's life commences. The object comes into being physically. During this stage the object obtains real physical properties—its characteristic features—which are usually a bit different from those described by the designers. At the end of production the object exists physically. It can go on sale. Machines, such as trucks, wheel loaders, bulldozers and scrapers, are examples of these objects. In mining, some objects are so huge that they are not mass produced. A client makes an order and the desired machine parts are manufactured and delivered to a location where the whole unit is erected. Sometimes this process is completed in weeks; sometimes it takes a few years. Enormous bucket wheel excavators, giant bridge conveyors and bucket chain excavators are examples of objects constructed in this way.

When a purchaser 'takes' possession of the object, it is delivered to the place where the object will begin to execute its duties and accomplish the formulated task. When the purchaser initiates use, the third stage of the object's life begins—exploitation of the object. During the utilisation of the object and its maintenance, its properties change. As the object executes its duties some changes in its properties can be positive (for example, when running some parts and/or assemblies cooperate better), but generally the changes are negative. The object begins to deteriorate, some parts become worn and malfunction, and failures occur. Maintenance is needed or, at least, some adjustments. Generally, the exploitation/operation process of an object is an interlacing process of utilisation and maintenance. It is worth noting that there are some mining objects that cannot be maintained or repaired—their properties cannot even be adjusted. Generally, the exploitation phase is finished when the physical or economic death of the object occurs. Fortunately, this *death* does not mean catastrophic failure, as a rule.

Today the majority of technical objects are well designed so that they fulfil the requirements of their owners. During both initial stages—design and production—there are many opportunities to improve an object's future achievements in order to make it more able to execute duties. In addition, the environment in which an object operates can have a strong influence on the course of the exploitation process. Generally, the environment of the object is everything that does not belong to the object but has a certain influence on the object. The object and its environment are in a mutual relationship. An armoured flight conveyor working at a coalface will have different values for its reliability parameters than those of the same conveyor operating as a braking element where the displacement of the mineral is done under a steep angle of inclination far from the working face. The mean durability of a head rope of a hoist operating in a dry shaft will very likely be longer than that of an identical rope with the same intensity of utilisation working in a wet upcast shaft.

---

[1] According to some researchers the life of an object has more phases. They would regard storage and the construction of the object from assemblies in the location where it will be utilised as separate phases. However, there is no need to introduce such phases into our examinations.

[2] A *need* is a primitive notion used, for example, in economics. One can find the definition of this term in psychology.

The selected method of object utilisation can also have a significant influence on the object's reliability and performance. Chapter 7.3 will focus on the methods of utilisation and maintenance of a pair of elements. The importance of selecting a proper method of service (repairs, preventive actions, etc.) in mining is easy to understand when one recalls that until the 1960s approximately 80% of machine failures resulted in production stoppages. After spectacular achievements in reliability theory and practice, by the 1980s about 80% of potential machine failures were cleared quickly without disturbing production—thereby creating real increases in mine production and drastically reducing the production time lost due to repairs.

An investigation of the properties of the object exploitation process sometimes concerns properties that can be identified immediately with regard to measure[3]—it is enough to observe the symptoms occurring during the process. However, there are many different properties that can be identified and to which measures can be attributed when handling appropriate data concerning the object and its process. Sometimes modelling is very useful here because working with a model allows interesting measures of the process to be constructed. Therefore, it is worth recalling the definition of an exploitation process formulated by Kaźmierczak (2000, p. 156) here: *the exploitation process of an object is everything that happens with the object from the moment of the end of its production till the moment of its final withdrawal from utilisation.*

There are two basic terms of exploitation theory associated with the term 'exploitation process'. These are the *state of the object* and the *exploitation event*.

During the object exploitation process, that is during the process of the object's utilisation and maintenance, the properties of the object change. For some features these changes will be continuous, sometimes slow, sometimes transitional, and sometimes drastic. Therefore, an object at a given moment in time is not identical to the object at a different moment in terms of its properties. In order to describe the process of these changes the term *state* is applied.

When defining a set of an object's essential properties $\mathfrak{C}$, $\mathfrak{C} = \{c_1, c_2, ..., c_m\}$, we can say that the state of the object at time t is determined by a certain function:

$$\mathfrak{G}(t) = f[\mathfrak{C}(t)] = f[c_1(t), c_2(t), ..., c_m(t)]$$

Kaźmierczak (2000, p. 119) gives a similar assessment of the term 'state': *under the term state of object we are going to understand here a 'photograph' of the values of object properties at a given moment in time.*

In practical applications this function is not considered to be a continuous one. Discretisation is regularly made, and the discrete states are named. These names are usually associated with the physical nature of the state, such as repair state, work state, stand-still state and so on. Notice that a simple conclusion can be made here: the exploitation process of an object can be understood as the sequence of the states of this object or—the usual formulation—as the process of changes of states.

As a result of this discretisation at each moment when a change of state occurs, an *exploitation event* has taken place. Sometimes these events are *visible* and to some extent *perceptible*, for example, a certain element of the object fails and the machine ceases operation. Sometimes events are defined by a convention or schedule—nothing physically happens apart from the fact that a certain object parameter exceeds its assumed limited value, such as a brake lining is worn excessively. At this moment it is assumed that the object is in a different state.

At the end of this short discussion concerning the fundamentals of the theory of exploitation, some brief general information should be given on the methods of investigating technical objects.

Scientific investigations can be divided into:

- theoretical examinations
- empirical examinations.

---

[3] 'Measure' in the sense of the theory of measure (see, for example, Fremlin 2000).

As a rule theoretical examinations are separated into:

- analytical study
- statistical experiments.

However, simulation almost always takes into account data gathered during field investigations.

Empirical examinations are frequently divided into two categories taking into account the way in which they are realised, namely:

- field investigation
- stand or lab investigations.

Statistical experiments will not be dealt with here for two main reasons. First, although simulation is effective to a certain degree for a given existing case, it does have some limitations. If some generalisation is made based on simulation outcomes concerning different cases, the conclusions formulated should be treated very carefully. Each statistical experiment is based on some short cuts and simplifications. Some stochastic phenomena are unknown and/or not taken into account. Each outcome is obtained with some accuracy and with some error that cannot be measured. All these factors mean that it is better to treat generalisations from statistical experiments as a hint not as a strong statement. Nevertheless, simulation is a unique method that works, and it allows for an analysis of systems that other methods fail to offer. It is quick and cheap.

Second, the library of simulation programs applied in mining is actually quite rich; these programs have their own literature that is also very rich. Therefore, it makes no sense to repeat it.

The models and descriptions of mining objects presented here are analytical in nature and are based on results of exploitation investigations (in the field), lab investigations and pure theoretical research based on information from mining practice. In addition, Chapter 3 concerns diagnostics; that part of diagnostics that is rarely covered in professional literature—statistical diagnostics. Because the scope of this book concerns the exploitation of technical objects, the diagnostics connected with this subject can be called exploitation diagnostics. This, in turn, can be divided into technical diagnostics and statistical diagnostics. Technical diagnostics, which have developed rapidly over the last thirty years, is well illustrated in many professional publications. Statistical diagnostics has remained unexplored and is rather unpopular.

2.2 EXPLOITATION THEORY IN MINING—PRELIMINARIES

Mining is a peculiar activity. There is no other engineering field that applies so many and such a variety of machines and other pieces of equipment. These machines are organised into systems; rarely does a machine work alone. This variety creates an immense challenge for mining engineers, who use this equipment in practice, and a great challenge to researchers attempting to describe the operation of systems analytically. This is the reason that the range of mathematical models employed is particularly rich (obviously depending on the case that is under investigation).

Models dealing with an object working until its first failure occurrence have found application, together with models examining objects that can be repaired. Due to the fact that the majority of machines operating in mines work in systems, the employment of models for systems of different natures have found wide application in the mining engineering world.

In the theory of mine machinery systems, four specific types are distinguished by their method of functioning (Czaplicki 2004b). These are:

- continuous systems (systems of continuous technological structures)[4]
- cyclic systems

---

[4]The description *systems of continuous technological structures* was probably formulated for the first time by Sajkiewicz at the beginning of 1970s in connection with an analysis of belt conveyor systems in surface mining.

- readiness systems
- mixed systems.

The theory and analytical description of continuous systems was developed mainly in the 1960s and 1970s and comprises both formal models and simulation models. These have been successfully verified by practice in mines, and one can say that the mining world is in possession of a wide range of confirmed mathematical tools to describe and analyse systems of this type. Further progress in this field can be observed in the development of new enhanced methods of simulation.

In the English mining engineering world, simulation programs are well developed and well known. Analytical models, in turn, are not so popular and for that reason their very convenient characteristic features are frequently overlooked. Therefore, these analytical models will be presented in this book.

The theory and analytical description of mine cyclic systems can be divided into four types:

- operation of hoisting installations
- operation of railway systems
- operation of loading machine systems—hauling units
- operation of systems involved in rock extraction by blasting.

These systems are modelled, analysed and calculated in different ways. Each needs completely different mathematical tools to describe its operation, however the general principles of reliability theory and exploitation theory are applied in all analyses. Two systems will be discussed in this book, a hoist system and a system for loading machines—haulers. The classic system—involving the shovel-truck—is a separate vast problem and the modelling, analysis and calculation of this system was presented in Czaplicki's monograph of 2009. The application of queue theory, which is applied in analysing some cyclic systems for the study of particular mining problems, is presented in Chapter 8.

The development of the theory and analytical description of readiness systems is strongly connected with military problems. The majority of military systems are simply readiness ones, that is systems with the main purpose of expectation and readiness for action. All rescue systems—fire brigades, police, medical emergency services, etc.—are examples of readiness systems. In mining, the rescue systems in place to release underground miners trapped by a roof collapse are extremely important. Generally, the mathematical models applied to these systems are well developed and have been confirmed in practice. The main point of consideration for these systems is the achievement of appropriate values for system parameters. This is a two-dimensional problem. Readiness systems are usually systems of people and the appropriate equipment. If the people are well trained in applying their tools in action, the system attains suitable values for its parameters. On the other hand, a problem can be posed in terms of the *correct* estimation of these parameters. These systems are peculiar ones and they will not be dealt with in this book.

In mining some systems have a mixed nature with, for example, one part of the system being continuous, another part cyclic. Three examples of this type of system will be taken into account:

- a shovel-truck system with in-pit crushing unit and conveyors
- a shovel-truck system with an inclined hoist
- a stream of rock extracted—shaft bin—hoist.

Discussions on these themes will begin with a presentation of some examples of formal descriptions of the exploitation processes for selected machines. This will allow readers to become familiar with the principles involved in the construction of models of the exploitation processes of machines. The first, and the most important, point is the purpose of such a construction, and the model should permit the identification of some regularities running during the object's exploitation and, further, the evaluation of the object's efficiency as a function of the object's parameters.

Two further fundamental terms of exploitation theory will be defined in regard to the first case study (see Chapter 4.1) concerning the exploitation investigation of an underground suspended locomotive used in coal mines in many countries today. These terms are *exploitation repertoire* and *exploitation graph*. It is better to define these expressions by giving practical examples because then their meaning is immediately well understood.

However, before discussing practical examples, one special topic is worth considering—statistical diagnostics. It is not very popular, especially where mining practice is concerned. However, it appears to be very useful and sometimes provides surprisingly vital information on the properties of the exploitation process under investigation. Moreover, importantly, statistical diagnostics allows much essential information to be obtained that should be at hand before the description and modelling of the exploitation process can be realised.[5]

[5]A recent publication by Corder and Foreman (2009) provides useful information on statistical analyses for mining engineers.

CHAPTER 3

Statistical diagnostics

For hundreds of years the term 'diagnostics' was associated exclusively with medicine as it is the field concerned with the methods of identifying sicknesses based on observed symptoms. This term is derived from Greek word *diagnosis* which means recognition, whereas *diagnosticós* represents the ability to recognise. In the second half of the twentieth century technical diagnostics began to be developed based on the same idea—to recognise the state of a technical object by means of objective methods and by means of tracing symptoms associated with the object's exploitation (Cholewa and Kościelny in Korbicz et al. 2004). More precisely, diagnostics evaluates the state of an object by investigating the properties of the operational process of the object together with processes associated with it and/or by studying the products of the object if any.

In regular technical diagnostics a primary problem is: *what kind of symptoms should be observed to obtain information about the state of the object?* This leads on to a second problem: *how can these symptoms be detected?* Usually, special diagnostic apparatus is needed, including highly sophisticated, precise procedures and well-trained personnel to apply diagnostic instruments and to read the recorded data. A similar situation is found in statistical diagnostics; the main difference is that the data gathered in statistical diagnostics are usually sequences of the times of the states of the object and, typically, no specific instrumentation is required. Diagnostic devices here are mathematical tools. A person doing these statistical investigations should possess the appropriate knowledge to conduct the research as well as to properly 'read' the outcomes of the statistical tools being applied. A point of consideration here is how to work with the data gathered during the investigations into the object's exploitation. The problems associated with the preparation, organisation and analysis of these investigations will not be considered here. These problems are described well in the reliability literature as well as in literature related to mathematical statistics. Keeping in mind that reliability literature is quite rich in this area (see, for example, Gnyedenko 1969, Melchers 1999, O'Connor 2005, Smith 2007), attention will be paid here to the first stage of research, immediately after the collection of data. This concerns the testing of the information gathered.

There is a lot of information in collected data that is not 'visible' at first glance. It is necessary to apply appropriate mathematical tools that allow many crucial questions to be answered and which indicate what should be taken into consideration. Sometimes these questions are related to what kind of decisions should be made in connection with further utilisation of the object, and sometimes they concern maintenance. These tools can be found in mathematical statistics.

Basically, much information can be obtained from the shape of the observed sequences of the times of the states. These sequences should be tested for:

a. randomness
b. possible outliers
c. stationarity of times of states in the sense of means
d. stationarity of times of states in the sense of variance
e. lack or existence of a cyclic component in the process
f. autocorrelation in the process
g. homogeneity of data
h. mutual independence of random variables
i. mutual dependence of random variables.

We will discuss these points in sequence and support our considerations with examples taken from mining practice. However, before we come to that, one term needs an explanation. This is *probability*. This expression will be used frequently throughout our considerations.

There is no doubt that probability is one of the cardinal terms in several theories, including probability theory, reliability theory, mathematical statistics and the theory of exploitation. (Other theories could be enumerated here but these are not the point of our interest.) If we examine the majority of books dealing with reliability, terotechnology, mathematical statistics or the theory of the exploitation of technical objects (see, for example, Gnyedenko et al. 1965, Ryabinin 1976, Migdalski 1982, Feller 1957, Melchers 1999, Smith 2007, O'Connor 2005, Bhandury and Basu 2003, Keller et al. 1987, Johnson 1984, Cramér 1999, Kaźmierczak 2000), there is no definition of probability. All these books present investigations that concern probability but there is no clear statement of precisely what the term means. In some of these books a certain idea is given, so readers may have some intuitive feeling of what it is all about. But all these volumes are strictly scientific, therefore there should be a clear explanation of probability. In wikipedia 2009 there are some incorrect and imprecise statements on this theme. Let us start to make the situation clear.

Until the beginning of the twentieth century the so-called *frequentive* approach was very popular and used almost exclusively. In this approach, the probability of a random event denotes the *relative frequency of the occurrence* of an experiment's outcome when the experiment is repeated. 'Frequentists' treated probability as the relative frequency of outcomes 'in the long run'. Speculatively, probability is the ratio between the number of favourable outcomes to the number of all possible events, provided that the outcomes have the same chance of occurring (for example, all faces of a die have an identical possibility of appearing). Notice that this definition makes no sense when the set of outcomes is infinite. Besides, at the end of the definition the statement *the same chance of occurring* means in fact *the same probability of occurring*, which immediately cancels out this definition. This situation existed until 1933 when a work by Kolmogorov was published. His approach to probability theory was completely different: '*this theory can and should be developed from axioms in exactly the same way as geometry and algebra.*' He presented a set that created the basis for the modern theory of probability. His original approach to probability was as follows:

'Let $\mathfrak{C}$ be a collection of elements which we shall call elementary events, and $\mathfrak{F}$ a set of subsets of $\mathfrak{C}$; the elements of set $\mathfrak{F}$ will be called random events.

Axioms:

I. $\mathfrak{F}$ is a field of sets.
II. $\mathfrak{F}$ contains the set $\mathfrak{C}$.
III. To each set A in $\mathfrak{F}$ a non-negative real number $P(A)$ is assigned. This number is called the probability of event A.
IV. $P(\mathfrak{C}) = 1$.
V. If A and B have no element in common, then $P(A \cap B) = P(A) + P(B)'$.

In this set of axioms only a finite number of events were taken into account. However, in the same publication, Kolmogorov extended the reasoning to an infinite number of events.

Based on the theory of sets it can easily be concluded that event $\mathfrak{C}$ is a sure event and $P(\Phi) = 0$, if Φ denotes an impossible event.

Resuming:
The probability is a **non-negative real number** supported at a [0, 1] interval.

Some mathematicians (Kopociński 1973, for example) are of the opinion that probability is the **function** in which arguments are random events and the values are real numbers, and this function fulfils the above axioms. For a fixed event, the probability function is a number. This approach is now more generally adopted than Kolmogorov's original one.

3.1   RANDOMNESS OF A SAMPLE

When a sample is taken it is in the form of a sequence of numbers and this information will be a point of comprehensive analysis. It is necessary to be sure that this sequence of numbers is random. A basic assumption that is formulated in mathematical statistics is that when a certain population is a point of interest with regard to a specified feature, we have to collect a representative sample (usually a small part of this population) in order to estimate the feature. Ignore the problem of why we only collect data from a certain part of the population as this is typically explained in a comprehensive manner in statistical literature. Let us focus our attention on the fact that we would like to be sure that the sample taken is a *good* representation of the whole population. A long time ago it was stated that a sample is a *good* representation if this sample is *random*.

We abandon here the definition of randomness. Investigations in this regard can be found in many mathematical statistics books and some other publications not directly associated with statistics (such as Wolfram 2002 p. 1067). We approach this issue from an engineering point of view—a sample is taken and we would like to be convinced that it has a randomness property. To solve such a problem the tools offered by the theory of verification of statistical hypotheses are very useful. Here a few tests are popular but the most frequently used are tests based on series, such as the Stevens test. We take into account that these tests make use of the median of the random variable[1] tested.

Recall, a median is the parameter of a random variable and it is such a value that the probability that the random variable value will be greater than the median equals the probability that the random variable value will be smaller than the median, and this probability is obviously ½. When a sample is given in the form of a sequence of numbers that occur one by one, it is necessary to order this sample monotonically. The middle number is an estimation of the unknown median of the population. This is true if the sequence has an odd number of values. For an even number of values, there are two middle numbers and we take the arithmetic mean of these two values as an estimate of the unknown median of the population.[2]

3.1.1   *The test procedure*

The sample collected is sorted monotonically and the median is found or calculated. Then the original sample is converted into a sequence of signs + and − ascribing to each element of the sample a sign. The sign + denotes an element which is greater than the median, the sign − denotes an element which is smaller than the median. Values of a variable equal to the median are rejected. Then the number s of the series (monomial signs) is calculated. This number is the value of the statistic used in the test. We also calculate the number n_+ of + signs and n_- of − signs.

The verified hypothesis H_0 proclaims that the elements of the sample were selected in a random manner, whereas an alternative hypothesis H_1 denies H_0. The level of significance α is selected and tables[3] with the critical values of the series are used to find the limits. There

---

[1] Bring to mind, a random variable X is a function, which has the domain 𝕰 and the values taken from the real axis, and this function fulfils two conditions:

-   the set $\{X \le x\}$ is the event for any real number x
-   the probability of events $\{X = +\infty\}$ and $\{X = -\infty\}$ equals zero.

[2] For samples given in different shapes, estimators for calculating the median are different (see, for example, Hogg and Craig 1995 or Jóźwiak and Podgórski 1997).

[3] There are many publications—books, monographs, leaflets of a statistical and reliability nature—with statistical tables (see, for example, Lindley and Scott 1995, Rohlf and Sokal 1981, Gnyedenko et al. 1965, Shor and Koozmeen 1968, Firkowicz 1970, Zieliński 1973, Keller et al. 1987, Maliński 2004). It will be presumed throughout this book that the appropriate statistical table is used to get the proper value required for a given test, ignoring from which set it has been taken.

are two: the minimum number of series—the left side of the critical region for $\alpha/2$; and the maximum number of series—the right side of the critical region for $1-\alpha/2$. If the number s falls between these two critical values, we have no ground to reject the null hypothesis. Otherwise, the hypothesis H_1 is the true one. This means that the rejection of the basic supposition is a consequence of the fact that either there are too many series in the sample or too few series. Let us consider an example.

■ **Example 3.1**
An LHD machine operating in an underground copper ore mine was observed and sequent repair times were noted. The following sequence was recorded:

$$0.9;\ 1.2;\ 2.4;\ 3.8;\ 1.8;\ 5.2;\ 1.6;\ 2.2;\ 2.9;\ 4.3;\ 0.4;\ 6.7;\ 1.7;\ 3.1;\ 0.6;\ 0.2\ \text{h.}$$

Verify the hypothesis that the observed sequence is random.

Calculate the median for the sample. Arranging the sequence monotonically we have:

$$0.2;\ 0.4;\ 0.6;\ 0.9;\ 1.2;\ 1.6;\ 1.7;\ 1.8;\ 2.2;\ 2.4;\ 2.9;\ 3.1;\ 3.8;\ 4.3;\ 5.2;\ 6.7$$

Because the sample has an even number of elements we select the two middle numbers and calculate the arithmetic mean:

$$\frac{1.8+2.2}{2} = 2.0$$

This value is the median of the sample.
Convert the original sample into a sequence of signs. We have:

$$0.9;\ 1.2;\ 2.4;\ 3.8;\ 1.8;\ 5.2;\ 1.6;\ 2.2;\ 2.9;\ 4.3;\ 0.4;\ 6.7;\ 1.7;\ 3.1;\ 0.6;\ 0.2$$

And:

$$-\ -\ +\ +\ -\ +\ -\ +\ +\ +\ -\ +\ -\ +\ -\ -$$

The number of series $s = 11$, the number of signs $n_+ = 8$, $n_- = 8$. Presuming the level of significance[4] $\alpha = 0.05$ and using the tables for critical values we have:

$\alpha = 0.025$ the minimum number if series: 4
$\alpha = 0.975$ the maximum number if series: 13.

The empirical number of series $s = 11$ falls between these critical values, thus we have no ground to reject hypothesis H_0—the sample has a randomness property.   ◄

---

[4]The most common level of significance presumed by engineers is 0.05.

Usually tables of the critical values for series comprise cases of up to 20 signs and so the problem arises of what to do when a sample size is greater than 20. It is a well-known fact that the series number distribution can be satisfactorily described by the normal distribution $N(m, \sigma)$ of parameters:

$$m = \frac{2n_+ n_-}{n_+ + n_-} + 1$$

and

$$\sigma = \sqrt{\frac{2n_+ n_- (2n_+ n_- - n_+ - n_-)}{(n_+ + n_-)^2 (n_+ + n_- - 1)}}$$

3.1.2 *Result of randomness investigation*

A finding that a given sample is non random does not occur frequently. However, it has to be stated clearly that if a given sample is not random, we cannot make any further statistical inference other than to trace why this situation has occurred.

There are many reasons for such a state of affairs. One possibility is the existence of a cyclic component in the exploitation process of the object being investigated (see Chapter 3.5). Another possibility is that during the repair of the object a certain assembly has been replaced by an assembly from a different machine and the intensity of failures of the piece of equipment being tested has changed significantly. In some cases, the 'modification' of the repaired machine may be even greater. All of these 'abnormal' events may generate non-randomness.

For a researcher investigating a given exploitation process it is a clear signal that something untypical has occurred and finding the source of this phenomenon is by all means recommended.

3.2 OUTLIERS ANALYSIS

When a sample is taken there is sometimes one outcome, extremely rarely two, that does not fit to the sample at first glance due to the fact that the observed value evidently differs from the others. The outcome either has a very high or very low value compared to the other data. In mathematical statistics such an outcome is termed an *outlier*. Immediately the question comes up as to whether this number belongs to the sample in a statistical sense or whether it has been put into the data due to a wrong decision or a mis-recording. Neglecting the problem that authors have given different definitions of an outlier (see, for example, Zeliaś 1996, Moore and McCabe 1999, Czekała 2004, www.mathwords.com/o/outlier.htm 2008), it is worth noting that mathematical statistics has a variety of tests allowing a hypothesis stating that such an outcome belongs to a given sample to be verified. Literature from this field is rich (cf. Barret and Lewis 1994) and particular problems of untypical observation have been considered for years by many authors (see, for example, Fisher 1929, Gnyedenko et al. 1965, Czaplicki 2006b).

In mining practice in this field one important issue has to be taken into account. In the engineering world in general, and especially in mining, there is a large class of technical objects that can be devastated during their operation. This means that a failure can occur that can destroy the object entirely or damage it very seriously. The environment of the object can also be devastated and the threat to the lives and health of personnel should be noted as well. For example, a hoist conveyance over-wound beyond its limited level could strike crush beams or a rupture could occur in a hoist head rope. This type of failure is called catastrophic and this type of event is a unique one. Its occurrence is very rare fortunately. The intensity of failures of this type is completely different from other regular failures of the object. In addition, the time of failure clearance is significantly longer. This means that these two types of exploitation events cannot be gathered together in a parameter analysis. The values of the intensity of failures and times of repair (if the repair is possible) are completely different but they are not outliers in a statistical sense in relation to the corresponding values noted in the regular exploitation of the object.

Returning to the main topic of this section, it should be stated that the analysis of data with an outlier (or outliers) collected from the exploitation of an object is more complicated when there is no indication in the records as to why the outlier has appeared.

Consider the following statistical scheme.

Assume there is a sample gathered from elements of two different populations. The variables connected with these populations differ from each other considerably in regard to the mean value. Moreover, one population is very large in number in relation to the second one. In such a situation we have a right to expect that in the sample taken a number may occur that is significantly different from the other elements of the sample. The occurrence of such an item allows a hypothesis stating inhomogeneity with regard to the feature that is the objective of investigation to be formulated. In order to formalise it, one can write it down.

There is a given sample:

$$\{X_1, X_2, \ldots X_k, \ldots, X_N\}$$

There is a supposition that the k-th element of the sample is untypical—an outlier.

Presume the mean value of the rest of elements is:

$$E(X_{i, j \neq k}) = \omega_i = \omega$$

Formulate a null hypothesis H_0: $E(X_i) = \omega$ which states that all elements are taken from one population versus the hypothesis H_1: $\omega_k \neq \omega \wedge \omega_{i, i \neq k} = \omega$ which states that the k-th element has a significantly different mean value.

A further part of the analysis depends on the kind of information that is at hand. In mining practice two probability distributions are most frequently applied, namely, exponential distribution and Erlang distribution. (Gaussian distribution is also often applied, especially when the times of the work cycle of machines operating periodically is concerned.)

3.2.1   *Exponential distribution*

If λ denotes the reciprocal of the expected value of a random variable exponentially distributed, the product $2\lambda \sum_{i=1, i \neq k}^{N} X_i$ has the probability distribution χ^2 with $2(N-1)$ degrees of freedom. It can conventionally be noted as:

$$2\lambda \sum_{i=1, i \neq k}^{N} X_i : \chi^2\big(2(N-1)\big) \tag{3.1}$$

Denoting:

$$\bar{X}_{N-1} = \frac{1}{N-1} \sum_{i=1, i \neq k}^{N} X_i \tag{3.2}$$

which is the arithmetic mean that is an estimator of the average value of a random variable based on an $(N-1)$ element sample, one can write down:

$$2\lambda(N-1)\bar{X}_{n-1} : \chi^2\big(2(N-1)\big) \tag{3.3}$$

Consider the ratio:

$$\frac{2\lambda N \bar{X}_n}{2N} : \frac{2\lambda(N-1)\bar{X}_{n-1}}{2(N-1)} : F\big(2N, 2(N-1)\big) \tag{3.4}$$

This quotient has an F-Snedecor probability distribution with $[2N, 2(N-1)]$ degrees of freedom, where $\bar{X}_N$ is the estimator of the mean including the outlier. Simplifying one can get:

$$\frac{\bar{X}_N}{\bar{X}_{N-1}} : F\big(2N, 2(N-1)\big) \tag{3.5}$$

Hypothesis H_0 should be rejected if—for a presumed level of significance α—the quotient of the left-hand side of this relationship is greater than the quintile of the order α of distribution F, that is

$$\frac{\bar{X}_N}{\bar{X}_{N-1}} > F_\alpha\big(2N, 2(N-1)\big) \tag{3.6}$$

A slightly different approach for the verification of an outliner was presented in 1929 by Fisher (the problem of outliers in those days was almost non-existent in the formal sense). In his paper he presented a test for the simultaneous verification of the hypothesis of the exponential distribution of a random variable tested together with the existence of an outliner significantly greater than the rest of the elements in the sample. Fisher proposed that the following formula be considered:

$$\eta = \frac{X_k}{\sum_{i=1}^{N} X_i} \tag{3.7}$$

In the cited paper the critical values for this function were given allowing for the verification of the formulated hypothesis (see also Gnyedenko et al. 1965 or Gnyedenko 1969).

■ **Example 3.2**
A belt conveyor hauling coal in an underground mine was investigated. The point of interest was the sequence of work times between two neighbouring failures. The following data was gathered [h]:

910, 170, 280, 790, 660, 1210, 510, 180, 6310, 140, 220, 440, 870, 380, 60, 1340.

Looking at this sequence one element seems strange: 6310 h.

Knowing that the times of work for conventional belt conveyors can almost always be satisfactorily described[5] by an exponential distribution, two means were calculated:

$$\bar{X}_{N-1} = 544 \text{ h} \quad \bar{X}_{N-1} = 904 \text{ h}$$

The first mean does not take into account the outlier, the second does. The difference between these average values is significant.

Calculate the ratio:

$$\frac{\bar{X}_N}{\bar{X}_{N-1}} = 1.66$$

The critical value taken from F-Snedecor tables for degrees of freedom for this example and presuming $\alpha = 0.05$ is:

$$F_{\alpha=0.05}(32, 30) = 1.82$$

Because the empirical value does not exceed the critical one, there is no objection to rejecting the hypothesis—there is no ground to remove the outlier from the sample. Consider the above problem by using Fisher's procedure. Applying formula 3.7:

$$\eta = \frac{X_k}{\sum_{i=1}^{N} X_i} = \frac{6310}{14470} = 0.44$$

The critical value taken from the appropriate table is:

$$\eta \, (\alpha = 0.05; \, N = 16) = 0.33$$

It can be stated that there is a ground to reject the hypothesis because this shows that the sample is taken from a population characterised by an exponential probability distribution and it is an element considerably greater than the rest of elements in the sample.

Both statistical procedures gave an identical result. ◄

3.2.2   *Erlang distribution*

The above considerations can easily be extended for a case where the values of the variable that is being investigated can be described by the Erlang probability distribution. It is easy to conclude that product $2\lambda \sum_{i=1, i \neq k}^{N} X_i$ has here the distribution χ^2 with $2k(N-1)$ degrees of freedom. Using analogical reasoning one can get:

$$\frac{\bar{X}_N}{\bar{X}_{N-1}} > F_\alpha \left(2kN, \, 2k(N-1)\right) \tag{3.8}$$

■ **Example 3.3**

The operation of a rack-and-pinion loco used in an underground coalmine was investigated. A sequence of repair times was noted giving the following sample:

---

[5]Very often in this book the phrase 'satisfactorily described' will be used in relation to a probability distribution—this should be taken to mean 'satisfactorily' in a statistical sense.

$$90, 45, 180, 90, 65, 55, 280, 245, 190, 140, 90, 270, 40,$$
$$125, 180, 310, 60, 1850, 130, 60, 235, 70, 120, 80. \quad [h]$$

The outlier in this sample is element number 18, which is 1850 min. There was no record of what had happened and why this time is so long.

An earlier reliability investigation gave grounds to conclude that the probability distribution of repair times can be satisfactorily described by a gamma distribution with the shape parameter $k \cong 2$. Thus, it can be assumed that Erlang distribution can be applied.

The two average times calculated based on the sample are:

$$\overline{X}_N = 208.3 \, \text{min} \quad \overline{X}_{N-1} = 137 \, \text{min}$$

The ratio is:

$$\frac{\overline{X}_N}{\overline{X}_{N-1}} = 1.52$$

The critical value is $F_{0.05}(96, 92) = 1.4$.

The empirical value is greater than the critical one. There are grounds to reject the verified hypothesis, stating the homogeneity of the sample. The outlier has to be rejected from further analysis.

Extra investigations showed that the surrounding rocks had been displaced and the loco under scrutiny had derailed. Several assemblies of the loco were destroyed causing the long repair time. It can be stated that a rare event was noted. ◀

3.2.3 *Normal distribution*

In mining practice Gaussian probability distribution also has a wide application. It is true that many empirical distributions of times of machines' work phases can be described satisfactorily by a normal distribution. This applies, for example, to wheel loaders, power shovels, trucks and drillers. A stream of broken rock transported by belt conveyors has a Gaussian character. The probability distribution of the total mass of mineral delivered daily to a shaft from an underground production level has a normal character. Distributions of measure errors are normal as a rule.

A random variable described by a Gaussian function is determined over the whole real number axis, from $-\infty$ to $+\infty$. For engineering practice this is not convenient. Physical magnitudes have their own natural limits. The approach that is presented by the theory can be taken into account as a margin model—an ideal one. The left-hand side (lower) natural limit is very often zero. The application of negative values mainly concerns load, stress, deceleration and dynamic moment. The right-hand side (upper) limit, in turn, when considering the capacity of mine transporting means say, is extremely important if likelihood results of considerations have to be obtained. All these clearly indicate that in engineering practice, especially in mining, the normal probability distribution should be truncated, usually from both sides. If—in spite of this—no truncation is applied, an estimation of the error made must be done. If the error generated by applying the ideal model is small and can be neglected, this model can be used (with care) for further analysis. If the error is significant, the results of applying the ideal model in terms of its inaccuracy have to be assessed.

Consider homogeneity analysis. A hypothesis that is verified is a statement that all elements of the sample are taken from one population in spite of the fact that one element differs considerably from the others. To check whether the supposition is true, this formula can be used:

$$T_k = \frac{X_k - \bar{X}_{N-1}}{S_{N-1}} \sqrt{\frac{N}{N-1}} \tag{3.9}$$

where:

$$S_{N-1} = \sqrt{\frac{1}{N-2} \Sigma_{i=1, i \neq k}^N (X_i - \bar{X}_{N-1})^2} \tag{3.10}$$

is the estimator of the standard deviation of the variable being observed without the outlier. If the verified hypothesis is true, the random variable T_k has Student's t-distribution of $N - 2$ degrees of freedom. If the alternative hypothesis H_1 is true, the random variable T_k has a non-central Student's t-distribution. The verified hypothesis should be rejected for the presumed level of significance if the following inequality holds:

$$T_k \geq t_\alpha(N-2) \tag{3.11}$$

where $t_\alpha(N-2)$ is the quintile of the order α of Student's t-distribution with $N-2$ degrees of freedom.

▪ Example 3.4

In an open-pit mine the times of *pure* loading (with no manoeuvrings of shovel and truck) were noted. This sequence was obtained:

2.4, 1.8, 1.6, 1.9, 2.6, 2.2, 2.1, 2.0, 1.8, 2.0, 5.3, 2.3, 2.1, 2.5, 2.1, 2.7,
1.9, 1.7, 1.6, 2.4, 2.3, 2.2, 2.9, 2.0, 1.8, 1.9, 2.7, 1.6, 2.3   min.

An outlier is the time of 5.3 min. Knowing that the probability distribution of the times of loading is normal, the mean and the standard deviation were calculated.

$$\bar{X}_{N-1} = 2.1 \, \text{min} \quad \text{and} \quad S_{N-1} = 0.4 \, \text{min}$$

and

$$T_k = 9.1$$

Presume now that $\alpha = 0.02$. The critical value for this level of significance is $t_\alpha(N-2) = 2.5$. Conclusion: the outlier in the sample has to be excluded from further considerations.

(By the way, after additional investigation the reason for the outlier was found—the shovel had loaded a large boulder.) ◀

Generally, the presence of an outlier generates some sort of problem. If further analysis is carried out and some statistical measures are estimated—or when, for example, performing least squares fitting to data—it is often the best solution to discard the outlier before computing. This should be done even when the appropriate statistical test gives no grounds to reject the hypothesis that the outlier belongs to the given sample characterised by a certain probability distribution. It is worth noting that if a sample is large, the rejection of one element does not cause a great loss. If however a sample is small, the value of the outlier has an enormous influence on the values of the calculated parameters; it is better to reject it.

3.2.4 *Result of an outlier analysis*

This provides univocal information about which part of the gathered data should be taken into further investigations.

In a case when an untypical element belongs to the sample, it is noteworthy information that during the operation of the object sometimes such an event can occur.

3.3 STATIONARITY TESTING OF SEQUENCES

Generally, stationarity means that something is fixed in a position or mode that is immobile or unchanging in condition or character. Stationarity in connection with the exploitation process of a technical object is associated with the way of realising the process. Therefore, it is a kind of property of a random process.

Let us neglect the subtleties connected with different kinds of stationarity considered in stochastic processes theory (see, for example, Kovalenko et al. 1983, Ross 1995). Instead, we will approach the problem of stationarity from a practical engineering point of view.

Each sequence of times of a given state should be tested to see whether observations increase or decrease on average over time. Because the parameter of the process is a time, verification concerns stationarity. Nevertheless, in engineering practice there are some operations for which the most important process parameter is not time. For a mine hoist installation, the number of winds executed is much more important. Similarly, for mine transport the number of tons of mass displaced is important. For many machines the number of work cycles performed is essential. For these reasons, the point of interest is the defined stochastic property of the sequence of the values being observed. However, in order to make further consideration more communicative, it will be presumed that the process parameter is a time.

There are several statistical tests that allow this property to be checked for a given sequence, but the one most frequently used in econometrics and technometrics seems to be the test applying Spearman's rank correlation coefficient (see, for example, Hollander and Wolfe 1973). Correlation is a kind of stochastic relationship between random variables. It is the statistical proportionality of the results of measures of different phenomena. It is sometimes stated that it refers to the departure of variables from independence.[6] A basic measure of correlation is a correlation coefficient, however several other coefficients are used in different situations.

The procedure for the Spearman rank correlation test is as follows. A natural number, following the sequence of the occurrence of the elements in time, is assigned to each element of the sample. These natural numbers are called ranks. Next, the sample is ordered monotonically. Natural numbers going up or down are assigned to all the elements of the new sequence. (For general purposes it does not matter if these numbers are assigned going up or down, it generates changes in the sign

---

[6] For more on correlation, see also section 3.9 in this chapter.

of the coefficient only.) In such a way a second set of ranks, v_i, $i = 1, 2, ..., N$ is obtained. These two sets of ranks create a matrix:

$$\begin{bmatrix} v_1 & v_2 & ... & v_N \\ 1 & 2 & & N \end{bmatrix}$$

where v_i is the rank of i-th element of the sample.

Then the value of the following function is calculated:

$$r^{(S)} = 1 - \frac{6R_N}{N(N^2 - 1)} \tag{3.12}$$

where:

$$R_N = \sum_{i=1}^{N} (v_i - i)^2 \tag{3.13}$$

The empirical value of Spearman's correlation is obtained in this way. This value is compared with the critical $r_\alpha(N)$ taken from appropriate tables.

The Spearman's rank correlation coefficient always fulfils the double inequality: $1 \geq r^{(S)} \geq -1$. If there is a full independence of values of the random variable with respect to time, $r^{(S)} = 0$. If the values of a variable grow strictly with time, $r^{(S)} = 1$. If the values of a variable also strictly decrease in time, $r^{(S)} = -1$. Empirical values lie between $[-1, 1]$.

To check whether the calculated value is significant, the statistical hypothesis H_0 is formulated stating that there is no dependence between the values of the variables and their positions in the sequence (that is, there is no dependence with respect to time). Formally, H_0: $\rho = 0$, where ρ is the correlation coefficient in the whole population. This hypothesis is set against H_1: $\rho \neq 0$, stating that the values of a variable depend on time. If the following inequality holds:

$$\left| r^{(S)} \right| > r_\alpha(N) \tag{3.14}$$

where $r_\alpha(N)$ is the critical value, the hypothesis H_0 must be rejected.

This test can be applied for $N \geq 4$ for $\alpha = 0.05$ and $N \geq 5$ for $\alpha = 0.01$. For larger N, practically for $N > 10$, this correlation coefficient has an approximately normal distribution $N(0, \sqrt{(N-1)^{-1}})$ that allows critical values to be calculated from the pattern:

$$r_\alpha(N) \approx \frac{u_{1-\alpha}}{\sqrt{N-1}} \tag{3.15}$$

where u_α is the quintile of the order α of the standardised normal distribution.

An alternative approach available for a sufficiently large sample size ($N > 20$) is an approximation to the Student's t-distribution with degrees of freedom $N - 2$. The variable:

$$t = r\sqrt{\frac{N-2}{1-r^2}} \tag{3.16}$$

has the Student's t-distribution in the null case (H_0). In the non-null case, tests are much less powerful, though the t-distribution can be used again.

The procedure for calculating Spearman's rank correlation coefficient needs adjusting if there are so-called *tied ranks*.

If there are two or more elements in a sample with the same value, you need a method to assign a rank to these elements when the sample is ordered monotonically. An arithmetic mean—calculated from all ranks belonging to these identical numbers—should be the rank assigned. Obviously, the ranks assigned have an influence on statistical inference. But tied ranks have a bad influence. The greater the number of tied ranks, the greater their influence and the more imprecise the calculation.

If there are tied ranks in a sample, the Spearman's correlation coefficient can be calculated using this formula (for example, Maliński 2004):

$$r^{(S)} = \frac{12\sum_{i=1}^{N} iv_i - 3N(N+1)^2}{\sqrt{(N^3 - N - T_x)(N^3 - N - T_y)}} \tag{3.17}$$

or:

$$r^{(S)} = \frac{(N^3 - N) - 6R_N - (\frac{1}{2})(T_x + T_y)}{\sqrt{(N^3 - N)^2 - (T_x + T_y)(N^3 - N) + T_x T_y}} \tag{3.18}$$

where:

$$T_x = \sum_{j=1}^{g}(t_j^3 - t_j) \quad \text{and/or} \quad T_y = \sum_{j=1}^{g}(t_j^3 - t_j) \tag{3.19}$$

and g is the number of groups of tied ranks whereas t_j is the number of tied ranks in j-th group.

■ **Example 3.5**

Reliability investigations concerned an AFC working at a longwall. Repair times were noted, obtaining:

25, 70, 50, 170, 20, 65, 40, 90, 210, 35, 60, 115, 130, 30, 355, 30, 30, 140, 20, 90, 55, 70, 125, 65, 155, 70, 20, 35, 30, 95, 100, 25, 15, 10, 270   min.

At first randomness of the sample was investigated. The median was found to be 65 min. Signs were assigned to the sample and the number of series computed, $s = 18$. The number of particular signs were calculated, $n_{+} = 17$ and $n_{-} = 16$. Presuming the level of significance $\alpha = 0.05$, critical values taken from appropriate tables are 11 and 23.

The number of series in the sample falls between the critical values and for this reason we have no grounds to reject the hypothesis proclaiming the randomness of the sample.

Next, the stationarity of this sequence was tested.

The Spearman's rank correlation coefficient was calculated by applying equation 3.12, getting

$$r^{(S)} = 1 - \frac{6\Sigma_{i=1}^{n}(v_i - i)^2}{N(N^2 - 1)} = 1 - \frac{6 \times 7808}{35(35^2 - 1)} = -0.094$$

A null hypothesis stating 'in the population there is no relationship between the time of an occurrence and the value observed' was verified presuming a level of significance $\alpha = 0.05$. The critical value for $N = 35$ is $r_{\alpha}(N = 35) \cong 0.282$. Because:

$$r^{(S)} = 0.094 < r_{\alpha}(N = 35) \cong 0.282$$

there are no grounds to reject the hypothesis. If approximation (equation 3.15) is applied the outcome is identical, namely:

$$r_{\alpha}(n) \approx \frac{u_{1-\alpha}}{\sqrt{N-1}} = \frac{1.645}{\sqrt{34}} = 0.282$$

We can assume that the sequence does not depend on time.

Perhaps readers have noticed that there are elements of the sample with identical values, and tied ranks should be applied. Therefore, the correlation coefficient should be estimated applying a corrected formula (say, equation 3.18).

By using equation 3.19, one obtains $T = 132$. For this reason the new value of the coefficient is:

$$r^{(S)} = -0.095$$

The difference is negligible in the case being considered.   ◄

3.3.1   *Result of stationarity testing with regard to mean*

Information on the stationarity or non-stationarity of a given sequence is very important. From a theoretical point of view it decides what kind of mathematical tools should be used in further investigations. If the statistical test applied gives no grounds to reject a hypothesis stating stationarity, further examinations can be conducted in the field of random variable theory. Conversely, if a statistical test rejects the hypothesis on stationarity, a further description of the data must be done using the tools of stochastic process theory. This is much more complicated than random variable theory.

In a physical sense, information on non-stationarity is very strong and significant—something is happening with the object. A certain process is running. There is a physical cause and—by all means—it should be identified, usually as quickly as possible. Sometimes the process is positive (time from failure to failure increases stochastically, caused, for example, by the running-in period for the machine) but more frequently it is negative. In the extreme, the object can be in danger, the environment of the object can be in danger, and/or the people working with the object can be in danger.

Generally, non-stationarity occurs rarely if the running-in period is excluded. More precisely, in the majority of cases the running-in period is now non-existent due to improvements in the design and production of objects. The rare occurrence of non-stationarity is also connected with the fact that sometimes equipment is withdrawn because of changes in the operation system in a mine. In some cases objects are withdrawn because of an *economic death*—a new machine appears on the market with much better parameters than the current machines of its kind. All these factors mean that the occurrences of non-stationarity are reduced. However, there are some mine machines and devices that have wearing processes that progress gradually in an irreversible way. An increase in the intensity of failures is observed when the object becomes older in terms of its worn elements, especially if a wearing process of a fatigue character is noted, such as in powered supports.

However, mining practice has shown that sometimes unusual situations arise. A failure occurs in a machine that is important for production and there is *time pressure*—this machine should be repaired as soon as possible. And, unfortunately, there is not a spare of the element that failed in the mine store. Miners sometimes try to solve this problem by making use of a *similar* element (directly or after a small modification) during the repair. The machine operates again but the intensity of failures changes. The modified element (sometimes it is an entire assembly) fulfils its duties but not entirely in the way that is expected. This period when the machine commences its duties and the intensity of failure increases is a non-stationary one, and this means that the work time between two neighbouring failures shortens and the intensity of the failures increases with time. This type of a repair practice is not recommended but unfortunately it can be observed in mining reality.

It should also be added that there are some mine technical objects, such as hoisting ropes, that have a non-stationary wear process almost from the beginning.

3.4   STATIONARITY TESTING OF VARIANCE IN SEQUENCES

If—after analysis—there are no grounds to reject the supposition that the analysed sequence of values is a stationary one, it does not mean that this sequence is totally free from any dependence on time. The relationship can be subtler—the dispersion of values of the random variable tested may be not stable. As a rule, in mining practice, if it is not constant, it increases over time.

A problem of this type can occur for instance when an examination of data is being carried out and a point of interest is to find the proper function that can satisfactorily describe the information obtained. Such a situation can be noticed when analysing records concerning the fatigue-wearing processes of the head rope of a hoist. An example plot is given in Figure 3.1, where the dotted line shows the empirical data.

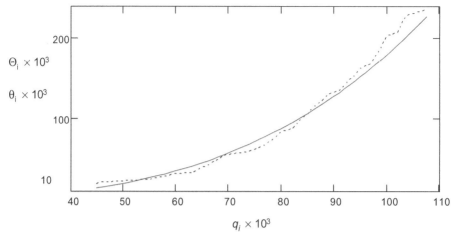

Figure 3.1.   Total number of cracks in the wires of a hoist head rope vs. the number q of winds executed; θ_i empirical plot, Θ_i theoretical plot.

There are a few different functions that can be used to describe this type of data but quite frequently a power function determined by this formula is applied:

$$\Theta_t = \delta t^{\chi} c^{\zeta_t} \tag{3.20}$$

where:

δ, χ – structural parameters of the function
Θ_t – total number of cracks of rope wires
t – time (or number q of winds)
c – constant
ζ_t – random component of the function.

It is presumed that the random component of the model comprises the whole stochastic nature of the process of the accumulation of cracks during the rope's utilisation. It exclusively explains the fact that the theoretical function does not cover empirical values precisely. This means that the right-hand side of equation is also a random variable, which means that the total number of cracks in rope wires is a random variable. This component is unobservable directly; however, it can be estimated by determining a sequence of residuals. A series of empirical values of θ_t is usually noted following rope utilisation and the creation of statistical data.

Considering function 3.20, the random component can be defined in a two ways:

a. as a sequence of the differences between the empirical values of θ_i and its theoretical counterpart; for formula (3.18) it is the sequence:

$$u_i = \theta_i - at_i^{b} \quad i = 1, 2, \ldots \tag{3.21}$$

b. as a sequence of residuals defined by pattern:

$$\hat{u}_i = \ln \frac{\theta_i}{at_i^{b}} \quad i = 1, 2, \ldots \tag{3.22}$$

which is the result of the appropriate conversion of formula (2.18). Notice, that the residual here is the index of power and $c = e$.

Both measures are correctly defined and they are two different measures of the random component for function 3.20.

The structural parameters are estimated by making a linearisation of the power function, that is:

$$\ln\theta_t = \chi\ln t + \ln\delta + \zeta_t \qquad (3.23)$$

and applying the method of least squares. Following this approach, one obtains for the data contained in Figure 3.1:

$$\Theta_i = 6.48\times10^{-5}q_i^{3.22}e^{u_i}$$

where q is the number of winds. This is shown in this Figure 3.1 as a continuous function.

Incidentally, many engineers use this way of reasoning and they are convinced that everything is correct. But this is not true. For more on this topic, see Chapter 5.3.2.

Figure 3.2 illustrates residuals u_i determined by function 3.21 for the sequence of data observed and Figure 3.3 shows residuals $\hat{u}_i$ (determined by function 3.22) against the number of winds executed by the rope.

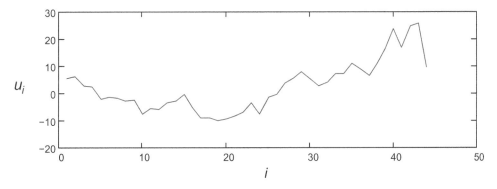

Figure 3.2.   Residuals determined by function 3.21 for the sequence of data noted.

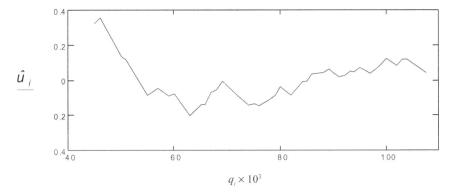

$$q_i\times10^3$$

Figure 3.3.   Residuals determined by function 3.22 vs. number of winds executed by a rope.

Both plots are different from the typical realisation of the pure stochastic process with a zero mean and a finite variance, as could be expected. A suggestion can be formulated that there has to be a factor making this difference. Looking at both figures a hypothesis can be created that the dispersion of values is not uniformly distributed in a stochastic sense. To verify this supposition the data was divided in half and for both sequences the standard error of estimation was calculated, giving 5.92 and 12.39 respectively. The problem now is to verify whether this difference is significant or only random. The proper statistical test must be applied to clear up these doubts.

Two tests can be considered:

a. a test applying F-Snedecor distribution
b. the Kruskal-Wallis one-way analysis of variance.

Reasoning based on (a) assumes a Gaussian distribution of the random variable being tested. This is the most frequently used method; however, this assumption must be taken into account before further analysis.

Reasoning based on (b) is a non-parametric method for testing the equality of population medians among groups. It is identical to a one-way analysis of variance with the data replaced by their ranks. Since it is a non-parametric method, this test does not assume a normal population (Kruskal and Wallis 1952, Siegel and Castellan 1988); therefore, this approach is more general.

In the case being considered an assumption can be made that the calculated sequence of differences has a normal probability distribution. So, case (a) can be analysed.

Formulate a null hypothesis H_0: $\sigma_1 = \sigma_2$, that the standard deviation in the first half of the data is the same as the standard deviation in the second half of the data. An alternative hypothesis can be H_1: $\sigma_1 \neq \sigma_2$. To verify the basic hypothesis calculate the ratio of empirical variances, which has an F-Snedecor distribution:

$$\frac{S_2^2}{S_1^2} : F(N-1, N-1) \tag{3.24}$$

where S^2 is the variance from the data and $S_2 > S_1$.

If the ratio is greater than the critical value $F_{\alpha/2}(N-1, N-1)$ taken from the table presuming a level of significance α, the hypothesis H_0 must be rejected.

For the example being considered:

$$\frac{S_2^2}{S_1^2} = 4.39 \quad \text{whereas} \quad F_{0.05}(24, 24) = 1.98$$

The null hypothesis must be rejected—the variance is not constant in time.

Further analysis can be conducted. By knowing the physical nature of the process being observed, it can be expected that the variance will increase in time. If so, the alternative hypothesis has to be modified to H_1: $\sigma_1 < \sigma_2$. For such an option the critical value is $F_\alpha(N-1, N-1)$ which gives 1.70. Conclusion: hypothesis H_0 must be rejected, hypothesis H_1 is the true one. Again, knowing the physical nature of the process being observed, one can formulate this statement: a characteristic feature of the process being observed is the stochastic increment of the variance.

3.4.1   *Result of stationarity testing with regard to variance*

Information on the non-stationarity of the variance of the process being observed is crucial for further analysis.

First, a clear hint is obtained that one of the basic assumptions that has to be fulfilled in order to allow the least squares method to be applied is not fulfilled.

Second, from a physical point of view, there must be a certain physical process running in time that generates the increment in the dispersion of the random variable being observed. The nature of this process should be recognised as soon as possible. A great deal of important information is associated with this process.

Third, a prediction interval for the prognosis of the future total number of cracks of wires of the rope will increase with time. Prognosis accuracy, in turn, will decrease with time. The exploitation situation becomes more dangerous over time.

Fourth, because no adjustment can be made in the operation of the rope, the only solution is to increase the frequency and accuracy of technical diagnostics performed on this rope.

During the regular exploitation of machines the increment in the variance of random variables connected with times of states is rarely observed. Sometimes this occurs in mechanical objects that are out of adjustment or out of order. A practical hint: an adjustment is required immediately if possible.

3.5   CYCLIC COMPONENT TRACING

Many processes in mining have a periodic character. This cyclic feature concerns the variety of courses of running action at different levels of generalisation.

The organisation of work in mines is cyclic. A shift is a period of the cycle. The exploitation of many machines operating in systems has a cyclic nature. Moreover, many machines have their own cycles of operation. The periods of all these cycles are more or less stochastic. A long time ago a hypothesis was formulated that this periodic operation can generate a periodic course of the exploitation process (Czaplicki 1974, 1975). In some periods of time—during exploitation of the object—the occurrence of some states is more probable than that of others. If so, the probability of the appearance of a given state is not constant in time but it is a function of time and this function is a cyclic one. In addition, a stream of rock extracted or hauled by transport also very often has a periodic character. These two functions are frequently correlated with each other. In some cases a stronger statement can be formulated: if the stream of rock being transported increases and usually has greater dispersion in value this means that the probability of the occurrence of failure in the transporting units increases. Therefore, the output of a hauling unit calculated as the product of a probability of the work of the unit and the mean mass of the rock being transported gives a wrong estimation because the higher the mass transported, the lower the probability of its transportation. Thus, this estimation gives higher values than it should.

Consider the problem of the existence of a cyclic component from a formal point of view. There are two cases to consider:

a. the period of cycle is known
b. the period of cycle is unidentified.

Let's consider case (a)—the period of the cycle is known—and turn our attention to the following scheme.

The exploitation cycle of the object is known and is marked by [0, *T*]. Presume that the object can be in any of *k* mutually excluding states, *j* = 1, 2, …, *k*. The object is observed *N* times—that is, records of what was going on with the object during *N* cycles are given. The cycle is divided in a discrete way into a certain elementary units of time. If we pay attention to one unit of cycle time (it can be any), we notice one out of *k* events. If so, the construction of the probability distribution of a random variable that a given state is observed *k* times in *N* independent trials ($N \geq k$) is now possible. This distribution is multinomial and is given by the formula:

$$P\{X_1 = b_1, X_2 = b_2, \ldots, X_k = b_k\} = \frac{N!}{b_1! b_2! \ldots b_k!} p_1^{b_1} p_2^{b_2} \ldots p_k^{b_{k1}} \qquad (3.25)$$

where obviously:

$$\sum_{j=1}^{k} p_j = 1 \quad \sum_{j=1}^{k} b_j = N$$

Consider one state of the object. Its realisation in time consists of realisations in N periods of time. If all of these realisations are put together, the frequency of the occurrence of this state versus the time cycle will be obtained for N independent trials. The relative frequency, in turn, is the j-th estimator unbiased, consistent and most efficient for parameter p of the distribution (formula 3.25).

The following hypothesis can be formulated: the method of the exploitation of an object and/or the conditions of the exploitation of the object can generate significant irregularity in the process of the changes of the states of the object. In other words, some states can occur more frequently in some periods of time during the cycle and some states will rarely be observed.

Having diagrams of the relative frequency against time in the cycle for all states, the above hypothesis can be verified. It is obvious that a certain irregularity of the process of the appearance of a given state will be visible due to the stochastic character of the process. However, a problem arises of whether the great changes observed can be connected with the stochastic nature exclusively. Reversing the problem a question can be formulated: how many times can a given state occur in a moment of cycle time in N trials that such an event can be assessed as very rare—so rare that a certain exploitation factor probably generated this irregularity? The factor is of a deterministic nature.

Consider the state of interest. Let us denote it by s. If it is the only one of interest, it can be specified in the following way:

$$P\{X_s = b_s\} = \frac{N!}{b_s! \Pi_{j \neq s} b_j!} p_s^{b_s} \prod_{j \neq s} p_j^{b_j}$$

However:

$$\prod_{j \neq s} b_j! = (N - b_s)! \quad \text{and} \quad \prod_{j \neq s} p_j^{b_j} = (1 - p_s)^{N - b_s}$$

thus:

$$P\{X_s = b_s\} = \binom{N}{b_s} p_s^{b_s} (1 - p_s)^{N - b_s} \tag{3.26}$$

The multinomial distribution is reduced to a binomial one. The problem of significant irregularity in the occurrence of state s is reduced to finding the number b_s which has a probability of appearance lower than presumed, a small level of probability, say υ (where $\upsilon \ll 1$), that is:

$$\binom{N}{b_s} p_s^{b_s} (1 - p_s)^{N - b_s} < \upsilon \tag{3.27}$$

Due to well-known properties of the binomial distribution there will be two values b_s that fulfil this inequality. They will be denoted by $b_s^{(l)}$ and $b_s^{(u)}$, where $b_s^{(u)} > b_s^{(l)}$ (see Figure 3.4). From this figure it is easy to observe that the critical area is determined by level υ; all events having a probability below this level should be comprehensively considered.

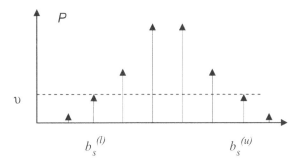

Figure 3.4.   Binomial probability density function.

Function 3.26 has a maximum for:

$$b = (n + 1)p \quad \text{if } (n + 1)p \notin \mathfrak{N} \qquad \mathfrak{N}\text{—set of natural numbers}$$

and it has two maximum values:

$$b_1 = (n + 1)p \quad \text{and} \quad b_2 = (n + 1)p - 1 \quad \text{if } (n + 1)p \in \mathfrak{N}.$$

Figure 3.5 shows the empirical probability of the occurrence of a repair state for a certain machinery system consisting of a shearer, two AFCs and a few belt conveyors operating in a series in an underground coal mine.

Looking at Figure 3.5, four events should be considered—all connected with the fact that the probability of the occurrence of a repair state for the system being analysed is above the critical level. Further analysis must comprise an examination of the records to find out whether the reasons for the appearance of this state are repeating or not. If the reasons are different in each, they were pure random events rarely occurring. If the reasons are repeating, they generate such a rare event. This rare event was not entirely random. These repeating reasons should be eliminated from the further operation of the system.

It is very important to understand that the *statistical procedure only indicates which events should be taken into further consideration*. And that is all. Further analysis must proceed outside of the mathematical area; physical aspects have to be taken into account and before final assessment can be made.

To complete these considerations it is necessary to construct an estimator of unknown probability p_s given in formulas 3.26 and 3.27. Following the relative frequency approach the number of favourable events is represented by the area of the histogram, whereas all possible events are

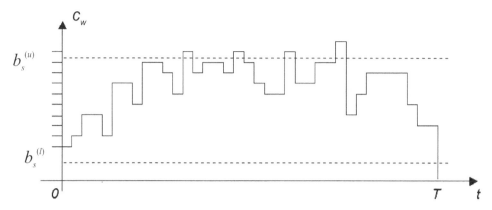

Figure 3.5.   Histogram of the relative frequency c_w of the occurrence of the repair state for a certain series system obtained by observing the system during N exploitation cycles.

represented by an extraordinary event when in all N trials only one state is observed. Therefore, the function:

$$\hat{p}_s = \frac{\Sigma_i t_{si}}{NT} \tag{3.28}$$

where t_{si} is i-th time of state s, is the estimator of interest.

It is worth mentioning that such regularity indicates that the probability of the occurrence of a repair state possesses a cyclic character and this has repercussions. But, let us start considerations from an empirical point of view.

For years in underground coal mines a steady increase in mine production has been observed; however, today production is decreasing in some countries. Extracting machines are being made with better parameters, which usually guarantee higher production. Hauling units must be able to take this increasing stream of extracted rock. But here is the problem. An increasing stream of rock flowing through conveyor systems has a great dispersion making it necessary to use wider belts. In surface mining this is not a problem. In underground mines using conveyors with wide belts, say more than 2000 mm, is inconvenient due to the necessity of constructing wider underground openings. This is very expensive. On the other hand, a stream of rock with high dispersion flowing on belt conveyors increases the probability of the occurrence of conveyor failures. These two reasons call for a reduction in the variation of the stream. The use of rock crushers/sizers is highly recommended. These devices can be found in some shearers and AFCs today.

Consider now the problem that the stream of rock being mined is of a high intensity with great dispersion. Because production is not steady during mine shifts, there are some periods of time when this stream will be of low intensity and a certain period of time when the stream will be of high intensity and high dispersion. This period with higher production causes an increase in the frequency of the occurrence of failures in all machines involved in production, particularly the extracting and hauling units. If so, it can be stated that two variables—the stream of flowing rock and the frequency of the occurrence of failures in machines—are mutually correlated.

Consider this problem a bit more expansively. For an assessment of what kind of changes can be expected due to the fact that these variables are correlated, two approaches can be considered. These two variables can be treated as periodic functions of time, or a Bayesian approach can be applied assuming that these variables are treated as mutually correlated random variables.

The first method is tedious and it needs rich statistical data and a long period of observation. Besides, when the function of interest is estimated it is very likely that the system being analysed will change its structure in a mine—some pieces of equipment will be withdrawn or added. Remember, one of cardinal characteristic features of mine machinery systems is its *changeability* due to the continuous displacement of working faces and incessant changes in transporting routes.

Thus, a simpler solution is to treat these variables as random and mutually correlated—moreover, correlated in a negative way. Because these variables are jointly correlated they create a two-dimensional probability distribution. Its marginal distributions are easy to recognise.

It is a well-known fact that a stream of extracted rock flowing through belt conveyors has a Gaussian character.

An example of the probability density function of stream q of extracted rock with a mean of $\eta_1 = 1200$ t/h and a standard deviation of $\sigma_1 = 300$ t/h is shown in Figure 3.6.

In light of previous considerations, the probability of the work of a series system when the process of exploitation is an alternative one, the work-repair type can be treated as a random variable supported on [0, 1] more or less symmetrically. Figure 3.7 illustrates the proposed model to describe this frequency of occurrence. This function is a Gaussian one with a mean of $\eta_2 = 0.75$ and a standard deviation of $\sigma_2 = 0.10$, however beta distributions fit better in many cases.

If the problem of support truncation is omitted, that is it is presumed that the lack of truncation will not generate a significant error, it can be assumed that the two-dimensional distribution can be approximated by the following probability density function:

$$f_{QC}(q,c_r)=\frac{1}{2\pi\sigma_1\sigma_2\sqrt{1-\rho^2}}\exp\left\{\frac{-1}{2(1-\rho^2)}\left[\frac{(q-\eta_1)^2}{\sigma_1^2}-\frac{2\rho(q-\eta_1)(c_r-\eta_2)}{\sigma_1\sigma_2}+\frac{(c_r-\eta_2)^2}{\sigma_2^2}\right]\right\} \quad (3.29)$$

The point of interest is now the product of two variables, the relative frequency of the occurrence of work state in the system and the stream of mineral transported that is:

$$E_w=E(QC)=\int\frac{1}{x}f_{QC}\left(x,\frac{w}{x}\right)dx \quad (3.30)$$

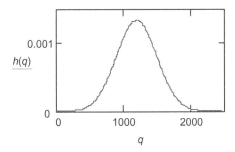

Figure 3.6.   Probability density function for stream of rock in t/h flowing through belt conveyors.

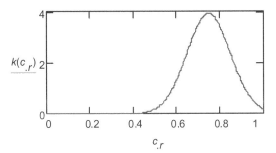

Figure 3.7.   Probability density function of the relative frequency of an occurrence of work state.

It is obvious that if these two variables are mutually independent (the correlation coefficient $\rho = 0$), then the above quotient is a product of both expected values:

$$\eta_1 \times \eta_2 = 900 \text{ t/h}$$

If these random variables are correlated and the coefficient is, say, $\rho = -0.80$, the expected value drops and:

$$E_w \cong 876 \text{ t/h}$$

This is not a great correction. However, it is worth noting that this value depends strongly on the standard deviations and in the above calculation no limits have been presumed. If, however, they are taken into account and dispersions are high, the difference can be considerable, reaching more than 20% in some cases.

3.5.1   *The exploitation cycle of object is unknown (case (b))*

Attention will now be placed on time series analysis although—as was stated before—the parameter of the process of interest in engineering practice is not always time.

In the theory of stochastic processes it has been proved that each stationary random process of a zero expected value can be expressed as the sum of two processes, mutually uncorrelated, stationary and with a zero mean. The properties of these processes are completely different—one is strictly periodic, whereas the second has no cyclic character. Both processes, when combined together, frequently create a process of a hardly recognised periodic character. The intensity of the obliteration of this character depends on the autocorrelation of the periodic process. Moreover, it increases when its variance increases. The methods that can be applied to trace the cyclic component in a discrete type random process are harmonic analysis and spectral analysis. This latter methodology has very rich literature (see, for example, Hannan 1960, Granger and Rosenblatt 1957, Granger and Hatanaka 1969, Hamilton 1994) and widely developed theory. Nevertheless, it has some disadvantages, such as the arbitrary selection of time lags in some cases or the disadvantageous properties of the estimators of the covariance processes. In addition, formulas to smooth out the processes are frequently very complicated. Harmonic analysis, in contrast, is very old but it is relatively simple. Let us pay attention to this method.

Consider a case when data is given as a time series $\{x_t; t = 1, 2, \ldots, T\}$. If a time series has no trend, it can be expanded in a Fourier series. In mathematics, a Fourier series decomposes a periodic function into the sum of simple oscillating functions, namely sines and cosines. Following this line of reasoning, we can write:

$$x_t = \frac{1}{2}a_0 + \sum_{i=1}^{T}\left(a_i \cos \omega_i t + b_i \sin \omega_i t\right) \tag{3.31}$$

where a_0, a_i, b_i are the Euler-Fourier coefficients.

Estimators of these coefficients are as follows:

$$\hat{a}_0 = \frac{1}{T}\sum_{i=1}^{T} x_t \tag{3.32}$$

$$\hat{a}_i = \frac{2}{T}\sum_{i=1}^{T} x_t \cos\frac{2\pi i t}{T} \tag{3.33}$$

$$\hat{b}_i = \frac{2}{T} \sum_{i=1}^{T} x_t \sin \frac{2\pi it}{T} \tag{3.34}$$

for $i = 1, 2, \ldots, T/2$.

The values of these statistics can be applied to verify (or reject) the hypothesis that a cyclic component exists in a given sample.

Consider the square amplitude of the process. It is given by the equation:

$$\hat{A}_i^2 = \hat{a}_i^2 + \hat{b}_i^2 \tag{3.35}$$

whereas its expected value is:

$$E(\hat{A}_i^2) = \hat{A}_E^2 = \frac{4\sigma^2}{T} \tag{3.36}$$

where σ^2 is variance of series x_t.

Replacing the unknown variance σ^2 by the unbiased estimator S^2 (the variance estimated from a sample) one obtains:

$$E(\hat{A}_i^2) = \hat{A}_E^2 \cong \frac{4}{T(i-1)} \sum_{i=1}^{T} (x_t - \bar{x})^2, \quad i > 1 \tag{3.37}$$

where $\bar{x}$ is the arithmetic mean of x_t.

A long time ago, in the theory of periodic oscillations it was proved (Schuster 1898, 1900) that the probability of an event that $\hat{A}_i^2$ will be α times greater than $\hat{A}_E^2$ is given by the formula:

$$P(\hat{A}_i^2 > \alpha \hat{A}_E^2) = \exp(-\alpha) \tag{3.38}$$

which gives:

$$P\left\{ \left(\frac{A_i}{A_E} \right)^2 > -\ln\alpha \right\} = \alpha \tag{3.39}$$

Now, it is possible to verify the hypothesis stating the significance of cycle of the process in a point i. The following hypothesis can be verified:

$$H_0 : \left(\frac{A_i}{A_E} \right)^2 > -\ln\alpha$$

This states that the value of quotient square to the amplitude to the mathematical hope is insignificant (random) and α is the level of significance.

■ **Example 3.6**

The main hoist output was noted in successive hours in one African mine. Data comprised $T = 64$ h. The average output was 560 t/h. A point of interest was the difference between the output in a given hour and the corresponding mean. An accuracy of recording was 10 t/h.

The data obtained is given in Figure 3.8.

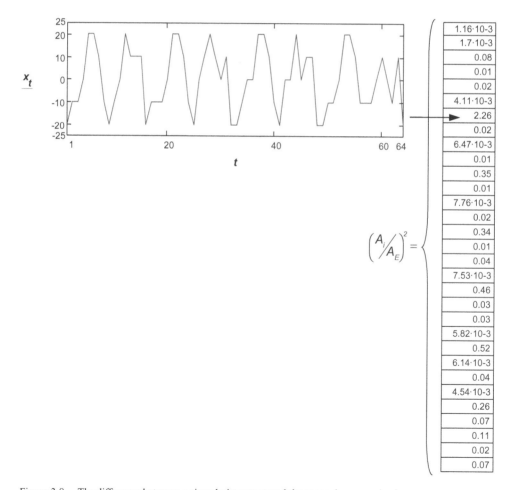

Figure 3.8.   The difference between a given hoist output and the mean in successive hours.

Calculations were conducted applying formulas 3.31 to 3.39. Outcomes of the computation of the square ratio are given in the column on the right.

It is easy to perceive that the only *jump* in value is at point $i = 8$. Therefore, a possible cycle of wavering is:

$$T/i = 64/4 = 8 \text{ hrs}$$

which is in accordance with intuition.

To close these considerations it is necessary to calculate the expression $-\ln \alpha$ for a presumed level of significance. Taking $\alpha = 0.10$ one obtains:

$$-\ln \alpha = 2.30$$

This value is a bit higher than 2.26, the value taken into account based on the data analysed. Thus, this gives no grounds to reject the hypothesis that the observed jump is a random one and no cycle can be identified. However, knowing mining reality and bearing in mind the weak accuracy of data, one tends to accept the hypothesis that the period observed is an 8-hours one.   ◀

3.6   AUTOCORRELATION ANALYSIS

The processes of the changes of the states of mine machines and other pieces of equipment characterise the mutual independence between states as a rule. Rarely does the time of an executed repair have an influence on the time of work after this repair. Dependent phenomena only occur in special circumstances, and are rather an exception to the rule.

The wearing processes of the elements of machines and some important mine devices have different characteristics depending on factors such as the properties of the object, the main character of the type of wearing process (fatigue, corrosion, abrasive wear, local mass reduction, for example) and so on. Mining engineers identify these processes for many reasons. A knowledge of the physical aspects of a given process together with a mathematical model of the course of run are very useful in making proper decisions about the exploitation process and in modifying the production process of the object and improving the object's performance. Very often the processes being investigated have interesting properties, and identifying these properties is advantageous. Some properties can even indicate what kind of mathematical tools should be applied for a suitable description of the process.

Consider the following example.

▪ Example 3.7

A point of interest was the wearing fatigue process of a hoist head rope 58 mm in diameter with triangular shaped strands. Data comprised $N = 33$ records of the total number of cracks of rope wires at the time of each inspection. Observations were made every 7 days ($\Delta t = 7$) of the rope's operation. The empirical data shown in Figure 3.9 (continuous line) was approximated using a power function (formula 3.20) obtaining:

$$\Theta_i = a t_i^b$$

where $a = 4.15 \times 10^{-4}$ and $b = 2.69$ after linearisation of the power function and application of the least squares method to get estimates of parameters of the function.

The approximation was tested through estimation of residuals $\hat{u}_i$ by applying formula 3.22:

$$\hat{u}_i = \ln \frac{\theta_i}{a t_i^b}$$

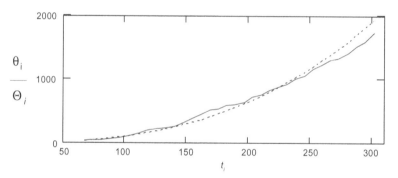

Figure 3.9.   Total number of cracks of wires of a hoist head rope vs. the number time t of the rope's operation; θ_i empirical plot, Θ_i theoretical plot.

The results of the calculations are shown in Figure 3.10.

One of the basic assumptions that has to be fulfilled in connection with the application of the least squares method is that residuals are not correlated—that is, that there is no autocorrelation in the random component of the model (see, for example, Goldberger 1966). This assumption can be written as:

$$E\{\zeta_i\zeta_j\}=0 \quad i\neq j \tag{3.39}$$

In order to verify the above assumption, four correlation coefficients were calculated for pairs (ζ_i, ζ_{i-1}); (ζ_i, ζ_{i-2}); (ζ_i, ζ_{i-3}); (ζ_i, ζ_{i-4}) and these coefficients were denoted by $r_1^{(a)}$, $r_2^{(a)}$, $r_3^{(a)}$ and $r_4^{(a)}$. A well-known formula for the calculation of the Pearson correlation coefficient was used and gave the following outcomes:

$$r_1^{(a)} = 0.83 \quad r_2^{(a)} = 0.62 \quad r_3^{(a)} = 0.39 \quad r_4^{(a)} = 0.17$$

It was of interest to check whether these values are important, and above the corresponding critical values.

For an answer to this question the Durbin-Watson test is usually the one that comes to mind (Durbin 1953, Durbin and Watson 1950, 1951). However, the Durbin-Watson statistic is only valid

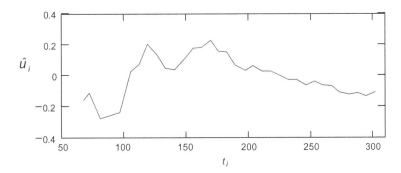

Figure 3.10.   Residuals determined by function 3.22 vs. number of days of rope work.

for stochastic regressors and first order autoregressive schemes (such as AR(1)). Furthermore, it is not relevant in many cases; for example, if the error distribution is not normal, or if it concerns the dependent variable in a lagged form as an independent variable. In these cases it is not an appropriate test for autocorrelation. A test that is suggested and does not have these limitations is the Breusch-Godfrey test (Breusch 1979, Godfrey 1978, 1988) as well as the Pawłowski test (Pawłowski 1973). Because the first of these tests is well-known and more simple to apply when compared to the latter test, further reasoning will be performed applying this kind of examination.

The Breusch-Godfrey statistic is determined by the following formula:

$$\chi^2(c) : (N - c)\,(r_c^{(a)})^2 \qquad\qquad (3.40)$$

where c is the autocorrelation order.

The hypothesis $H_0 : \rho_c = 0$ proclaims that there is no autocorrelation of the order c. If this is true the following inequality holds:

$$\chi_\alpha^2(c) > (N - c)(r_c^{(a)})^2 \qquad\qquad (3.41)$$

otherwise we have the ground to reject the hypothesis, that is we can presume that there is autocorrelation of the order c.

In the case considered we have for the first, second, third and fourth order:

$$22.04\ (3.84)\quad 11.92\ (5.99)\quad 2.56\ (7.82)\quad 0.84\ (9.49)$$

where the numbers in brackets are corresponding critical values.

The Breusch-Godfrey test can also be supported by the F-Snedecor statistic by making use of the well-known relationship between the χ^2 statistic and the F-Snedecor one. It has been proven by a simulation technique that for small samples such an approach is better than basing them on a χ^2 statistic.

In the case being analysed it does not matter which statistic is applied (either χ^2 or F-Snedecor), the result of verification is identical: the critical value is greater than the empirical one in the third and fourth cases. Conclusion: autocorrelation up to second order was traced in the data for the applied approximation function.   ◀

Often in processes of rope wearing in a fatigue manner autocorrelation exists and it can be traced for both cases, that is when the residuals are defined as differences as well as when residuals are defined in a different way.

The term 'autocorrelation' has been introduced here presuming that it is sufficient that readers have an idea about correlation. For further considerations, this term should understood as the correlation of a variable with itself over successive time intervals.[7]

Four important conclusions can be formulated in connection with autocorrelation.

- Autocorrelation traced in a physical process is very rarely connected with a purely random event (stochastic set up of numbers) that has no association with the physical nature of the process.
- Autocorrelation means that there is a 'memory' in the process and a future state depends—as a rule—on the state just before, sometimes on some states that happened earlier.
- Autocorrelation indicates that an autoregression function should be considered as an approximation function of the process.
- If autocorrelation has been found, further investigation should be done in order to recognise the physical grounds that are the source of this autocorrelation. This can provide important knowledge on the nature of the process being observed, which may allow some counteraction if this phenomenon is disadvantageous (if possible) or allow engineers to make use of this course if it is useful.

3.7   HOMOGENEITY OF DATA

The statement that something is homogeneous means that it is alike, similar or uniform from a certain point of view. In the statistical sense that is considered here, homogeneity is a property of a data set and relates to the validity of the very advantageous assumption that the statistical properties of samples taken are identical to the whole population. In mining engineering practice the problem of data homogeneity is extremely crucial because, as a rule, the sample size is not very large. However, in many cases it is possible to examine a certain number of similar objects operating in similar conditions and it can be expected that the data taken will be homogeneous; all observations can be gathered together creating a large sample. It has to be stated that in such a situation engineers are 'guilty'. They design objects of high reliability and failures seldom occur. Even a good quality belt conveyor operating in tough underground conditions only needs to be repaired a few times each year on average. To get an appropriately large sample concerning conveyor failures a long period of observation is required theoretically. However, frequently a given conveyor does not work intact for such a long period of time. It will either be changed in terms of construction (through a significant increase in its length, for instance) or it will be dismantled and erected in a new place. This 'old' device can have different values of reliability indices in a new location. To get large sample size, the only solution is to observe a number of the same belt conveyors operating in the same mine. The lengths of these conveyors will be different. However, fortunately, reliability investigations conducted in the middle of 1970s in several underground coal mines as well as in a few opencast mines in Poland proved that for conventional belt conveyors reliability indices do not depend on conveyor length. This information was extremely important for conveyor system designers. It allowed them to formulate a practical recommendation: if the rock mass is to be hauled by belt conveyors, and if two points are given—one, where the stream of extracted rock is loaded and a second, where this stream should be delivered—it is advisable to connect these two points by a minimum number of conveyors. The reliability of a system of these units connected in a series depends on the number of units involved (this will be proved in Chapter 7.2). The more conveyors in a series, the lower the reliability of the series.

Consider the following exploitation situation. The time of observations is given along with a certain number of the same technical objects. The term 'the same' can be understood in different ways, but here it means machines made by the same producer having the same technical parameters

---

[7]We neglect here the problem of some differences in definitions of autocorrelation presented by different authors.

and operating in similar exploitation conditions. The point of interest is a certain feature of these objects that can be observed during their exploitation. The appropriate data is gathered—that is, samples from each object that is being examined. The problem is whether these samples can be grouped together into one large sample or, in other words, whether the collected data is homogeneous from the given point of interest—the common feature of objects.

From a statistical point of view the problem concerns equality in probability distributions. There are a number of statistical tests that allow the hypothesis that the data is homogenous to be verified. One of the most frequently applied in engineering practice is the Kruskal-Wallis test based on the sum of ranks (Kruskal-Wallis 1952). A model considered for this test is as follows.

Presume that k same objects are observed with regard to a certain feature and therefore k samples are obtained. A convenient aspect of the test is that samples can have different sizes. Let us denote them by N_i, $i = 1, 2, ..., k$. Assume that random values of a measure of the feature can be described by a certain probability distribution $F(x)$ and a statistical hypothesis H_0 is formulated that all probability distributions are identical that is:

$$H_0 : F_1(x) = F_2(x) = ... = F_k(x)$$

An alternative hypothesis H_1 denies the null hypothesis.

3.7.1 *The test procedure*

All elements of all samples are gathered together and ranks are assigned for the monotonically ordered set—from 1 to N, where N is the total number of elements in all samples, $\Sigma_i N_i = N$. The average of ranks must be assigned to tied values, if tied values exist. Next, the following statistic is calculated:

$$K_N = \frac{12}{N(N+1)} \sum_{i=1}^{k} N_i \left(\bar{V}_i - \frac{N+1}{2} \right)^2 = \frac{12}{N(N+1)} \sum_{i=1}^{k} \frac{V_i^{(\Sigma)}}{N_i} - 3(N+1) \qquad (3.42)$$

where:

$V_i^{(\Sigma)}$ is the sum of ranks in i-th sample

$\bar{V}_i = \frac{1}{N_i} V_i^{(\Sigma)}$ is the average rank for i-th sample.

Looking at formula 3.42, it is easy to notice that if there are more differences between average sample ranks and the general mean rank, the statistic K_N is larger. A low dispersion in this regard, in turn, will be favourable for hypothesis H_0—providing no ground to reject it.

Kruskal and Wallis noticed that if k grows and if sizes N_i increase, the random variable K_N has an asymptotic probability distribution of χ^2 with $k - 1$ degrees of freedom.

Thus, if the following inequality holds:

$$K_N \geq \chi_\alpha^2(k-1) \qquad (3.43)$$

where $\chi_\alpha^2(k-1)$ is the critical value for the assumed level of significance α, the verified hypothesis H_0 should be rejected.

For a large k, the random variable $\sqrt{2\chi_{k-1}^2} - \sqrt{2(k-1)-1}$ has approximately a normal distribution $N(0, 1)$. Accordingly, for large k the following approximations can be used:

$$\chi_\alpha^2(k-1) \cong (k-1) \left(1 - \frac{2}{9(k-1)} + u_{1-\alpha} \sqrt{\frac{2}{9(k-1)}} \right)^3 \qquad (3.44)$$

or:

$$\chi_\alpha^2(k-1) \cong \frac{1}{2}(\sqrt{2(k-1)} + u_{1-\alpha})^2 \tag{3.45}$$

If some of N_i values are small (that is, less than 5) the probability distribution of K_N can be quite different from this chi-square distribution.

In order to obtain a more precise reasoning when tied ranks are in the samples, a correction should be done by first calculating:

$$(3.46)$$

$$t = \left(1 - \frac{\Sigma_{j=1}^g (t_j^3 - t_j)}{N^3 - N}\right)^{-1}$$

where g is the number of groups of tied ranks, and t_j is the number of tied ranks in j-th group.

Then multiply t by K_N. It can be proved that always $t > 1$ and for this reason the new value of statistic 3.42 will be greater than that calculated without the correction. This means that by taking the correction into account the chance for a rejection of the verified hypothesis increases.

■ **Example 3.8**

Four hoist installations of the friction type having identical electric movers of 1600 KW and the same wheel diameter of 6000 mm were investigated. Work times (h) between failures were noted obtaining:

Hoist (1):   752, 304, 82, 1288, 601, 476
Hoist (2):   88, 576, 48, 1308, 711, 373, 191, 870
Hoist (3):   118, 534, 39, 1410, 660, 399, 199
Hoist (4):   228, 596, 66, 1099, 801, 370.

A hypothesis was formulated that this data is homogeneous and can be gathered together. Calculations are shown in Table 3.1.

Based on the calculations shown in Table 3.1, the statistic K_N was estimated by this formula:

Table 3.1.   Calculations for the evaluation of homogeneity of data.

	Hoist (1)		Hoist (2)		Hoist (3)		Hoist (4)	
	Time (hrs)	Rank	Time (hrs)	Rank	Time (hrs)	Rank	Time (hrs)	Rank
1	752	21	88	5	118	6	228	9
2	304	10	576	16	534	15	596	17
3	82	4	48	2	39	1	66	3
4	1288	25	1308	26	1410	27	1099	24
5	601	18	711	20	660	19	801	22
6	476	14	373	12	399	13	370	11
7			191	7	199	8		
8			870	23				
		92		111		89		86

$$K_N = \frac{12}{27(27-1)}\left(\frac{92^2}{6}+\frac{111^2}{8}+\frac{89^2}{7}+\frac{86^2}{6}\right)-3(27+1)=6.86$$

For the case being considered, the critical value for the presumed level of significance $\alpha = 0.05$ and $k-1 = 3$ degrees of freedom is $x_\alpha^2(K-1) = 7.81$.

Looking at these values there is no doubt that there are no grounds to reject the verified hypothesis stating that the data can be treated as homogeneous.   ◄

Results of considerations on homogeneity of data are important for at least two reasons.

- If there is no ground to reject the hypothesis that the investigated data is homogeneous, there is the possibility to gather all data and to create a large sample. This is important because unit samples are usually small in mining practice. Working with a large sample, stronger statistical inferences can be made with a greater likelihood that the findings are true.
- If there is a basis to reject the supposition on homogeneity in data, further investigation should be done to trace the reason why data is inhomogeneous and which object has 'made' it. To find of the reason behind this 'unfitness' is valuable information from an exploitation point of view.

3.8   MUTUAL INDEPENDENCE OF RANDOM VARIABLES

When two features of an object are being investigated, their mutual relationship is very important. In the majority of reliability models as well as in exploitation models, the usual assumption is that states are mutually independent. A measure is assigned to each state—a time or another parameter important from an engineering point of view (such as tons of mass being transported)—and independence means freedom from any stochastic relationship between these variables. The assumption that two or more states are mutually independent must be verified to check whether a given model can be applied or—if not—a different model should be found (which is usually more complicated and difficult to apply).

An investigation concerning independence can be treated as a generalisation of the problem of stationarity. Sometimes the point of interest is not only the independence of some states between each other but also independence in respect of some accompanying processes—processes that run concurrently to the main exploitation process of the object.

In mathematical statistics there are several different tests to verify the independence of random variables. The most frequently used are tests based on contingency[8] tables 2×2 and 2×3 as well as the test of independency based on Pearson statistics χ^2.

Consider a case where elements of a certain population are classified taking into consideration two characteristic features, say X and Y. The problem is whether these features are mutually independent. The range of possible values of X is divided into w groups, and the range of possible values of Y into k groups. Thus, the sample that was collected can be shown in a so-called contingency table as presented below.

A hypothesis H_0 stating the mutual independence of features X and Y is:

$$H_0 : P\{X = x_i, Y = y_j\} = P\{X = x_i\}P\{Y = y_j\} \tag{3.47}$$

The following relationships hold:

$$n_{i.} = \sum_{j=1}^{k} n_{ij} \qquad n_{.j} = \sum_{i=1}^{w} n_{ij} \tag{3.48}$$

$$N = \sum_{i=1}^{w} \sum_{j=1}^{k} n_{ij} = \sum_{i=1}^{w} n_{i.} = \sum_{j=1}^{k} n_{.j} \tag{3.49}$$

Making use of the above, the sizes of the marginal probabilities are estimated:

$$p_{i.} = \frac{n_{i.}}{N} \quad p_{.j} = \frac{n_{.j}}{N} \tag{3.50}$$

and they are used to calculate the value of the statistic:

$$\chi^2 = \sum_{i=1}^{w} \sum_{j=1}^{k} \frac{(n_{ij} - Np_{ij})}{Np_{ij}} \tag{3.51}$$

Table 3.2. Contingency.

		Feature Y							
		y_1	y_2	...	y_j	...	y_k	$n_{i.}$	$p_{i.}$
Feature X	x_1	n_{11}	n_{12}	...	n_{1j}	...	n_{1k}	$n_{1.}$	$p_{1.}$
	x_2	n_{21}	n_{22}	...	n_{2j}	...	n_{2k}	$n_{2.}$	$p_{2.}$
	...	...	...	...	...	...	...	...	...
	x_i	n_{i1}	n_{i2}	...	n_{ij}	...	n_{ik}	$n_{i.}$	$p_{i.}$
	...	...	...	...	...	...	...	...	...
	x_w	n_{w1}	n_{w2}	...	n_{wj}	...	n_{wk}	$n_{w.}$	$p_{w.}$
	$n_{.j}$	$n_{.1}$	$n_{.2}$	...	$n_{.j}$	...	$n_{.k}$	N	
	$p_{.j}$	$p_{.1}$	$p_{.2}$	...	$p_{.j}$	...	$p_{.k}$		1

---

[8]Contingency is the state of being contingent on something.

This statistic—assuming that the hypothesis H_0 is a true one—has an asymptotical χ distribution with $(w-1)(k-1)$ degrees of freedom. The critical area in this test is always right-sided. The value obtained from formula 3.51 should be compared with the critical one for the presumed level of significance α.

3.8.1 *Results of independence analysis*

Basically, there are two important pieces of information obtained from the independence analysis.

As has been stated, in many exploitation models used to describe the operation of technical objects an assumption is made that the objects being analysed are independent from each other in terms of time. Therefore, it is compulsory to get verification of this assumption from the empirical data in order to get information as to whether a given model with this condition can be applied.

If the independency investigation gives information that the independency hypothesis should be rejected, then it is highly advisable to conduct further research to find the reasons for this situation. The scope of consideration should be located outside statistics in the physical nature of the observed process. If the process has such a property, a simple model should be discarded. As a rule, a more complicated and difficult analysis will need to be carried out to trace the regularities existing in the investigated process.

Remark: Independence is a stronger feature than non-correlation.

3.9 MUTUAL DEPENDENCE OF RANDOM VARIABLES

As a result of reliability or exploitation investigations of technical objects, a great deal of data has been gathered that gives information on the process of changes of properties of the object for different methods of the utilisation and maintenance. The object alone has a set of characteristic parameters that are important from an engineering point of view. Very often it is of great interest to recognise how the exploitation process of an object depends on some features of the object, some characteristics of the method of its utilisation and maintenance expressed by appropriate parameters. Because these parameters can be treated as random variables from a mathematical point of view, a suitable mathematical tool to trace and analyse existing relationships in this regard is correlation analysis.

Correlation is a kind of statistical relation between two or more random variables such that stochastically systematic changes in the value of one variable are accompanied by stochastically systematic changes in the other variable(s). Sometimes this term is defined as the degree to which two or more attributes or measurements on the same group of elements show a tendency to vary together.

There are a number of statistical measures of this kind of relation, starting from linear instruments—correlation coefficients of different types (see Chapter 3.3 for example)—to non-linear measures, partial correlation, multiple and cross correlation, etc. Many of these measures have been successfully applied in mining engineering investigations for years.

There is no doubt that the one most frequently used is the Pearson correlation coefficient (the linear one), and this is defined in cases where there are two random variables X and Y by the formula:

$$\rho^{(P)}_{X,Y} = \text{cov}(X, Y)/\sigma_X \sigma_Y \tag{3.52}$$

where:

$$\text{cov}(X, Y) = E(XY) - E(X)\,E(Y) \tag{3.53}$$

Note, $\text{cov}(X, Y)$ is the covariance of both random variables, and σ_X and σ_Y are standard deviations, appropriately, of X and Y.

The value of the linear correlation coefficient is 1 in the case of a strict increasing linear relationship between random variables, −1 in the case of a strict decreasing linear relationship, and a value in between −1 or 1 in all other cases, indicating the degree of linear dependence between the variables. The closer the coefficient is to either −1 or 1, the stronger the correlation between the variables.

Remember these points:
If the empirical correlation coefficient between random variables is near 1 or −1, this does not necessarily mean that there is a really strong linear relationship between these variables. This coefficient shows only the relationship between numbers, these two variables *could be* totally unrelated to each other in a physical sense.

If the empirical correlation coefficient between random variables is near 0, this does not necessarily mean that these variables are not really related to each other. The relationship between these variables *could be* quite strong but not a linear one.

A correlation between two variables is diluted by the presence of measurement error around the estimates of one or both variables, in which case disattenuation[9] provides a more accurate coefficient.

If a sample is taken, the following estimator[10] can be applied to get an assessment of the unknown real value of correlation coefficient:

$$R_{XY} = \frac{\sum_{i=1}^{N}(x_i - \overline{x})(y_i - \overline{y})}{\sqrt{\sum_{i=1}^{N}(x_i - \overline{x})^2 \sum_{i=1}^{N}(y_i - \overline{y})^2}} \tag{3.54}$$

where:

$\overline{x}, \overline{y}$ are the arithmetic means of x_i and y_i, $i = 1, 2, ..., N$ respectively.

Note that because the correlation coefficient given by equation 3.54 is linear, $R_{XY} = R_{YX}$.

If for a given sample an estimate of the unknown correlation coefficient in the population is obtained, a point of interest is whether this value is significant or whether this value is so small that it indicates that there is no linear relationship. To solve this problem, appropriate mathematical tools must be applied using the theory of verification of statistical hypotheses.

The statistical hypothesis to be verified is H: $\rho = 0$ stating that there is no linear correlation between the random variables under investigation. To check whether this premise is true, several different schemes can be used.

If $N \geq 3$ and hypothesis H holds than the statistic:

$$t_{N-2} = \frac{R_{X,Y}}{\sqrt{1 - R_{X,Y}^2}}\sqrt{N-2} \tag{3.55}$$

has Student's *t*-distribution with $N - 2$ degrees of freedom.
If the sample size is larger, say $N \geq 50$, then the statistic:

$$\chi_{N-1}^2 = NR^2 \tag{3.56}$$

has χ^2 distribution with $N - 1$ degrees of freedom.

---

[9]Disattenuation of a correlation between two sets of parameters or measures is the estimation of the correlation in a manner that accounts for the measurement error contained within the estimates of those parameters.
[10]There are different varieties of this estimator.

If the sample size is very large, $N \geq 100$, then the statistic:

$$U = \frac{R}{1 - R^2} \sqrt{N} \tag{3.57}$$

has an approximately normal distribution $N(0, 1)$.

Of course, we can obviously use tables with critical values for the Pearson correlation coefficient directly.

A further procedure, associated with the construction of a critical region for the given random variable, can be applied in a similar way to the procedures described in the preceding sections.

Let us now consider the application of these relationships to some problems taken from mining engineering practice. The correlation analysis that will be presented here is much more penetrative in order to acquaint readers with its possibilities.

During the vast number of reliability investigations of main hoists in Poland carried out at the late seventies of the twentieth century, their exploitation processes were identified as a process of the changes of four states:

– work
– repair
– planned standstill (out of disposal time)
– unplanned standstill during disposal time.

Obviously, the occurrence of the states repair, unplanned standstill and work are random, except in the case when a new disposal time in a day commences. In connection with these states, four random variables were of interest:

– time of repair (R)
– time of work between two neighbouring repairs (W)
– time of unplanned standstill between two neighbouring repairs (U)
– time of planned standstill between two neighbouring repairs (P).

The mean times were taken as a representation of these variables, so four average times were ascribed to each hoist. The question was whether these parameters are in some way associated with each other, or to be more precise, answers to these questions were sought:

a. Is there any correlation between the mean time of a repair and the mean time of work between two neighbouring repairs?
b. Is there any correlation between the mean time of a planned standstill between two neighbouring repairs and the mean time of work?
c. Is there any correlation between the mean time of an unplanned standstill between two neighbouring repairs and the mean time of work?
d. Is there any correlation between the mean time of a planned standstill between two neighbouring repairs and the mean time of repair?

Table 3.3a. Correlation coefficients between selected random variables.

State	W	R	P	U
W	R_{WW}	R_{WR}	R_{WP}	R_{WU}
R	R_{WR}	R_{RR}	R_{RP}	R_{RU}
P	R_{PW}	R_{PR}	R_{PP}	R_{PU}
U	R_{UW}	R_{UR}	R_{UP}	R_{UU}

Table 3.3b. Correlation coefficients between selected random variables.

State	W	R	P	U
W	1	0.486	0.648	0.554
R	0.486	1	0.076	0.197
P	0.648	0.076	1	0.588
U	0.554	0.197	0.588	1

Notice, that there is difference between the 'independence of two random variables (times of work and repair) for a given exploitation process of object' and the 'correlation between mean time of repair and mean time of work between two neighbouring repairs' for a set of homogeneous objects—homogeneous in a sense that they execute the same type of job.

Knowing that the correlation analysis will be done in a penetrative way, all correlation coefficients were calculated using formula 3.54. Table 3.3a illustrates the meaning of the notation used, and the results of the calculations are presented in Table 3.3b.

The results of the calculations can also be presented in a matrix form that is easy to construct from Table 3.3b. Let us denote this matrix conventionally by C.

We want to verify statistical hypothesis H stating that there is no correlation in the general population. To check whether this supposition is true, we have to compare the empirical correlation coefficient with its critical counterpart. The sample size is only one for all correlation coefficients, and for this reason the critical value is also only one.

Therefore, for the presumed level of significance $\alpha = 0.05$ and for an alternative hypothesis H_1 that negates the basic H hypothesis, the critical value is $R_{cr}(N - 2) = 0.325$. This value can be obtained directly from critical values table or it can be obtained by applying formula 3.55 and using appropriate critical value from the Student's table.

Looking at the calculated correlation coefficients and comparing them with the critical value, we can say that in the four cases there are reasons to reject the verified hypothesis H; in two cases we have no grounds. This means that the following statements can be formulated in connection with the set of hoists under investigation:

- the mean times of work are positively correlated with the corresponding mean times of repair
- the mean times of work are positively correlated with the corresponding mean times of planned standstill
- the mean times of work are positively correlated with the corresponding mean times of unplanned standstill
- the mean times of planned standstill are positively correlated with the corresponding mean times of unplanned standstill
- the mean times of repair are not correlated with the mean times of planned standstill
- the mean times of repair are not correlated with the corresponding mean times of unplanned standstill.

The last two conclusions are obvious and intuitively expected. The first four conclusions need further investigation.

There was *great pressure* to get mine output as high as possible (because of the possibility to sell coal from Poland to Western countries) at the time in which the hoists were investigated. For this reason, several mines extended their disposal time to achieve higher productivity. Thus, the time available for maintenance of hoisting installations and equipment in shafts was reduced. Researchers in mine faculties suspected that the reliability of hoisting installations would therefore be reduced (although, fortunately, a good safety level in respect of hoists was maintained).

Keeping this mind, it was a supposition that a longer standstill time is good for hoists. A longer planned standstill should have a positive influence on the reliability of a hoist, whereas an unplanned standstill should be neutral. For this reason partial correlation coefficients were included in the analysis and estimated.

If several (say, k) variables have an influence on the variable being investigated and a point of interest is the stochastic relationship between two variables excluding the influence of the remaining variables, a partial correlation coefficient is calculated that can be expressed by the pattern:

$$\rho_{XY.1...(i-1)(i+1)......(j-1)(j+1)...k} = -C_{ij}(C_{ii}\,C_{jj})^{-1/2} \tag{3.58}$$

where C_{ij} is the algebraic complement of the element ρ_{ij} of the determinant of the matrix C.

For three random variables, say, W and R excluding the influence of P, the correlation coefficient is determined by the equation:

$$\rho_{WR.P} = \frac{R_{WR} - R_{RP} R_{WP}}{\sqrt{(1 - R_{RP}^2)(1 - R_{WP}^2)}}$$ (3.59)

and, for the data in hand, $\rho_{WR.P} = 0.575$.

Further calculations of correlation coefficients gave the following outcomes:

$$\rho_{WP.U} = 0.478 \quad \rho_{WR.U} = 0.462 \quad \rho_{RU.P} = 0.189$$
$$\rho_{WU.P} = 0.280 \quad \rho_{RP.U} = -0.05 \quad \rho_{WP.R} = 0.701$$

Due to application of three random variables at this stage of the reasoning, the critical value changes and it can be determined from this formula:

$$R_{cr}(N - 3) = \frac{t_{N-3}}{\sqrt{t_{N-3}^2 + N - k}}$$ (3.60)

This is the appropriate modification of formula 3.55 and t_{N-3} has a Student's distribution with $N - 3$ degrees of freedom. Obviously, all partial correlation coefficients have one critical value which is 0.329. Thus, three correlation coefficients are significant, and three are not significant.

It looks as though major comments in connection with this part of the analysis can be expressed as:

- if *either* the influence of the mean times of a planned standstill between two neighbouring repairs is excluded, *or* the influence of the mean times of unplanned standstill between two neighbouring repairs is excluded, the mean times of work are positively correlated with the mean times of repair
- if *either* the influence of the mean times of an unplanned standstill between two neighbouring repairs is excluded, *or* the influence of the mean times of repair is excluded, the mean times of work are positively correlated with the mean times of a planned standstill between two neighbouring repairs
- knowing that the mean times of work are positively correlated with the mean times of repair and the influence of the mean times of repair is eliminated, then the longer the planned standstill, the longer the mean time of work; the correlation coefficient is very high, reaching 0.701.

To get stronger statements a partial correlation coefficient was applied again, but this time engaging four random variables.

The following partial correlation coefficients were calculated:

$$\rho_{WR.UP} = \frac{R_{WR.U} - R_{RP.U} R_{WP.U}}{\sqrt{(1 - R_{RP.U}^2)(1 - R_{WP.U}^2)}} = 0.554$$

$$\rho_{WR.UR} = \frac{R_{WP.U} - R_{RP.U} R_{WR.U}}{\sqrt{(1 - R_{RP.U}^2)(1 - R_{WR.U}^2)}} = 0.503$$ (3.61)

$$\rho_{WU.RP} = \frac{R_{WU.P} - R_{RU.P} R_{WR.P}}{\sqrt{(1 - R_{RU.P}^2)(1 - R_{WR.P}^2)}} = 0.214$$

The corresponding critical value is now 0.334.

Looking at these figures, in turn, a short engineering summary can be made stating that:

- the longer the mean time of a hoist repair, the longer the mean time of hoist work and *vice versa*
- the longer the mean time of a planned hoist standstill, the longer the mean time of hoist work.

These statements need some elucidation. However, before we come to that, one point requires clarification: the impact of an unplanned standstill on the reliability of hoists. During correlation analysis statements have sometimes been made that this state has an influence, and sometimes it does not. To sort out this problem the multiple correlation coefficient that is directly associated with the coefficient of determination[11] was applied.

At first, the point of interest was the answer to the question: how strong an influence do changes of variables R, P, U have on variable W? The multiple correlation coefficient was calculated using the formula:

$$R_m = \sqrt{1 - \frac{|C|}{C_{WW}}} \quad 0 \leq R_m \leq 1 \tag{3.62}$$

where:

$|C|$ is the determinant of the matrix C
C_{WW} is the algebraic complement of element ρ_{WW} of the determinant of the matrix C.

In the case being analysed we have:

$$C = \begin{vmatrix} 1 & 0.486 & 0.648 & 0.554 \\ 0.486 & 1 & 0.076 & 0.197 \\ 0.648 & 0.076 & 1 & 0.588 \\ 0.554 & 0.197 & 0.588 & 1 \end{vmatrix} = 0.232 \quad C_{WW} = \begin{vmatrix} 1 & 0.076 & 0.197 \\ 0.076 & 1 & 0.588 \\ 0.197 & 0.588 & 1 \end{vmatrix} = 0.627$$

and:

$$R_m = \sqrt{1 - \frac{0.232}{0.627}} = 0.794$$

The coefficient of determination is the square of the coefficient of correlation R_m, that is:

$$R_m^2 = 0.630.$$

Now, we can translate the result obtained into practical terms. It means that changes of variables R, P, U generate 63% of the changes of variable W.

Following this way of thinking, we remove variable U from the reasoning and repeat the calculations. In this case we have:

$$\begin{vmatrix} 1 & 0.486 & 0.648 \\ 0.486 & 1 & 0.076 \\ 0.648 & 0.076 & 1 \end{vmatrix} = 0.386 \quad \begin{vmatrix} 1 & 0.076 \\ 0.076 & 1 \end{vmatrix} = 0.994$$

---

[11] We ignore the fact that different definitions of this coefficient can be found in professional literature here (see, for example, Draper and Smith 1998, Nagelkerke 1991, Everitt 2002).

and:

$$R_m = \sqrt{1 - \frac{0.386}{0.994}} = 0.782$$

which gives $R_m^2 = 0.612$.

Therefore, a conclusion can be formulated stating that changes of variables **R** and **P** generate up to 61% of changes of variable **W**. This allows a final decision to made that the variable **U** should be excluded from further considerations.

Obviously, the correlation coefficients R_m that have been just calculated have corresponding critical values and their significance can be verified formulating appropriate statistical hypotheses.

If so, we are able to state that a longer mean time of a planned standstill between hoist failures gives a longer mean time of work between hoist failures. The opposite affirmation obviously makes no sense. A practical suggestion was formulated advocating that it was a wrong decision to shorten hoist disposal time, as this affects hoist reliability.

A separate problem was how to explain why the mean time of hoist work is positively correlated with its mean time of repair and *vice versa*. Here the points of interest were hoist repairs exclusively and their occurrence in time. Failures were divided into two groups—mechanical and electrical. Failures that did not fit into these categories were small in number and they were neglected. Analysing failures belonging to these two groups, a simple and obvious conclusion was drawn that mechanical repairs took significantly longer than electrical repairs. However, mechanical repairs were rarely required compared to electrical ones. Additionally, the period of hoist operation that was considered in the investigation comprised a period of rapid development in the construction of winders. More electrical and, especially, electronic parts were introduced. Because it was the initial period of their operation, failures occurred more frequently. Hoists of older construction had longer repair periods, but the frequency of their occurrence was lower compared to new hoists with more electronics. These were the reasons for the observed stochastic regularity.

An interesting regularity was noticed while analysing both the mean time of hoist repair versus the date of putting this hoist into service and the mean time of hoist work between two neighbouring repairs versus this date. In both cases the calculated correlation coefficients were significantly negative. This indicated that the frequency of hoist failures was increasing with time, however the mean time of repair was decreasing. All these regularities were generated by the changes in hoist construction connected with electronics.

Reliability investigations that were carried out several years later observed a decrease in the frequency of failures and found that generally the reliability of hoisting installations had increased.

3.9.1 *To summarise*

In the first part of this book, the theory of exploitation has been described with a focus on mine equipment problems. To illustrate the theory in a more penetrative way, some examples seem desirable. However, before the presentation of a description of a few examples of exploitation processes for pieces of mining equipment, the problem of statistical diagnostics has been considered in this chapter. There were two basic reasons to show these considerations.

First, examples of exploitation processes describe mine reality and the models used to analyse these processes. Models always have a certain number of assumptions that must be fulfilled in order to make use of them and, for this reason, we have to verify whether they correspond with mining practice. Statistical tools are extremely useful in this regard.

Second, during this statistical analysis, interesting information can be obtained on the exploitation processes being investigated. Much crucial information is frequently gathered that creates a good basis for making proper decisions in connection with the object. Some further proofs in this regard will be given by considering the wearing processes of hoist head ropes. Taking decisions based on statistical information can have an influence on the design and production of the object.

CHAPTER 4

Exploitation processes—case studies

Recall that the exploitation process of a technical object—according to the classical definition—is the process of changes in its properties during operation. In a wider sense, it is everything that happens to the object from its first application until the moment of the object's final withdrawal from utilisation.

It is obvious that not 'everything' is important from an exploitation point of view. What is essential and what is the object of interest depends on the purpose of the research that will be conducted. Usually, researchers are interested in changes in an object's properties that influence its efficiency and performance, as understood in a broad sense. In mining, two main problems are considered. They are the profitability of the object's exploitation and safety dilemmas associated with this exploitation.

Safety problems have their own specific nature. They must be considered in two different ways.

The first approach is based on statistical data. It has to be rich, in the sense of the number of objects investigated, because the probability of an occurrence of a catastrophic failure in particular object is low, often very low. Therefore, that part of probability theory which deals with rare events must be taken into consideration and the theory of extreme values must be applied (Gumbel 1958, Castillo 1988).

Generally, all failures that can occur in a given piece of equipment can be divided into one of three categories:

- catastrophic events
- potentially catastrophic events
- 'safe' events.

Usually, a probability of the occurrence of catastrophic consequences can be assigned to each failure that may appear in an object. The space of all possible events is divided into three sub-spaces that have this probability of occurrence as the criterion of division. These three types of events can be defined as follows.

A catastrophic failure is an event that causes the loss of human lives, affects human health and/or causes serious damage to the object or its total devastation, and is usually accompanied by disaster in the object's surroundings.

A potentially catastrophic failure is an event which fortunately does not cause any of these consequences, but one for which the probability of the occurrence of fatal results is significant.

A 'safe' failure is one for which the probability of a catastrophe related to the object is non-existent or very low and can be neglected.

The second approach to this theme is based on considerations of particular failures of precise parts or object assemblies that may generate catastrophic events. The construction of these parts/assemblies are analysed comprehensively, together with their possible operation with regard to safety problems.

Safety problems have their own specificity. There are specific mathematical models to deal with them. Consideration of this area lies outside the scope of this book.

Four examples of exploitation processes of mine machines will now be presented to acquaint readers with the way such processes are described. It should be emphasised that each technical object has its own identity and, for this reason, the description of each exploitation process differs from those for other objects. The construction of a model not only allows regularities in the process to be recognised, but also basic exploitation measures of the process to be determined and, further, the efficiency of the object operation to be evaluated.

4.1   AN EXPLOITATION PROCESS OF AN UNDERGROUND MONORAIL SUSPENDED LOCO

The application of monorail suspended systems in underground mines, especially coal mines, has become more and more popular in many countries. Equipment in mines has become heavier and its displacement has created serious problems.

As a solution to this problem, a variety of supplementary transportation systems have been constructed to cope with it successfully. Paths of transportation in these systems are:

- rope
- rail
- floor.

The part that a rope plays in these systems can be described as follows:

- a rope is the path on which small containers with loads are moved
- containers (skips, scrapers, buckets, etc.) fixed to this path are drawn by means of a rope
- different carriers are coupled to a rope and they move together with the rope.

In a drive unit, four different types of propulsions are applied:

- fluid drive (pneumatic or hydraulic)
- winch
- electric engine
- diesel engine.

There are two main transportation systems in use underground, namely:

- rail transportation
- tyre transportation.

There are auxiliary transport systems are associated with each of these main transportation systems.

- Tyre back-up systems are applied in those mines where vehicles with tyres are in wider use.
- Rail transportation is more usually used to deliver heavy units to the loading station of production sections, but transport over the section from the loading station to working faces is done using a different type of supplementary rail transportation. Here two solutions are available depending on the place of the system's construction:

  o roof-mounted systems
  o floor-mounted systems.

A roof-mounted monorail system has many merits, including the option to construct a branched track, a theoretically unlimited range of transportation—the system can cope with steep inclines, the possibility to mechanise reloading works—and the fact that the loco operator is in the front of the train. The basic disadvantages are its dependence on roof conditions—the applied loco is heavy as a rule—and the application of a diesel engine in underground mines. Diesel units generate problems, especially where underground coal mines are concerned. Roof-mounted monorail systems are mainly used to move heavy equipment, such as shearers, ploughs, armoured flight conveyors and powered support units, for longwall coal faces.

Let us examine the exploitation process of a diesel loco applied in this kind of system.

Usually, the crucial questions connected with its operation are:

– *What does the everyday practice of its usage and maintenance look like?*
– *Is it being deployed according to a planned scheme?*
– *Does the loco meet the requirements of a mine?*
– *Are the many problems associated with its exploitation?*

The answers to these questions are important for a mine where this equipment is already in use, for mines where the application of these units is planned, for producers and for constructors of locos.

To get the answers to these questions, field investigations are the best solution. However, they must be well prepared and preceded by a preliminary analysis of the exploitation processes of locos.

Let us consider the exploitation process of a loco as a process of the changes of states. First, an *exploitation repertoire* $\mathfrak{E}$ must be determined.

An **exploitation repertoire** $\mathfrak{E}$ is a defined set of the possible states of a given object (that is, states in which object can be). In a general case, several different exploitation repertoires can be constructed depending on the set of states that are the points of interest. Therefore, if the point of interest is an object's reliability, the set constructed must contain all of the reliability states of the object. A different exploitation repertoire is built if the accomplishment of a given task by the piece of equipment being considered is taken into account, such as a loading task, a hauling task, a dumping task, etc.

A loco, as an object that can be repaired, can be in at least two states. That is a **work state** $\mathfrak{s}_w$, when the object executes its duties, and a **repair state** $\mathfrak{s}_r$, when failure occurs and restoring its ability to function is required. But a loco does not operate continuously. There are some periods when the loco is waiting for a job. This third possible state is called a **standstill** $\mathfrak{s}_s$. The object is able to work but there is no need for its services. It appears that all possible states have been enumerated. However, mining practice gives additional information—a fourth state sometimes occurs. In some mines, this fourth state happens when there is a shortage of spare parts—there is a failure in the loco and repair is required, but it cannot be done because of the lack of the necessary spare parts. The loco must wait until the required part is delivered to the mine. Let us call this state **idleness** $\mathfrak{s}_i$. Therefore, the exploitation repertoire $\mathfrak{E}_1$ of the loco in a general case can be determined as:

$$\mathfrak{E}_1: \; < \mathfrak{s}_w, \mathfrak{s}_r, \mathfrak{s}_s, \mathfrak{s}_i >$$

An example course of an exploitation process of a loco of this type is shown in Figure 4.1.

All states enumerated here—which is important—are separable.

Once information on all the feasible states is available, the next step is a determination of the possible transitions between these states.

When a real course of an exploitation process is examined, it is easy to notice all the possible transitions between states that can occur in practice. These are shown in Figure 4.2. It illustrates a so-called *exploitation graph*. The term '*graph*'[1] used here does not mean any conventional picture.

---

[1] To get some idea on the theory of graphs look to Gould 1988 or Agnarsson and Greenlaw 2006, for example.

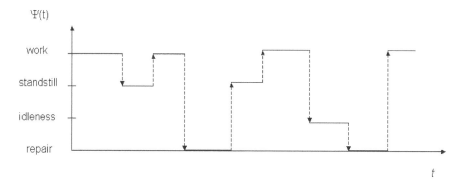

Figure 4.1.   An example course of the exploitation process of a suspended loco.

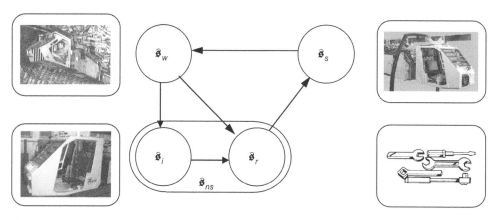

Figure 4.2.   An exploitation graph of a suspended loco.

This term is taken from mathematical graph theory in which the subjects of study are graphs, mathematical structures used to model the pair-wise relationships between 'objects' from a certain collection. A 'graph' in this context refers to a collection of 'vertices' or 'nodes' and a collection of 'edges' that connect pairs of vertices. Graphs used in the theory of reliability and the theory of exploitation are 'directed', which means it is important to know from which vertex it is possible to go to another vertex.

An **exploitation graph** is a graphical representation of the principle of the transitions between states in the exploitation process of an object understood as a process of changes of states (see also Chapter 7.1). Note that Figure 4.1 shows an example course whereas Figure 4.2 points out the principles followed in the process. It contains information on:

- all possible transitions between states
- the directions of these transitions.

If it is required, possible intensities are assigned to each transition to give information about how often a given transition takes place.

There is also a third mathematical way to illustrate process transition principles, namely in the shape of a matrix, although in the case being considered here, there is no need to present it. However, if the process of changes of states is of the Markov type (this will be defined in Chapter 7), construction of a matrix demonstrating a possible changeover at given intensities of these passages is very helpful. By using data contained in the matrix, the construction of equations to determine the limited probability of states can be done. This method will be presented in Chapter 7.1.

At the end of these considerations about the possible states of the object, it should be added that if the point of interest is the loco's ability to fulfil its duties then two states can be joined together, or lumped.[2] These are the idleness state $\mathfrak{s}_i$ and the repair state $\mathfrak{s}_r$—for in both of these states the machine is in failure and it is not able to work.

After lumping:

$$\mathfrak{s}_i \cup \mathfrak{s}_r = \mathfrak{s}_{ns}$$

we have a new state, $\mathfrak{s}_{ns}$ called the *state of the unserviceability of the loco*.

All **states** of the process have their own **properties** that can be expressed by selected measures. Some of them are crucial from a practical point of view but two are most important. They are the:

- frequency of the occurrence of a given state in the process
- relative frequency of the times of a given state.

This first measure is an estimate of the corresponding probability, whereas the second is an estimate of the probability density function.

The rationale for using a loco is stronger for a high probability of the occurrence of the work state, for frequent calls to use a loco and when the probability of a standstill is low. Sometimes, by analysing the exploitation parameters, it is easy to conclude that an additional loco is required.

If the mean time of the work state is long in relation to the mean time of repair, we are inclined to say that the reliability of the machine is high. Going further with our analysis, the standard deviations of the random variables work time and repair time should be analysed. Generally, it is preferable that the dispersions of these variables are low. This is more convenient for engineering planning. However, these dispersions are functions of the types of parts and assemblies used in the construction of the loco. For an object with many electrical and/or electronic parts, both variables can be successfully described by exponential distributions.[3] If there are a large number of mechanical parts and their reliability is important, the probability distribution of repair times does not fit the exponential function. Weibull or gamma distributions are usually applied, and in some cases other functions.

Mines using a great number of locos of the same type from the same producer should pay attention to the values of the reliability parameters estimated from data taken from mining practice. If the dispersion of the values of a given parameter is low and the mean value is high, it can be concluded that the production of locos of this type is stabilised and the machines will be of high quality. For a producer of 'poor' equipment, the dispersion of the values of reliability parameters is high as a rule. However, there are some exceptions to these rules. If a 'good' producer commences to manufacture a certain type of machine, reliability indices of this machine could initially have a wide spread, but usually in a short time the production process improves and the values of parameters are *stabilised* and their dispersion becomes reduced. A similar situation can be noted when there is a modification to an existing type of machine currently in production.

If the fourth state occurs in the exploitation process of a loco, it is a clear indication that the management of spare parts is poor. If this situation is incidental, it can be neglected. If an idleness state appears more frequently, improvement in spare parts management is compulsory—and should be undertaken at once. If no action is taken, the parameters of the machine's performance will decrease significantly.

Generally, the occurrence of an idleness state is connected with poor management in a mine and poor cooperation between the mine and the producer or machine dealer.

---

[2] In addition to lumping, one can find the terms merging, aggregation and consolidation (see, for example, Limnios and Oprişan 2001 p. 46).

[3] The relationships and figures presented here are taken from outcomes of field investigations that are described by Czaplicki 2006a. Giza 2008 obtained similar results.

The values of the exploitation parameters of a given machine depend on the policy on machine usage and servicing and maintenance. This policy has a great influence on the frequency of the occurrence of some states and on the mean values of the times of some states. Every machine producer gives clear indications and recommendations about how the machine should be utilised, what kinds of service should be performed and what kinds of maintenance regime should be provided in order to keep the machine in the appropriate general state. Mine management is responsible for implementing these recommendations and if they are not followed, the efficiency of the machine will decrease.

The so-called *exploitation conditions* (operational constrains) have a great influence in the course of the exploitation process of a technical object.

Everything that is 'around' the object and has some influence on it can be named the exploitation conditions. There is a certain interaction between the object and its environment. In the case being considered, the exploitation conditions include the mine atmosphere underground, the composition and parameters of the air, such as temperature, humidity, the dust content, corrosivity, etc. The stability of the roof has an immense influence on the optimal operation of the suspended loco. Some further components could be listed here but this general principle always holds true: the exploitation conditions of a technical object operating underground are very difficult and that the environment of the object is problematic. All these components are taken into account at the design stage of the object. For these reasons, many of these objects have a variety of redundancies in a reliability sense. These excesses can be parametric, structural, informatics, etc. Their existence (or their lack) frequently determines the degree to which the object fulfils its duties both from the operational and safety points of view. They have a strong influence on the frequency of the occurrence of the object's failures, including those connected with safety. In the case being considered here, the observation of locos was done over a period of 1500 days. During this time 130 repairs were noted along with the corresponding times of repair. All elements that failed were identified, which allowed for a comprehensive failure analysis. All instances where there was a lack of spare parts were also noted. All this information provided the basis on which to make both a reliability analysis and synthesis.

At first, a preliminary statistical analysis was done which comprised assessing the randomness of sequences, outliers tracing, stationarity analysis and autocorrelation analysis. Outliers tracing comprised investigations into whether extremely large values were observed in sequences of work times and repair times. There was no evidence in this regard. The stationarity analysis examined whether the observed sequences of work times and repair times, as well as times when the loco waited for a call to work, increased or decreased stochastically in time. In all cases there were no grounds to reject the hypothesis of stationarity. The autocorrelation analysis tested whether there was a memory of a first, second and third order in the sequence of repair times. In all of the cases investigated there were no reasons to speak of memory in the observed process. Finally, mutual independence was examined, which was a verification of the supposition that the time of a given repair has an influence on the time of work just after this repair. Again, there was no data to validate such an hypothesis.

This analysis permits a statistical synthesis to begin.

First, the essential statistical parameters were estimated for all states (see Table 4.1). These are:

- average times for all states
- mean deviation for all states.

Table 4.1.   Estimates of the basic reliability parameters of a suspended loco.

State	Mean time min	Mean deviation min
Repair	128	87
Work	4540	2990
Standstill	121650	146620

The next step comprised the identification of probability distributions that satisfactorily describe times for given states, one by one, except for the idleness state. Histograms were constructed showing the empirical functions of probability density. The shapes of these histograms indicated that Weibull or gamma distributions should be taken into consideration. Gamma distributions were chosen and their parameters were estimated using Pearson's method of moments. The outcomes obtained were used to improve the estimations by applying the method of maximum likelihood.

Estimates of the parameters of probability density functions were applied to test the hypothesis that probability distribution functions describe empirical data well. The Kolmogorov-Smirnov test was used as a test of the goodness of the fit. In all cases there were no reasons to reject the statistical hypothesis that the observed distances between the empirical and theoretical distributions were non significant. The probability density function of repair times is shown in Figure 4.3 and Figure 4.4 illustrates the probability density function of work times for the machine being investigated.

Up to this point the parameters and functions that have been estimated characterise states individually. Now the characteristics that describe the process of changes of states will be examined. The most important for repairable objects is steady-state availability. According to the classical definition, this is the probability of an event that at any given moment in time the object is in a work state provided that its process of changes of states is of the work-repair type. Therefore, for the case being considered it is necessary to modify this availability measure and consider it in a wider sense.

First, if this probability of work for the loco is to be calculated it is necessary to lump two states: repair and idleness. In both of these states the machine is out of order and is not able to fulfil its

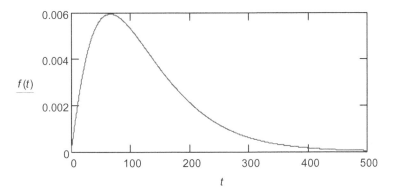

Figure 4.3. The probability density function of repair times of the loco.

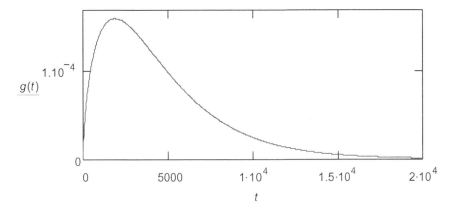

Figure 4.4. The probability density function of work times of the loco.

duties. Therefore, a parameter υ should be determined to indicate how often the idleness state appears in relation to the state of repair. For the investigated loco $\upsilon \cong 21$, which means that every 21 downtimes is connected with a lack of spare parts on average. Thus, the exploitation steady-state availability A_l of the loco defines formula:

$$A_l = \frac{T_w}{T_w + T_r + (T_i / \upsilon)} \qquad (4.1)$$

where:

T_w – the mean time of work
T_r – the mean time of repair
T_i – the mean time of idleness.

For the case analysed, $A_l = 0.434$ which is very low.

Suppose that the producer of the loco assures delivery of the lacking spare part, say, in 48 hours. In this case the exploitation steady-state availability becomes:

$$A_l = \frac{4540}{4540 + 128 + \frac{48 \times 60}{21}} = 0.945$$

The difference is great.

An interesting case arises when spare parts are available without any problem. The steady-state availability reaches $A_l = 0.973$.

The next attractive exploitation parameter is loco utilisation.[4] This can be defined as a ratio:

$$U_l = \frac{T_w^{(\Sigma)}}{T_f} \qquad (4.2)$$

where:

$T_w^{(\Sigma)}$ – total time of loco work in presumed time T_f
T_f – presumed time of loco operation, say, a month.

This measure informs to what extent the loco is needed. The higher its value, the greater the rationale to use this machine. For a high value of U_l, practically for $U_l > 0.85$, the mine should consider the application of an additional unit.

Mines that use locos of a given type should compare the values of the parameters obtained after exploitation investigations with the values of corresponding parameters for locos produced by different manufacturers. This will provide clear information on whether the decision to use the particular units in the mine is a good one.

In the case being considered here the main conclusions were as follows.

- The process of changes of states is a stationary one, which means that the object is properly manufactured in production and its exploitation process is a standard one.
- The states of work and repair are mutually independent and the sequence of repair times has no autocorrelation. This means that there is no unusual process accompanying the process of changes of states; the loco wearing process runs in a standard manner.

---

[4]Note that availability is a property of the object and steady-state availability is the main measure of it (one of many). Utilisation, in the theory of exploitation, is a measure of the process of the exploitation of an object and refers to the degree to which the object has been utilised to execute its duties. The term 'utilisation' is also applied in some other scientific disciplines and has a similar definition. For example, in the theory of queues it is the proportion of a system's resources which is used by the traffic which arrives at it.

- The average time of loco repair is approximately 2 hours and this value is similar to the values of locos made by other producers.
- The average work time of the loco, which is about 76 hours, is a good value; some locos made by competitors have lower work time values.
- The coefficients of variation for both work and repair times, at around 60%, are similar to the coefficients characterising products of competitors.

A special comment, with appropriate recommendations, has been made in connection with the curious situation of lack of spares.

The results of the comprehensive failure analysis which was undertaken at the same time have been omitted in this account because this is a separate issue of an almost purely reliability nature. By the way, the highest frequency of failures was connected with the propulsion mechanism and the main engine, causing about 80% of all failures and accounting for approximately 90% of total time of repair.

4.2   AN EXPLOITATION PROCESS OF A TRUCK IN A SHOVEL-TRUCK SYSTEM

In open-pit mining a truck is an essential piece of equipment but shovel-truck systems are also applied in some opencast enterprises. These systems are common in earth-moving projects. There are thousands of these machines working all over the world.

As a repairable object, a truck can be in two reliability states. These are work state $\mathbf{s}_w$ and repair state $\mathbf{s}_r$. A third state is reserve $\mathbf{s}_0$—the truck waits for a call. It is presumed, as a rule, that in reserve the intensity of truck failures is zero or, to be more precise, the intensity of failures of a truck in reserve is negligible. Because trucks operate in real systems one more state must be considered. Each repair shop for trucks has a certain capacity and its ability to operate is constrained by this fact. Therefore, a queue of failed trucks waiting for repair can sometimes occur. Denote this state by $\mathbf{s}_q$. A queue can occur due to two reasons. First, the number of repair stands is finite and usually not very great compared to the number of units to be serviced. Second, there can be a shortage of some spare parts—this type of case was analysed in Chapter 4.1.

If these four states are considered, an exploitation graph can be presented as in Figure 4.5. Notice that this set of states and their mutual relationships are of interest, first of all, from a reliability point of view. It is worth noting that:

- state $\mathbf{s}_q$ occurs because trucks operate in a system, so state $\mathbf{s}_q$ is not the *own state* of this machine
- when evaluating a truck's steady-state availability and later a truck's system efficiency, state $\mathbf{s}_q$ must be lumped with the state of repair $\mathbf{s}_r$ (in these two states a truck is not able to execute its duties—it is in failure), thus:

$$\mathbf{s}_q \cup \mathbf{s}_r = \mathbf{s}_{ns}$$

and this new state, $\mathbf{s}_{ns}$ is called the *state of unserviceability of truck*.

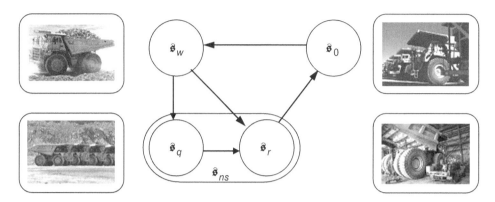

Figure 4.5.   Truck exploitation graph with reliability states.

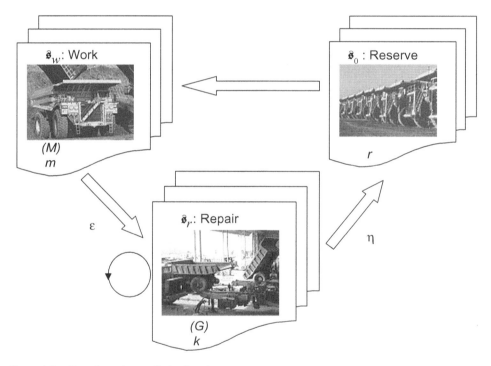

Figure 4.6.   Operating scheme of a truck system.

Truck exploitation has a periodic character and for this reason its operation can be modelled by queue theory.[5] Knowing the range of models that are the subject of analysis in this field, it is easy to observe that the classical repairman problem considered by Palm (1947) fits well in this case. It can be presented in a communicative way by Figure 4.6. In a general case it is assumed that

---

[5] This theory, born at the beginning of the twentieth century, was termed as the theory of queues or waiting line theory. In Central and Eastern Europe this theory has been termed as the mass servicing theory. The latter term seems more appropriate, expressing the essence of matter.

there are many probability distributions of work and repair times and the intensities of transitions between states are determined, say, ε and η. In this picture Kendall's notation is introduced (this system is described comprehensively in Chapter 8.3). To properly read the notation presented in this figure, the letter M denotes exponential distribution. It has been used here because reliability investigations have shown (Czaplicki 1989, Temeng 1988) that the probability distribution of work times can be described satisfactorily by an exponential law. A circle on the left of the repair state denotes a queue; a queue of failed trucks waiting for repair. Such a queue is almost permanent in a badly organised truck system without a back-up facility (sometimes this is the result of a shortage of money). In a well-organised system, a queue can occur from time to time, but does not last long and does not comprise many units.

The exploitation process of truck will now be examined from a functional point of view. A truck's purpose is to transport material and in mining—first of all—to haul the extracted rock. Therefore, it is necessary to consider a truck's accessibility for conducting transportation as well as its inaccessibility. A list of possible states is as follows:

$\hat{\mathbf{s}}_r$ – repair state (failure clearing)
$\hat{\mathbf{s}}_{nd}$ – state of inaccessibility for transporting
$\hat{\mathbf{s}}_{nz}$ – state of incapability for transporting
$\hat{\mathbf{s}}_w$ – work state
$\hat{\mathbf{s}}_0$ – reserve state
$\hat{\mathbf{s}}_{zd}$ – state of ability (and also accessibility) for transporting

and an exploitation repertoire $\mathfrak{E}_t$ is:

$$\mathfrak{E}_t : \; < \hat{\mathbf{s}}_w, \hat{\mathbf{s}}_r, \hat{\mathbf{s}}_{nd}, \hat{\mathbf{s}}_{nz}, \hat{\mathbf{s}}_0, \hat{\mathbf{s}}_{zd} >$$

One state enumerated here needs some more explanation—state $\hat{\mathbf{s}}_{nd}$. It is a well-known fact that a truck does not execute its duties continuously. It needs fuel and having a tank of, say, a few thousands litres capacity, it will be unable to transport for some time while its empty tank is being filled. Changes of drivers also need some time. A driver is not continuously behind the steering wheel for eight or ten hours. Coffee breaks must be provided, especially for safety reasons. Generally, a truck is in an upstate, ready to execute its duties but it is not hauling. From a reliability point of view the truck is in a work state.

The list of possible states set out above are not separable. Their mutual relationships are shown in Figure 4.7. An alternative way of illustrating these states with their relationships is to present an appropriate system of equations in which states are understood as sets. Such a set will be presented when considering the exploitation process for a shovel in Chapter 4.3.

A corresponding exploitation graph for a truck taking into account only separable states will now be constructed. It can be depicted as in Figure 4.8. All transitions between states are possible in practice and this is shown in this figure.

Let us now look at the exploitation process by considering the *truck work cycle*. This is an ordered set of truck working phases with assigned times that are random variables. Each phase characterises a different type of job to be done by the machine. Usually a truck work cycle is understood as a four-stage cycle, load-haul-dump-return (Figure 4.9), or a two-stage cycle, a particular type of service such as load-travel (travel means haul-dump-return) (Figure 4.10).

Field investigations have shown that distributions of the times of work cycle phases can be described satisfactorily by normal probability distributions (cf. Czaplicki 2009).

Examining Figures 4.9 and 4.10 more carefully and comparing them with mining practice, some doubts may arise in connection with truck queues. Theoretically, a queue of trucks can occur at any point on a truck route (neglecting here a queue of failed units waiting for repair). First, note that a queue is a *by-product* of the operation of trucks in the system. When observing shovel-truck systems during their exploitation, one can perceive that a queue may be seen at dumping points, at a crusher, at the point where trucks connect themselves to an overhead wire if a trolley assistance system is applied in a mine and, obviously, at loading machines.

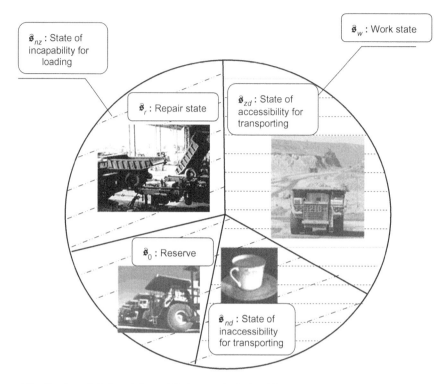

Figure 4.7.   Truck exploitation states and their mutual relationships.

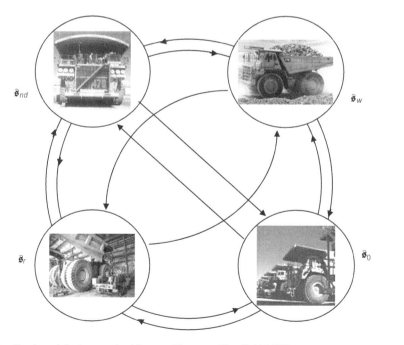

Figure 4.8.   Truck exploitation graph with separable states (Czaplicki 2009).

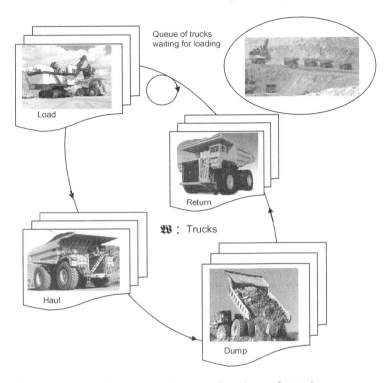

Figure 4.9. Operating scheme of a shovel-truck system—four phases of operation.

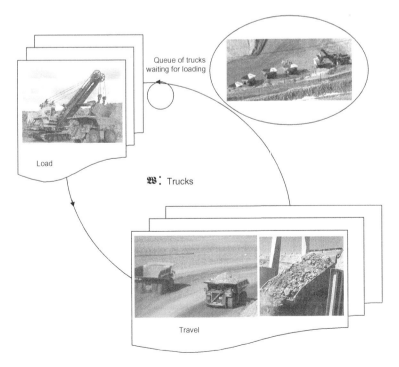

Figure 4.10. Operating scheme of a shovel-truck system—two phases of operation.

Knowing that an analytical description of queuing systems with many points of possible queue occurrence is usually complicated, it can be assumed for simplicity that a queue of haulers will occur before the repair shop and at loading machines. However, there is the option to include an additional queue at a crusher, for instance. This requires the application of a stepped-chain procedure.[6] At first the queue at the crusher is considered, and its effect on the elongation of the mean time of the truck work cycle is estimated. Then further considerations are shifted to loading machines. Here a short queue is advisable because loading machines should operate continuously. Again, this queue extends the mean truck time work cycle. This must be taken into account. Examples of this kind of procedure are presented in Czaplicki's textbook (2004b).

A truck is a repairable machine and therefore the main reliability measure is its steady-state availability. As we know, there is a probability that at any moment in time a truck is in work state. To determine its value the exploitation graph from Figure 4.1 must be taken into consideration.

If there is no queue of failed trucks before a repair shop, that is either the number of repair stands is equal to the number of trucks in the system or the number of repair stands is smaller but large enough to make the average truck queue length negligible (this is the recommended principle to determine the number of repair stands in the repair shop for a given shovel-truck system), steady-state availability is determined by a well-known simple formula:

$$A_{ts} = A_w = \frac{E(T_p)}{E(T_p) + E(T_n)} = \frac{\eta}{\eta + \varepsilon} \tag{4.3}$$

where:

$E(T_p)$ – the expected mean time of truck work
$E(T_n)$ – the expected mean time of truck repair
A_{ts} – the steady-state availability of a truck in the system
A_w – the steady-state availability of a truck.

If a truck operates in a system and a queue of failed haulers waiting for repair must be taken into account, formula 4.3 must be modified to:

$$A'_{ts} = \frac{\varepsilon^{-1}}{\varepsilon^{-1} + E(T_{ns})} \tag{4.4}$$

where A'_{ts} is called the *adjusted steady-state availability* and $E(T_{ns})$ is the expected mean time of the unserviceability of the truck. The mean time T_{ns} of state $\mathbf{S}_{ns}$ is the sum of two components: the mean time of truck repair plus the unconditional mean time spent by the truck in a queue waiting for repair. For the shovel-truck system described by the Sivazlian and Wang $G/G/k/r$ model, both mean values are determined in Chapter 7.2 of Czaplicki's monograph (2009).

The last topic that should be discussed here is the level of a truck's steady-state availability. Producers of trucks assess their products in a way that differs from that of truck users but it is obvious that the latter's evaluation is more important. It seems that following a short assessment a determination may be presented. This is a hint, rather than a strong statement.

Trucks with a steady-state availability higher than 0.75 can be assessed as good. When the value of this measure reaches 0.80, this is very good and above 0.85 it is excellent; however, unfortunately such assessments rare.

Values of availability measures are not consistent with the number of tons transported and kilometres driven. During the first two or three years of operation they should be relatively stable.

---

[6] A similar procedure is well-known in the theory of prediction, see Wold 1964.

Later, a slow decrease in the value of steady-state availability can be observed. For trucks of low quality this decrement can be significant, and it also starts earlier. There are a number of factors that have an influence on the course of the values of reliability and availability measures and usually managers of operation control in mines have good information in this regard, information based on observations of the machines operating in their mines (see, for example, Crawford 1979). It is really vital to obtain this information for a given mine, because this then takes account of the influence of the specific exploitation conditions which exist in this mine on the values of reliability indices. Readers can get some idea about how great an influence reliability indices values have on the efficiency of a machinery system and on the number of repair stands required by studying Czaplicki's monograph (2009).

4.3 AN EXPLOITATION PROCESS OF A POWER SHOVEL

Power shovels like trucks are in common use, and not only in mining. They are reparable objects and for this reason their exploitation process from a reliability point of view is understood as a process of changes of states: work ($\mathbf{\hat{s}}_w$)—repair ($\mathbf{\hat{s}}_r$) type.

The process of their functioning is more complicated.

For shovels to perform their loading task it is important to know whether the shovel is available for loading or not. It is obvious that during the repair state loading is impossible. But there are some periods during the work state when the machine is able to load but does not do so because it is being used to carry out a different operation, such as moving to a new loading point or moving away from the working face because of a large blasting. We can say that in these cases the shovel is in a state of inaccessibility for loading ($\mathbf{\hat{s}}_{nd}$). If we consider the problem of a shovel accomplishing a loading task, we can lump this state and the repair into one—a state of incapability for loading ($\mathbf{\hat{s}}_{nz}$),

$$\mathbf{\hat{s}}_{nz} = \mathbf{\hat{s}}_r \cup \mathbf{\hat{s}}_{nd}$$

Thus, a power shovel exploitation repertoire $\mathbf{\mathfrak{C}}_s$ and the list of all states are:

$$\mathbf{\mathfrak{C}}_s : < \mathbf{\hat{s}}_w, \mathbf{\hat{s}}_r, \mathbf{\hat{s}}_{nd}, \mathbf{\hat{s}}_{nz}, \mathbf{\hat{s}}_{zd} >$$

$\mathbf{\hat{s}}_r$ – repair state
$\mathbf{\hat{s}}_{nd}$ – state of inaccessibility for loading

$\mathbf{s}_{nz}$ – state of incapability for loading
$\mathbf{s}_{w}$ – work state
$\mathbf{s}_{zd}$ – state of ability (and also accessibility) for loading.

These states are not separable. The relationships between the states can be described in terms of set theory and can be presented as:

$$\mathbf{s}_r \cup \mathbf{s}_{nd} = \mathbf{s}_{nz} \quad \mathbf{s}_{zd} \cup \mathbf{s}_{nd} = \mathbf{s}_w$$

$$\sim \mathbf{s}_r = \mathbf{s}_w \quad \sim \mathbf{s}_w = \mathbf{s}_r$$

$$\mathbf{s}_{zd} = \sim(\mathbf{s}_r \cup \mathbf{s}_{nd}) \quad \mathbf{s}_r \cup \mathbf{s}_w = \mathbf{s}_{nz} \cup \mathbf{s}_{zd}$$

In a graphical form they can be illustrated as shown in Figure 4.11.

It appears at first glance that all possible shovel states are enumerated. However, this is not true. One unwanted state is likely to occur when a shovel starts working in tandem with a means of haulage, that is when it becomes an element of a particular system. This state is a stand-still state $\mathbf{s}_s$,—when the machine is waiting idly for a truck. The shovel is in a work state (reliability state) and also in a state of accessibility for loading (exploitation state), but it does not load because of the lack of a transporting machine. The frequency of the occurrence of this state depends on the organisation of the whole machinery system, the number of elements of a particular type in the

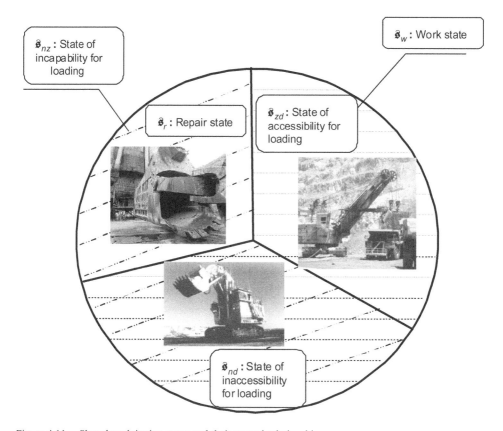

Figure 4.11.   Shovel exploitation states and their mutual relationships.

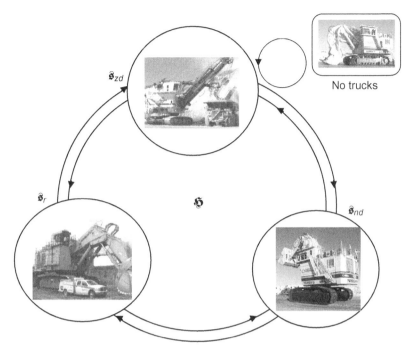

No trucks

Figure 4.12. Process of changes of states of a shovel in graph form: accessibility for loading—repair—inaccessibility for loading.

system, their reliability and the geological and mining conditions of the mine. Thus, an exploitation graph for a shovel can be depicted as in Figure 4.12.

The expression *shovel work cycle* is also vital. This is a measure of shovel actions made in time that are associated with loading, actions that are repeated periodically.

In a system of loading machines spare shovels do not exist. They are too costly, especially the huge units. They should operate 24 hours a day, 7 days a week. However, if one of the loading power shovels operating in an open pit or in a large machinery system of this kind fails, it begins to generate a loss of mine production. For this reason spare loading machines are kept by mines, and as a rule spare wheel loaders are kept. The changes in the machinery system parameters and the improved reliability of the system that results from existence of spare loading units is discussed in Chapter 12 of Czaplicki's monograph (2009). The method of calculating the efficiency of a machinery system is also given.

Power shovels are machines with a complicated construction, however designers have introduced several kinds of redundancies into their structure that have significantly improved their availability. It may be presumed (Sense 1968, Crawford 1979, Czaplicki and Temeng 1989, Sargent 1990) that the steady-state availability for good quality machines is 0.80 and above, and sometimes even exceeds 0.90 (excellent availability). Units below 0.75 can be assessed as machines of low steady-state availability.

4.4   EXPLOITATION PROCESSES OF A CONTINUOUS MINER AND A SHUTTLE CAR

Having established the basic ideas about the construction of exploitation processes for different pieces of mine equipment, let us add a short, slightly simplified consideration for the processes of a continuous miner and a shuttle car. These two machines cooperate with each other.

Consider a system consisting of a certain number of continuous miners operating on a production level and a certain number of shuttle cars hauling the extracted coal and serving the winning machines. This type of system is in frequent use in underground coal mining in the USA.

This system consists of a two subsystems:

a. winning machines
b. hauling machines.

The first subsystem (a) is made up of, say, n extracting machines operating in parallel, that is independently of each other. We may presume in advance that all the machines are identical, so for this reason their technical and exploitation parameters differ only in a random way. From an operational standpoint, two parameters are the most important: the steady-state availability A_{cm} and the accessibility coefficient B_{cm}. Knowing that states of work and repair are independent of each other, we can determine a basic reliability function that is the probability distribution of the number of machines in a state of accessibility to be able to execute their duties. This distribution is given by:

$$P_d^{(zd)} = \binom{n}{d} G_{cm}^d (1 - G_{cm})^{n-d} \quad d = 1,2,\dots,n \quad G_{cm} = A_{cm} B_{cm}$$

where $P_d^{(zd)}$ is the probability that d machines are in a state of accessibility to be able to execute their duties, that is d machines are able to accomplish their task. For more on a system of machines working in parallel[7] see Chapter 6.

Let us now look at a single continuous miner operation. Its regular work cycle consists of two alternative phases: **loading ($\mathbf{\hat{s}}_L$)** and **waiting**. The second phase is the sum of two stages: the continuous miner waits ($\mathbf{\hat{s}}_1$) because a full shuttle car just loaded moves away to a change point plus the continuous miner waits ($\mathbf{\hat{s}}_2$) because an empty shuttle car approaches to the winning machine when a free path has been made by the full hauler; the car moves from the change point to the continuous miner (only one car can be at the winning machine at any one time). If we would like to construct the exploitation repertoire $\mathbf{\mathfrak{E}}$ for the miner three additional states must be taken into account.

First, there will be some periods of time when the machine is able to execute its duties (that is in the up state) but winning action is not being performed because something has happened at the working area or some different actions are being done in this area and, therefore, extraction is stopped. Denote this state as a **standstill, $\mathbf{\hat{s}}_x$**.

The next state that has to be enumerated is a state of **idleness $\mathbf{\hat{s}}_i$**—the continuous miner is in an up state and extraction can be done but there is no shuttle car to load. Obviously, this is an unwanted state but, frankly speaking, unavoidable in practice. The problem is to reduce the frequency of the occurrence of this state and time spent in it. The occurrence of an idleness state can be generated by:

– a failure in a different unit amongst the shuttle cars serving a given continuous miner; a failure that obstructs the flow of cars for a certain period of time
– a shortage of cars because a haulage unit is in repair
– the occurrence of a delay state during shuttle car travelling.

Obviously, the stochastic nature of the process of the operation of the system may generate an idleness state as well.

From a reliability standpoint there is no difference between a state of standstill and a state of idleness, because in both states the intensity of failures of the machine is negligible and can be presumed to be zero. Conversely, from the operational standpoint, a standstill state must exist

---

[7] Notice that exploitation of continuous miners is in parallel but a miner can serve two machines. Sometimes he can walk from the recently completed cut to the next machine, which is set up to mine the next cut. By the way, this walking time must be included in the evaluation of the system's efficiency.

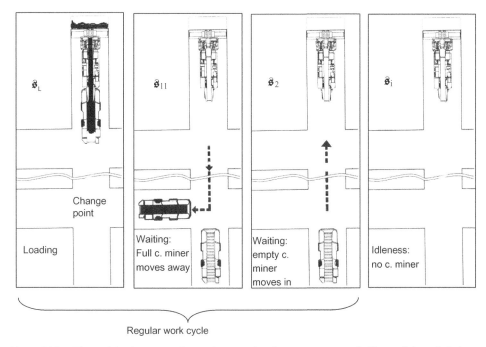

Figure 4.13. The exploitation states of a continuous miner in an up state; standstill state $\hat{\mathbf{s}}_s$ is excluded.

because of the sequence of elemental operations whereas an idleness state should be avoided. Its occurrence means losses.

All these above states are illustrated in Figure 4.13.

The last state that must be taken into account is a **repair** state, $\hat{\mathbf{s}}_r$. Failures sometimes occur in the machine and therefore a down state is inevitable. We neglect here a case where a continuous miner is in failure and it is necessary to wait for a spare part for a couple of days (if such a situation occurs it means we have a new state).

Thus, the exploitation repertoire $\mathfrak{E}_{cm}$ of a continuous miner in a general case can be determined as:

$$\mathfrak{E}_{cm}: < \hat{\mathbf{s}}_L, \hat{\mathbf{s}}_1, \hat{\mathbf{s}}_2, \hat{\mathbf{s}}_r, \hat{\mathbf{s}}_s, \hat{\mathbf{s}}_t, >$$

Possible transitions between these states are shown in Figure 4.14. The regular work cycle states of the machine are indicated by the triangle.

Let us a make few brief comments.

From a reliability standpoint the states idleness ($\hat{\mathbf{s}}_t$), waiting ($\hat{\mathbf{s}}_1 \cup \hat{\mathbf{s}}_2$) and standstill ($\hat{\mathbf{s}}_s$) have the same meaning—the intensity of the machine failures is null.

When a continuous miner is executing its duties, that is extracting and loading (state $\hat{\mathbf{s}}_L$), it can fail—the intensity of failures is positive.

The times of a continuous miner work cycle without any losses, that is ($\hat{\mathbf{s}}_L \rightarrow \hat{\mathbf{s}}_1 \rightarrow \hat{\mathbf{s}}_2$) can be satisfactorily described by a Gaussian probability distribution.

The majority of repairs of a continuous miner involve a regeneration in the exploitation process of the whole machinery system.

Remember to take into consideration the fact that after a certain period of time a cut is completed and mining will be restarted after the necessary and appropriate mining actions are performed.

This last remark is significant. It means that the exploitation process of a continuous miner can be identified as an intermittent stochastic process, similar to the process associated with mine hoists, for example. When a cut is completed the exploitation process of the extracting machine is terminated

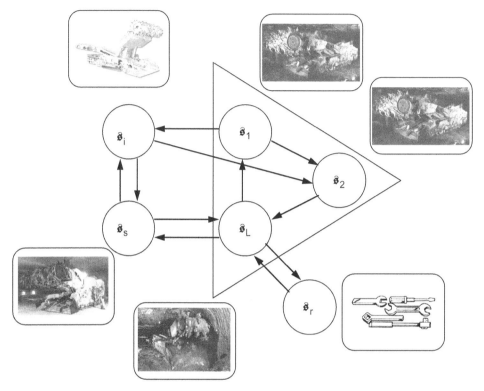

Figure 4.14.   An exploitation graph of a continuous miner.

for the time being and the process regenerates itself. A simple immediate conclusion can be drawn— the exploitation process of a shuttle car also has an intermittent character with regeneration.

Let us look now at a single shuttle car. Generally, its regular duties consist of four stages repeated cyclically similar to that of the majority of haulers, namely: load—haul—dump—return. However, it is advisable to look at this work cycle more attentively.

Let us start from the loading phase. The next step is the withdrawal of the full machine from the working face area and hauling the load to the dumping point through the change point (**CP**). But now, unlike the previous machine, it is not necessary to distinguish this point during the phase being considered. The haulage should be treated as one phase. The next phase is obviously unload-ing and subsequently it is going return to the machine, but only up to the change point. Now, it is essential to discern this fact. Depending on the size of the system, but quite often just before this point, the machine will be waiting for a free pass to join the continuous miner. Notice, the situation here is slightly different from the majority of systems of the loader-hauler type. In a shovel-truck system, for example, a queue of dumpers is formed at the loading machine. In the case being ana-lysed a queue of haulers can be formed before a change point. Thus, it looks like the operation of a shuttle car can be presented in a graphical form as shown in Figure 4.15.

An exploitation repertoire for this machine consists of 8 states:

$$\mathfrak{C}_{sc} : < \mathfrak{s}_L, \mathfrak{s}_r, \mathfrak{s}_h, \mathfrak{s}_u, \mathfrak{s}_3, \mathfrak{s}_d, \mathfrak{s}_4, \mathfrak{s}_5 >$$

where all states are explained in Figure 4.15 except for the state $\mathfrak{s}_r$ of the car's repair.

Looking at Figure 4.15 it is easy to conclude that a shuttle car's time work cycle T_c is the sum the following components:

$$T_c = T_l + T_h + T_u + T_3 + T_{cp} + p_d T_d + p_w T_4 \tag{4.5}$$

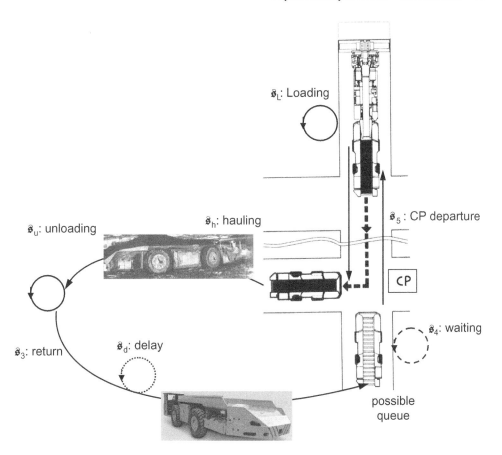

Figure 4.15.   Operating scheme of a shuttle car.

where:

T_l – time of loading
T_h – hauling time
T_u – time of unloading
T_3 – time of return from the dumping point to the change point
T_{cp} – tram time from the change point to the face
T_d – time of delay
p_d – probability that wasting time will occur during a work cycle
T_4 – time of waiting at the change point
p_W – probability that waiting time will occur during a work cycle.

Some explanations connected with these parameters are necessary.

Let us commence our discussion from a general statement that a shuttle car's time work cycle is a random variable that is the sum of several different random variables; two of them have their own probability of occurrence. Let us analyse the components of the sum.

Time of loading. This random variable has a probability distribution whose shape is a symmetric bell form similar to the normal distribution. Moreover, this distribution is double-sided truncated. The left side truncation is clearly specified at zero, but the right side truncation is hard to define univocally. If the dispersion of this variable is relatively low (this usually holds in practice), the regular Gaussian distribution can be applied.

Hauling time from the winning machine to the unloading point. This is a homogenous random variable; its probability distribution is a symmetric bell form similar to the normal distribution. Obviously, this distribution is double-sided truncated; however, the right side truncation is again hard to estimate precisely. The left side truncation—at zero—is usually not so important because the mass of probability is shifted away from the inception point. This distribution can be approximated by a Gaussian function.

Time of unloading. This random variable is again similar to the normal distribution. It can be approximated by this type of function.

Time of return from the dumping point to the change point. This random variable has an identical character to the time of haulage from the change point to the dumping point.

Sometimes a state $\hat{s}_d$ may occur associated with time of delay. We neglect here the problem of the car failures (state of repair, $\hat{s}_r$). It often happens that a car's work cycle is disturbed by miscellaneous factors (Bise 2003 Chapter 9). All events generating car stoppage for a certain period of time or causing the speed of the car's motion to be reduced increase the time of the work cycle. Here two problems must be taken into account. First, frequently a car cycle is executed without any disturbance but during some other cycles disturbances occur and the time of the work cycle lasts longer. Thus, it is necessary to estimate the probability of the occurrence of a disturbed cycle p_d—the probability that wasting time will occur during a work cycle. Second, we have to add this elongation to the time of the cycle—the time of the total delay during the work cycle. Estimation of the probability p_d is simple if there is some exploitation data at hand but an important problem is the identification of the probability distribution that may be associated with this time of delay. Having no operational information on this issue, we may consider this problem by looking at it from theoretical standpoint. It looks as though very often the time of the delay will be short, rarely will it be long. Thus, it seems that a correct proposition may be the application of exponential distribution.

After the arrival of the empty shuttle car to the change point an exploitation situation can be twofold: either the continuous miner is free and there is no other shuttle car in the working area or there is no way to approach the continuous miner because this area is occupied. The probability of each event must be evaluated and for this reason we have probability p_W in the pattern, which is the probability that waiting time will occur during the car's work cycle. Similar to the previous reasoning, we need information on the probability distribution of the time of waiting.

The last state to discuss is the tram time from the change point to the face. This time is usually short, its variation is not so great and even the corresponding probability distribution can be assumed as uniform supported on a certain time interval. However, to make our analysis more practical and simplified, it can be assumed that this probability distribution is normal.

If we look at the work cycle of a shuttle car more thoughtfully, we can come to the conclusion that a *sensitive point* is the change point and the whole investigation of the machine work cycle should be focused on this point. Thus, the cycle sequence of states without any losses is:

$$\hat{s}_5 \rightarrow \hat{s}_L \rightarrow \hat{s}_h \rightarrow \hat{s}_u \rightarrow \hat{s}_3$$

Consider formula 4.5. If it is assumed that the times that are the components of the right-hand side of the equation are average ones, the left-hand side of the equation determines the average time of the shuttle car's work cycle. Otherwise, the right-hand side of the equation shows the sum of seven random variables, which means that the left-hand side of the equation also defines a random variable. The question now is: what can we say about the probability distribution of this random variable?

Try to make use of the knowledge about the probability distributions of the components of the sum. The first five random variables have probability distributions of a normal type, thus their sums are random variables also having a normal probability distribution of appropriately defined parameters. The next component is the random variable of exponential distribution.

Let us rewrite formula 4.5 in the following form:

$$T_c = T_n + p_d T_d + p_w T_4$$

where T_n is the normal random variable of the mean, which is the sum of all the average values of all five times (T_l, T_h, T_u, T_3, T_{cp}), and the variance, which is the sum of all the variances. Let us denote the probability density function of this variable by $f_n(t)$.

If the next component of the sum is added to the variable T_n, the component exponentially distributed and weighted by the probability p_d of the probability density function:

$$f_d(t) = \frac{\varphi}{p_d} e^{-\frac{\varphi}{p_d} t}$$ (4.6)

then the probability density function of this sum is determined by:

$$f_s(t) = \int_0^\infty f_d(t - y) f_n(y) dy$$ (4.7)

provided that the normal component function is specified over the positive interval only.

It is hard to show this function in an explicit form. However, in practical terms, this additional random variable makes a certain small displacement away from zero of the mass of probability determined by $f_n(t)$—see Figure 4.16 in which a long travel time was presumed. This fact of a small displacement will be useful in further analysis.

The last component to identify is the time of waiting at the change point. This random variable is a function of the following components:

– the reliability of the equipment involved
– the number of haulers used to serve the shuttle car
– times of phases of the work cycle of the shuttle car.

For a given machinery system, a sample can be collected and the probability distribution of interest can be identified. However, when workings are planned on a given production area, it is necessary to evaluate the system which will be used in this area and the number of machines that will be in operation.

We can look at the system of a continuous miner and shuttle cars as a cyclic system and we can try to describe its operation by making use of one of the models taken from queue theory (see Chapter 8). Before beginning to search for an appropriate model it is important to remember that:

– there is only one service point (loading machine) in this system
  the car queue is formed facing the change point and queue length is limited

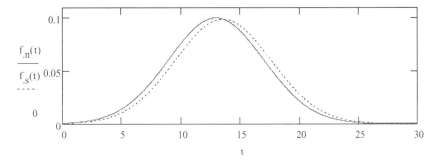

Figure 4.16.   An example course of the probability density functions: $f_n(t)$ and $f_s(t)$.

– machines are loaded in the sequence of their arrival
– the shuttle car repair shop is of a limited capacity, however the number of haulers to repair is relatively small
– there are no spare hauling machines
– shuttle cars are driven electrically via a power cable and this must be kept in mind because it has an influence on the sequence of their elemental operations
– battery-powered and diesel-powered rubber tyre haulage vehicles are not constrained in their movement.

These features of a cyclic mine system must have their corresponding features in the selected model. By applying a *good* model, information on the probability distribution of the number of machines waiting for loading can be obtained. The situation is more practical here than for the large machinery systems of this kind in surface mining because there is only one service point (for many service points of a given cyclic system we only have information on average characteristics). Knowing this distribution we can convert it into the probability distribution of the time of waiting in the queue.

At the very beginning of the investigation in this section (Chapter 4.4), it was stated that the analysis was to be simplified. Two issues have not been discussed.

First, in reasoning connected with the fact that with shuttle cars sometimes a situation occurs in which a given car is directed to the next cut to serve a different continuous miner. This has not been taken into account.

Second, some machinery systems of this kind can be reduced when the cut to be mined is near the belt entry. For such a case, the haulage 'system' may consist of two (or even one) cars and the elemental distances driven are very short.

Let us construct—at the end of these investigations—an exploitation graph for a shuttle car in a general case. This is shown in Figure 4.17. By analysing the exploitation graph presented in Figure 4.14 and the graph in Figure 4.17, and having identified the measures describing the transitions between states, several essential characteristics of the operation of these machines can be constructed.

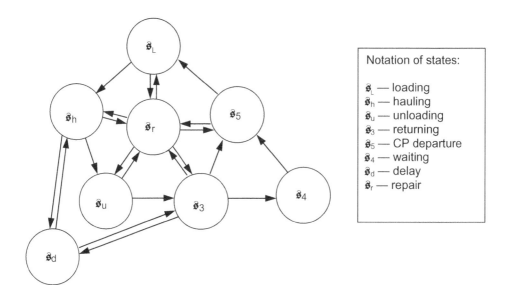

Notation of states:

s_L — loading
s_h — hauling
s_u — unloading
s_3 — returning
s_5 — CP departure
s_4 — waiting
s_d — delay
s_r — repair

Figure 4.17.   A shuttle car exploitation graph.

CHAPTER 5

Reliability models applied for a single piece of equipment

As has been stated, mining engineering deploys a wide range of different technical objects of various constructions which are applied for diverse purposes. For these reasons a large variety of reliability models are used. However, some models have common application.

In this chapter three reliability models of mine equipment will be presented, covering equipment that is in widespread use in mining. The models cover belt conveyors, hoisting installations and the head ropes of a hoist. However, the models presented in connection with these units can also be applied to describe the reliability of different pieces of mine equipment. It is easy to recognise that the first two are repairable objects, whereas head ropes are unrepairable.

5.1 BELT CONVEYORS

There are some pieces of equipment that are heavily used in mining, both in underground and surface mines. There is no doubt that in addition to loaders and trucks, belt conveyors can be counted in this class of objects. Their 'popularity' is directly connected with their wide range of merits. The main ones are a small transverse section area, low resistance to motion, electrical driving force, possibility of automation, very high output, low operating cost, high safety, light construction and usually the very high reliability of a single unit.

It has to be added that approximately 80% of the energy consumed by a conventional conveyor is used to move the material being transported, and only 20% is used to overcome resistance to motion. By comparison a haul truck, for example, consumes almost 60% of its energy to move itself—so a belt conveyor is really valuable transport mean. Its main weaknesses are associated with the belt itself: it is very expensive and sensitive to the properties of the bulk being transported. These disadvantages reduce the application of this type of conveyor. Belt conveyors are also elements of some compound machines used in mining such as bucket wheel excavators, bucket chain excavators, stackers, etc. For this reason, in some cases they have difficult exploitation conditions—for example, in receiving a recently excavated and extracted stream of rock. The intensity of failures of these 'compound' conveyors is considerably higher (Aiken 1996, Wolski and Golosinski 1986) than the intensity of failures of conveyors operating in a series as the main haulage system in both underground and surface mines.

A conventional belt conveyor usually operates 24 hours a day carrying extracted rock. It is still in motion even when the stream of bulk being hauled is zero during a certain period of time. In a majority of cases, failures rarely occur—they are as low as approximately a few times per year on average. Therefore, there are two reliability models used to describe the operation of these units from a reliability standpoint. They are a stream of failures model and a process of changes of states of the work-repair type.

5.1.1   *Stream of failures model*

Work time between two neighbouring failures of belt conveyors is usually very long. Repair time, compared to work time, is quite short as a rule. Therefore, if repair times of a conveyor are not important in a given investigation, these can be neglected. It can also be presumed that after repair the unit is as good as new. The process of object exploitation from a reliability standpoint is as follows.

At a certain moment in time, a conveyor starts its duties. After time t_1 of work a failure occurs. This first work time is t_{p1}, $t_{p1} \equiv t_1$. Renewal is instantaneous in this model and it is assumed that its length is zero. The conveyor commences its duties for a second time. In time t_2, $t_2 > t_1$, a new failure occurs. The work time between the first failure and the second one is t_{p2}, $t_{p2} = t_2 - t_1$. The conveyor again starts to work. It can be assumed that random variables t_{p1}, t_{p2}, ... have the same probability distributions. Let us denote this by $F(t)$ and obviously:

$$F(t) = P\{t_{pn} < t\}.$$

The moments of failure occurrence (see Figure 5.1) are:

$$t_1 = t_{p1}, \quad t_2 = t_{p1} + t_{p2}, \dots, t_n = \sum_{i=1}^{n} t_{pi}$$

They illustrate a random process called a stream of failures[1] (Smith 1958, Gnyedenko et al. 1965).

Usually, the main point of interest for this stream is a random variable $v(t)$—the number of failures that occurred up to the moment t, $t_{v(t)} \leq t < t_{v(t)+1}$. It is easy to see that the following equations hold:

$$P\{v(t) \geq n\} = P\{t_n < t\} = P\{t_{p1} + t_{p2} + t_{p3} + \cdots + t_{pn}\} = F_n(t) \tag{5.1}$$

$$F_n(t) = \int_0^t F_{n-1}(t - t_w) dF(t_w) \quad F_1(t) = F(t) \tag{5.2}$$

$$P_n(t) = P\{v(t) = n\} = F_n(t) - F_{n+1}(t) \quad P_0(t) = 1 - F(t) \tag{5.3}$$

These equations determine the random variable $v(t)$ well. A very important statistical function that is of great value in practice is the renewal function $H(t)$, which determines the expected number of failures up to the moment t. It can easily be proved (see, for example, Gnyedenko et al. 1965) that

---

[1] If a stochastic process has a continuous parameter, say, time and a discrete space of events (here failures), such a process is termed a stochastic stream.

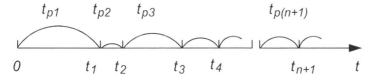

Figure 5.1.   A stream of failures.

$$H(t) = \sum_{n=1}^{\infty} F_n(t) \tag{5.4}$$

It should be added that all basic probabilistic characteristics of the stream are expressed by means of this function. For example, the variance of the number of failures up to the moment *t* is given by the formula:

$$\sigma^2 [\nu(t)] = 2\int_0^t H(t - t_p)dH(t_p) + H(t) - H^2(t) \tag{5.5}$$

The average number of failures in a time interval (t_1, t_2) is obviously

$$\bar{\nu}(t_2, t_1) = H(t_2) - H(t_1) \tag{5.6}$$

For reliability researchers who are interested, a function called the density of renewal *h(t)* is determined by the equation:

$$h(t) = H'(t)$$

which yields the following equations:

$$h(t) = \sum_{n=1}^{\infty} f_n(t) \quad f_n(t) = F_n'(t) \tag{5.7}$$

This way of modelling is good if the exploitation process of an object is characterised by seldom failures, which means that the object is of high reliability. Thus, in many cases this model is applied to investigate a series of failures of cable belt conveyors. This modelling approach is also excellent when rare events are observed, such as catastrophic or potentially catastrophic events, and the object of interest is a set of single units working in parallel. In the late 1970s this model was applied to describe a series of catastrophic events involving hoisting installations in the Polish mining industry (Czaplicki and Lutyński 1976).

From a theoretical standpoint the convolution of functions $F(t)$ can only be expressed in an explicit form in a few cases. Fortunately, in all the above cases exponential distributions are applied and all functions connected with the stochastic stream have a precise open form. Thus, if

$$f(t) = \lambda e^{-\lambda t}$$

The main functions are determined by the equations:

$$P_n(t) = P\{\nu(t) = n\} = \frac{(\lambda t)^n}{n!} e^{-\lambda t}$$
$$H(t) = \lambda t \quad h(t) = \lambda \tag{5.8}$$

Amongst all series of failures only the above process, called the Poisson process, is a stationary one and is memoryless.[2]

Sometimes the point of interest is only a particular type of failure. In such a case different types of failures are neglected. These failures are 'dummy' failures. A new random variable appears in this problem—a number w of dummy failures up to the moment of the occurrence of the particular failure.

Generally, a sum of w random variables exponentially distributed with parameter λ has an Erlang distribution of the order w for which a probability density function is given by:

$$f(t) = \frac{\lambda^w t^{w-1}}{(w-1)!} e^{-\lambda t} \tag{5.9}$$

In the case being considered parameter w, $w \in \mathfrak{R}$, is a random variable of a certain distribution $p(w)$. To get the unconditional probability function, it is necessary to calculate:

$$g(t) = e^{-\lambda t} \sum_w \frac{\lambda^w t^{w-1}}{(w-1)!} p(w) \tag{5.10}$$

It is easy to observe that from a theoretical standpoint the problem of finding the probability density function is trivial one. Conversely, from a practical point of view it is a serious problem. In order to estimate the function $p(w)$, data is needed, which in practical terms means that a given conveyor should be operated for a very long time (many years) to observe the appropriate number of failures of a given type. In many cases there is no way to fulfil this condition, as mine systems do not operate for such a long time. For this reason instead of function $p(w)$ the corresponding expected value is taken into account. Data gathered to estimate this average value is based on recording the numbers of failures of several conveyors. Obviously, their exploitation processes must be homogeneous.

Because the renewal function $H(t)$ is of great importance for reliability calculations, it is advantageous to have some approximations of this function.

- If probability distribution function $F(t)$ is known, then:

$$F(t) \le H(t) \le \frac{F(t)}{1 - F(t)} \tag{5.11}$$

- If mean time of work T_p is known, then:

$$\frac{t}{T_p} - 1 \le H(t) \le \frac{t}{T_p} \tag{5.12}$$

- For longer time t:

$$H(t) \approx \frac{t}{T_p} + \frac{\sigma_p^2}{2T_p^2} - \frac{1}{2} \tag{5.13}$$

where σ_p^2 is the variance of work time.

---

[2] A process is memoryless if the probability of the occurrence of k failures in the interval $(T, T + t)$ does not depend on how many failures and in which way they have occurred up to the moment T (Gnyedenko and Kovalenko 1966).

Figure 5.2. A stream of failures and residual work time.

Information that a random variable $v(t)$ has an asymptotically normal distribution[3] is of practical use:

$$N\left(\frac{t}{T_p}, \sqrt{\frac{\sigma_p^2 t}{T_p^3}}\right)$$

(5.14)

and an estimation of the number of failures over a long time interval can be made.

For large t the following equation holds:

$$P\left\{\frac{t}{T_p} - u_{\alpha/2}\frac{\sigma_p\sqrt{t}}{T_p^{3/2}} < v(t) < \frac{t}{T_p} + u_{\alpha/2}\frac{\sigma_p\sqrt{t}}{T_p^{3/2}}\right\} = 1 - \alpha$$

(5.15)

where u_α is a quintile of the order α of the standardised normal distribution.

One side assessment can also be made by appropriately modifying the inequality in brackets.

For some technical objects for which the exploitation process is described by a series of failures, an interesting characteristic is the residual work time ζ (see Figure 5.2). If an exploitation process runs over a certain period of time, and the object is in an up state at a particular moment in time, then a point of interest is the remaining work time up to the nearest failure occurrence. Usually this characteristic concerns the process of object exploitation after a long time, which means that the stochastic properties of the process have been stabilised.

The probability function of this residual work time is determined by the equation:

$$P\{\zeta > \tau\} = \frac{1}{T_p}\int_\tau^\infty [1 - F(t)]dt$$

(5.16)

The expected mean of residual work time is determined by equation:

$$T_{res} = \frac{T_p}{2} + \frac{\sigma_p^2}{2T_p}$$

(5.17)

One can find further interesting approximations by studying these publications: Smith (1959), Gnyedenko et al. (1965) and Kopociński (1973).

5.1.2 *Process of changes of states: work-repair type*

This type of process is well-known in the theory of exploitation; however, following the nomenclature used in the previous section, this process should be called a renewal process with a finite time of repair. States are alternative; nevertheless, a long time ago some investigations were oriented

---

[3] Notation $N(m, \sigma)$ denotes normal distribution where the mean equals m and the standard deviation σ.

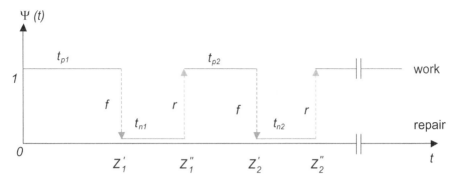

Figure 5.3.   Process $\Psi(t)$ of changes of states: work—repair.
Key words: *f*—failure, *r*—renewal, t_p—work time, t_n—repair time.

toward a non-alternative process, with regeneration resulting from object surveys done periodi-cally (see, for example, Czaplicki[4] 1985).

Almost all machines and technical devices applied in mining are repairable objects and, if repair time is not to be neglected, the process of changes of states of the work-repair type finds applica-tion to describe the exploitation process of these units. It is important to be aware of the fact that this type of process describes the reliability properties of the object well, but very often more states are of interest from the exploitation standpoint (from the object functioning standpoint, for example) and the whole process is more complicated.

This process can be determined in the following way.

A certain object starts its duties. It is good and able to fulfil its requirements. It commences work and after a certain period of time t_{p1} a failure occurs. A repair is needed. It is done in time t_{r1} and after the repair the object is reinstated—it starts its duties again. After time t_{w2} a new failure occurs and the repair takes time t_{n2}. A new reinstatement is noted and work commences once more.

It is assumed that random variables t_{p1}, t_{p2}, ... have one probability distribution:

$$P\{t_p < t\} = F(t) \quad i = 1, 2, ... \tag{5.18a}$$

and all random variables t_{n1}, t_{n2}, ... also have one probability distribution:

$$P\{t_n < t\} = G(t) \quad i = 1, 2, ... \tag{5.18b}$$

A further assumption is that these random variables are mutually independent and their basic statistical parameters are known:

$$E(t_p) = T_p = \lambda^{-1} \quad E(t_n) = T_n = \mu^{-1}$$
$$D^2(t_p) = \sigma_p^2 \qquad D^2(t_n) = \sigma_n^2 \tag{5.19}$$

The process can be defined as:

$$\Psi(t) = \begin{cases} 1 & Z_n'' < t \leq Z_{n+1}' \\ 0 & Z_{n+1}' < t \leq Z_{n+1}'' \quad n = 0, 1, ... \end{cases} \tag{5.20}$$

---

[4] The model described in this paper was a trial of the depiction of a certain exploitation process of mine hoists.

Look at equations 5.19 more carefully. Parameter T_w is the average work time, whereas its reciprocal λ determines the intensity of the object's failures (failure rate). Parameter T_n is the average repair time, whereas its reciprocal μ determines the intensity of the object's repair. It is important to remember that all these parameters have dimensions. The dimensions of both intensities are a time unit to the power -1. Let us read that if λ is, say, $1/100$ it means that one failure occurs after 100 hours of the object's work on average. If μ equals, say, $1/2$ it means that a failure is cleared up after 2 hours of repair on average.

Remember that formulas 5.18 and 5.19 determine the states quite well separately. Now it is time to construct some formulas that describe the properties of the whole process.

The most important is the probability:

$$A(t) = P\{\Psi(t) = 1\}, \tag{5.21}$$

which is the probability of an event that the object is in a work state (up state). This probability is a function of time.

Measure 5.21 is called *pointwise availability* (see Kodama and Sawa 1986 or Malada 2006); this term has been used since the late 1950s (Hosford 1960). According to some authors an alternative term for this probability is *instantaneous availability* (Elsayed 1996, www.weibull.com 2007).

It can be proved (see, for example, Gnyedenko et al. 1965, Kopociński 1973) that:

$$A(t) = 1 - F(t) + \int_0^t \left[1 - F(t-x)\right] dH_\Phi(x) \tag{5.22}$$

where:

$$H_\Phi(t) = \sum_{n=1}^\infty \Phi_n(t) = \sum_{n=1}^\infty P\left(Z_n'' < t\right) = \sum_{n=1}^\infty \int_0^t F_n(x-u) dG_n(u)$$

These formulas are almost never used in regular exploitation practice. The only exception is their application in the operation of the readiness systems—rescue units—employed in mining when a roof collapse calls for instant emergency action to release miners trapped by this dramatic event.

The limited value of the function $A(t)$ called **steady-state availability**, **long-run availability** or **limiting availability** has found a very wide application in engineering practice (see, for example, Gnyedenko et al. 1965, Ryabinin 1976, Kiliński 1976, Adachi et al. 1979, Beichelt and Fischer 1979, wikipedia 2007). This parameter is defined by the well-known formula:

$$A = \lim_{t \to \infty} A(t) = \frac{E(t_w)}{E(t_w) + E(t_r)} \tag{5.23}$$

In many English-language engineering publications and elaborations this measure is known in short—unfortunately—as availability.

Availability is the property (ability, feature) of objects that can be repaired, that is the exploitation process from a reliability standpoint is a two-state one: work—repair (up state—down state). However, some authors (see Bobrowski 1985 p. 77) extend this idea to unrepairable objects stating that the availability of an object that cannot be repaired at a given moment in time $t > 0$ is equal to the reliability of this object in time interval $[0, t]$. In this way the idea of availability can be enlarged to include systems consisting of unrepairable elements. This reasoning can be helpful when considering some of the electronic systems being applied in mining.

It looks interesting to translate the mathematical notation $t \to \infty$ into engineering reality. A point of interest here is the answer to the question: how quick does the function $A(t)$ tend to its limited value A?

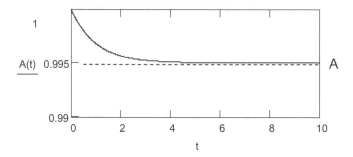

Figure 5.4.   Function $A(t)$ and its limited value A in relation to time.

A reliability investigation carried out in the mid-1970s in the Polish mining industry (Antoniak et al. 1973–75) showed that for the majority of machines the time of stabilisation is short—a few hours, as a rule (see Figure 5.4). Therefore, the use of a limited value is completely rational.

As in the case of reliability, several different measures for availability have been defined. Notice, that if we discuss and analyse a certain idea (term) and several different measures of it have been defined, it means—as a rule—that this idea determines a certain object property (ability, capability). Do not identify the measure with the property that it concerns. These two things are different; however, measures define this ability to a certain degree.

Generally, from a mathematical standpoint, **all measures of availability are probabilities**.

The first papers on availability appeared in the early 1960s together with reliability considerations (Hosford 1960, Bielka 1960, Malikov et al. 1960). Nagy (1963), Bailey and Mikhail (1963), Gnyedenko et al. (1965), Thompson (1966), and Gray and Lewis (1967) dealt with the problem of the estimation of availability from statistical data. Brender's (1968) two papers gave a comprehensive lecture on the definition of availability by defining steady-state availability, point availability, steady-state mission availability, steady-state repair availability, transient point availability, steady-state availability of the second kind, transient mission availability and steady-state repeated demand availability. Problems connected with availability were seen as so vital that in 1971 alone several publications considered this issue, such as Martz, Kodama et al., Nakagawa and Goel, Das, and McNichols and Messer. A few years later the term *availability theory* was formulated (see Baxter 1985). Bobrowski's extension (1985) supports Baxter's proposition.

Equation 5.23 determines **the probability of an event that the object is in a work state at any moment of time** but it concerns the process of work-repair exclusively. This measure is not connected with a precise moment t at all.

There is no doubt that this is one of the most important reliability/availability parameters of repairable technical objects. Its estimation is usually given by the producer in the list of the product's main parameters. This information is very useful for an estimation of an object's production cycles and, later, an estimation of an object's effective output and other measures of an object's efficiency of exploitation.

Now we introduce a non-dimensional parameter, the ratio of expected values and obviously a ratio of intensities:

$$\kappa = \frac{E(T_n)}{E(T_p)} = \frac{\lambda}{\mu} \tag{5.24}$$

This quotient is called a repair rate or (rarely) a fault coefficient (see, for example, Ryabinin 1976). Its relationship with steady-state availability is a strict one because:

$$A = \frac{\mu}{\lambda + \mu} = \frac{1}{1 + \kappa} \tag{5.25}$$

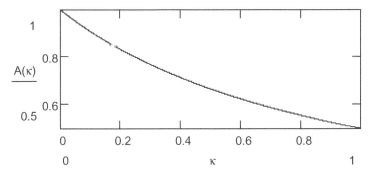

Figure 5.5. An example relationship between repair rate κ and its corresponding steady-state availability $A(\kappa)$.

Very often an interesting parameter is the number of failures (repairs executed) $v(t)$ within a given time interval $(0, t)$. If this interval is relatively long, a random variable $v(t)$ has an asymptotically normal distribution:

$$N\left(\frac{t}{T_n + T_p}, \sqrt{\frac{\left(\sigma_p^2 + \sigma_n^2\right)t}{\left(T_p + T_n\right)^3}}\right) \tag{5.26}$$

This statement enables a similar interval estimation to be built for the random variable $v(t)$ as was done in equation 5.15.

One statement requires explanation here: if this interval is relatively long. For technical objects of high reliability like belt conveyors operating as main haulage units, this interval should be many months (at least a year or so). For an AFC working at a coal face this interval is long enough if it is only a few weeks.

Another interesting characteristic of the process is the total work time $T_\Sigma^{(w)}$ of the object over a time interval $(0, t)$. It was proved (see Gnyedenko et al. 1965, Kopociński 1973) that this random variable has an asymptotically normal distribution:

$$N\left(At, \sqrt{\frac{\left(\sigma_p^2\mu^2 + \sigma_n^2\lambda^2\right)t}{\left(T_p + T_n\right)^3}}\right) \tag{5.27}$$

An interval estimation for $T_\Sigma^{(W)}$ is easy to construct based on expression 5.27.

In the mid-1970s some researchers were interested in a better evaluation of the random variable $T_\Sigma^{(W)}$. A new normalised random variable $T_\Sigma^{(w)}/t$ was introduced and its essential statistical parameters were found. A proposition was made to apply a beta distribution, which is also defined over a [0, 1] interval, and to presume that this is the distribution of the parameters found from this new random variable.[5] This way of reasoning made a slight improvement in the estimation of the probability distribution of random variable $T_\Sigma^{(W)}$ (Czaplicki and Lutyński 1987).

Where conventional belt conveyors are concerned, comprehensive reliability investigations carried out in the Polish mining industry for both underground and opencast mines showed that in more than 95% of cases both work time and repair time probability distributions were satisfactorily described by exponential functions. Therefore, it can be assumed that an exploitation process

---

[5] The results were published in IEEE Transactions on Reliability more than thirty years ago.

of conventional belt conveyors can be modelled satisfactorily by a two-state Markov process (for the definition of the Markov process, see Chapter 7).

A different situation in this regard can be noted in connection with belt conveyors that are constructional elements of large mine machines. These haulage units that operate close to an excavating device receive streams of extracted rock with a high dispersion. This affects the durability of some conveyor elements and assemblies, thereby generating a higher intensity of failure. However, these conveyors do not operate continuously. Their run is stopped when the whole machine is in failure, which is not very rare, and some repairs take a long time. Moreover, these machines require technical surveys and they are usually long-lasting. For these reasons, the intensity of usage of these conveyors is lower than that of belt conveyors operating in conventional conveyor haulage lines.

5.2   HOISTING INSTALLATIONS

Almost all underground mines all over the world have hoisting installations. A small number of these installations can also be found in opencast mines. Generally, they are mounted on the surface, but some hoisting installations are located underground in very deep mines.

Hoisting installations are an organised set of machines and other pieces of equipment that are designed to accomplish basic and auxiliary transportation tasks in a mine shaft or on the slope of an open pit. In the majority of cases a hoist's importance is hard to overestimate. Almost everything that should be delivered underground and that should be transported out of a mine must be done by a hoist.

A hoist is a system consisting of two subsystems:

- driving subsystem
- transporting subsystem.

The driving subsystem is a winder. The transporting subsystem consists of conveyances, ropes, rope attachments and a guiding system.

Each hoist has an assigned number of hours per day in which transportation tasks are to be realised. This time is termed as *disposal time*, T_d—a mine has it 'at disposal' to accomplish its transportation duty. The remaining time, usually night-time hours, is devoted to prophylactic actions in a shaft, such as repairs if needed, diagnostic actions, etc. Let us call this time *disposal standstill* or *planned standstill*. This time is a deterministic value. Planned standstill is obligatory for a mine and during this time a hoist is idle or takes on supplementary duties different from those realised during its disposal time. A hoist is neither heavily loaded nor heavily used. For these reasons, as a rule, when the process of hoist operation is the point of interest for engineers and researchers everything that goes on with this system during disposal time is important. Each day a hoist commences its regular duties early in the morning when the standard exploitation process starts and every day late at night

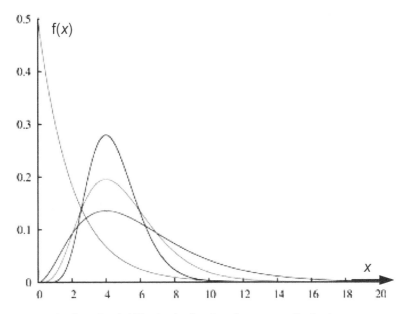

Figure 5.6.   Example plots of probability density functions for a gamma distribution.

the process is stopped (neglecting large breaks in production during scheduled holidays). This type of operation process belongs to the class of *intermittent stochastic processes*[6] (Czaplicki 1981).

During disposal time a hoist can be in three separable exploitation states: work, repair and standstill.

As comprehensive exploitation investigations carried out in Polish mining industry showed (Antoniak et al. 1973–75), repair times of hoists can be satisfactorily described by a gamma probability distribution with a mean of a bit more than one hour. The shape parameter of the distribution was in some cases below one, usually was just about one and in some cases was more than one. Work times were also described by a gamma probability distribution with the shape parameter less than one in some cases, just about one in many cases and sometimes more than one. The average mean work time was about 200 hours (Czaplicki and Lutyński 1982). Based on the outcomes of these investigations an important statement was formulated: in approximately 70% of cases the exploitation process of hoisting installations of a work-repair type was a Markov one.

Therefore, the following probability density function is used here:

$$\begin{cases} f(t) = 0 & \text{for } t < 0 \\ f(t) = \dfrac{\lambda^p}{\Gamma(p)} t^{p-1} e^{-\lambda t} & \text{for } \lambda > 0, p > 0, t \geq 0 \end{cases} \tag{5.28}$$

For $p \in \mathfrak{R}$ the gamma distribution becomes an Erlang distribution. Obviously, for $p = 1$ this distribution is the same as an exponential distribution. For p being a multiple of ½ the gamma distribution is identical to the χ^2 distribution.

---

[6] This term was defined by Buslenko et al. (1973, Chapter 5.4).

The two main statistical parameters are defined by the formulas:

$$E(T) = p/\lambda \quad \sigma^2(T) = p/\lambda^2 \tag{5.29}$$

An important problem, crucial from a practical point of view, is the analysis of a standstill state. Its occurrence sometimes affects mine productivity to great extent, especially where the main output shafts are concerned.

During a hoist operation two factors have an identical effect on hoist efficiency. They are the unreliability of a hoist and the stoppage of hoisting due to the occurrence of a standstill state. Unreliability of a hoist means the occurrence of a failure in the system such that further transportation is halted.[7] In engineering calculations these situations are represented by the steady-state availability of hoist, A_h. But hoisting can be stopped for different reasons—because, for example, there is a lack of rock to transport, there is a lack of power, failure appears in a piece of equipment connected with the hoist in a series, etc. These situations have a stochastic character and unfortunately they are inevitable. The problem is to reduce the occurrence of these events. In engineering calculations this second component is included in hoist utilisation, U_h.

An investigation of hoist standstill done by Czaplicki (1977) comprised the observation of hoisting installations during disposal time over many days. Every day the total work time $T_{\Sigma W}^{(d)}$ of a given hoist was noted and the following ratio calculated:

$$u = \frac{T_{\Sigma w}^{(d)}}{T_d} \tag{5.30}$$

Each day the outcome was different, which indicates that this quotient is a random variable. Because it is normalised to support [0, 1], a beta probability distribution was applied to describe the random variable of interest. A test of goodness of fit gave no grounds to reject the statistical hypothesis stating that this data can be satisfactorily described by this theoretical distribution. If so, we can write down:

$$U : Be(a,b) \quad a > 0 \quad b > 0 \quad 0 \leq u \leq 1$$
$$f_u(u) = \frac{\Gamma(a+b)}{\Gamma(a)\Gamma(b)} u^{a-1}(1-u)^{b-1} \tag{5.31}$$

The expected value and the variance of this random variable are obviously given by the formulas:

$$E(U) = \frac{a}{a+b} \quad \sigma^2(U) = \frac{ab}{(a+b)^2(a+b+1)} \tag{5.32}$$

Plots of function $f_U(u)$ can be quite different (see wikipedia) but in the case being considered the expected value of U is above 0.5 and for this reason they have shapes like those shown in Figure 5.7.

How many hours per day are lost due to standstill state occurrence depends on many factors. The main ones are:

– number of working faces
– reliability of equipment involved

---

[7] Sometimes hoist failures occur that do not cause hoisting stoppage. These concern failures of some electronic parts as a rule.

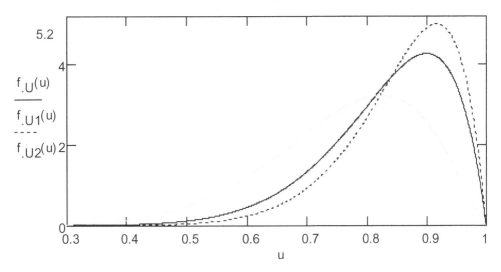

Figure 5.7.   Plots of the probability density function of hoist utilisation for different parameters.

– numer of pieces of equipment engaged in the operation
– structure of excavation and haulage systems
– management of the machinery system involved
– application of rock accumulators.

Losses connected with the existence of a standstill state can be counted as between 5% of disposal time (an excellent result) to up to 25% (or even more) of disposal time (a poor result) for skip systems (cf. Albert et al. 1975, Czaplicki 1977). For cage systems losses are usually greater for obvious reasons.

Compared to the steady-state availability of hoists, which is normally above 0.95, losses due to their poor utilisation are greater.

For further analysis of the exploitation process and reliability of hoists it is valuable to know whether the states in the process are independent. In addition, the answer to the question of what has a significant influence on the values of the basic reliability indices of these installations is crucial.

During the many cited reliability investigations of hoisting installations (Antoniak et al. 1976–80) correlation analysis was done (Czaplicki 1978) taking into account factors such as the type of conveyance applied, type of driving engine, travel speed, depth of wind, conveyance payload, mean repair time, mean work time, disposal standstill time and failure rate. The mathematical tools applied in these analyses were Pearson's correlation coefficient, Spearman's rank correlation coefficient, the partial correlation coefficient and non-linear correlation measures. The most interesting results are as follows.

– For skip hoists the longer the disposal standstill time per day, the higher the hoist's reliability.
  Comment: skip hoists are the main haulage units transporting minerals on the surface. They usually have a high intensity of usage and obviously they need proper service, including repair, adjustments and other prophylactic actions, as well as suitable diagnostics. All these require time to do in a proper way with high accuracy. Having a longer time to do them means that these actions are more likely to be performed more correctly. This creates a higher reliability of the installation. Moreover, in some cases a significant correlation was traced between the time of hoist repair and the time of work time after repair—a longer repair time led to a longer work time. It looks as though this finding can be explained in the same way.

– Hoists having different types of drives have a different reliability; both mean measures of work time and repair time are significantly different.
– The type of conveyance applied also has an influence on the mean repair time of hoist.

A more in-depth analysis along with an application of correlation measures was given in Chapter 3.9.

Generally, mine hoisting installations are systems of a high reliability. Mine winders have few failures per year that cause a stoppage of winding. The remaining hoist elements are also of a high durability, their wear processes are of a mechanical nature in the majority of cases. They are usually mended or replaced if needed during the disposal standstill time. Rarely is a hoist utilisation poor where the main transporting units are concerned.

5.3   HOIST HEAD ROPES

These are extremely important technical structures of hoists. Their reliability determines to a great extent the production of an underground mine and their reliability is also strongly connected with the problem of safety.

They are of various constructions and of various properties; however, one aspect has remained stable for years—they are made of steel wire. (Although, it seems that the day when steel will be replaced by compound plastics is not too distant).

From a reliability point of view they are unrepairable objects operating continuously from the moment of their mounting in a hoist till the moment of their withdrawal from further exploitation. Therefore, all reliability models considered in this book up to this point are useless. A new approach has to be presented.[8]

---

[8] The considerations presented in Chapter 5.3 are different from those offered by professionals in their elaborations (see Tytko's 2003 monograph or papers of the OIPEEC).

5.3.1 *A model of the hoist head rope wear process*

First, a definition of failure must be decided upon. Failure is one of cardinal terms in reliability and exploitation theories and it looks at first glance as though it is commonly understood. But, the reality is—as usual—more complicated.

It is true that in some fundamental publications concerning reliability, failure is not defined as it is assumed that the meaning of this term is obvious (see, for example, Gnyedenko et al. 1965, Gertzbah and Kordonsky 1966, Melchers 1999). But in a majority of publications some definitions are given that appear to mean approximately the same thing at first glance, but they do not entirely. Let us cite a few definitions.

The temporary or permanent termination of the ability of an entity to perform its required function (Atis Telecom Glossary).

Failure in general refers to the state or condition of not meeting a desirable or intended objective (wikipedia).

Failure is any event that impacts a system in a way that adversely affects the system criteria ... failure definition can change on a given system over time (www.asknumbers.com/WhatisReliability.aspx).

Failure is non-conformance to some defined performance criterion (Smith 2007).

The termination of the ability of an item to perform a required function (O'Connor 2005, BS 4778).

Failure is an event after which a system is incapacitated (Ryabinin 1976).

Failure is an event or passage from a state of ability to perform determined functions to a down state (Migdalski 1982, Kaźmierczak 2000).

Failure: the inability of an item/piece of equipment/system to operate within specified guidelines (Dhillon 2008).

Notice, that according to some authors failure is a kind of an event (Atis, www.asknumbers, O'Connor, Ryabinin, Migdalski, Kaźmierczak) but some researchers are of the opinion that it is a state (wikipedia, Smith, Dhillon). These two terms—event and state—have completely different meanings. Because reliability has a stochastic nature, one must look to probability theory for definitions of these two terms. The difference is clearly visible: an event has a zero time length, a state a non-zero duration. An event is equivalent to an impulse, while a state is associated with a random variable that determines its time span.

In our considerations we assume the following approach. **Failure is an event** of a stochastic nature during the operation of a technical object. It can be presumed that this event involves the passage from one state to another, from better to worse. In some cases, this passage means the termination of an object's life. In some cases, this transition means that the object is not able to fulfil its duties but after some actions this capability will be restored. Sometimes it means that the capability of the object is only a little worse, and many similar failures can occur in the object—the object fulfils its duties but with lower efficiency. In some cases the occurrence of a failure is associated with the object's safety. The object still performs its duties at full capacity but with lower safety. This last case applies to mine hoisting ropes. It seems that this issue is absent from many publications in this field. Nonetheless in some advanced considerations such a case is included (see Anderson and Randell 1979) but in a slightly different way to that presented in this book.

There is no doubt that the definition of failure that has been presumed has considerable consequences for further reliability and exploitation considerations. In many publications one can find proof of the truth of this statement (see, for example, Fashandi and Umberg 2003).

In reliability studies of unrepairable objects the most frequently applied model is called the *reliability of an item working up to the first failure occurrence*. However, mine head ropes operating as hoisting elements do not work up to the first failure occurrence. If the term *failure* is understood to mean a rope rupture, then fortunately this event occurs very rarely and it can only be considered in the context of the study of catastrophic events when applying the theory of extremes. If the term *failure* is understood to mean a single crack of a single wire, then there can be many such events, sometimes counted in hundreds or even more. For these reasons, failure is not the proper word to use here. It is necessary to look at the rope wear process in a more in-depth way.

The rope wear and tear process runs continuously from first day of a rope's operation. It is hard to find another example of a technical object which operates under such a great number of various loads that are constantly changing over time; and changing dramatically in a short time. In each work cycle a rope is suddenly heavily loaded when a huge portion of broken rock falls into a skip container from a measure pocket underground. Then the conveyance motion commences and a dynamic load occurs when acceleration is applied from a winder. While the skip is running rope, vibrations accompany this motion and waves run in the rope and are dissipated in the rope's attachment. In the meantime a torque moment that exists in the rope due to its construction changes its value, as it is a function of the rope length in a shaft. This torque moment tries to rotate the conveyance but this cannot be done due to the existence of a guiding system. Therefore, the rope is further twisted (or untwisted) depending on the stage of the skip's motion. For these reasons, the rope length is shortened or an elongation of this rope is noticed. And what is more, during rope transportation in a shaft there is a small mutual displacement of wires and friction force, and during this displacement friction gradually destroys the wire's surface. When the skip is near the surface, the winder begins to decelerate and a new dynamic load occurs. When dumping is finished the rope is suddenly unloaded from this great weight. The conveyance jumps up. Apart from these loads and forces, the rope is also bent over pulleys and over a drum during the work cycle (if this type of rope carrier is applied). This bending can be one-directional or it can have a double direction. Very often the mine air flowing around the rope carries many constituents, and some of them create acid or base mist. Rope rust is likely to occur. Dirt and rock dust are agglutinated to the wires, which increases the negative effects during friction between the wires themselves.

Often the number of rope work cycles is counted in the hundreds in a single day. Even if a certain component in the work cycle has a very small influence on the rope wear process, when the rope executes thousands of cycles this small component becomes important and its significance can be great.

Physically, a rope wear process is first of all a fatigue process in character. Wire cracks are observed and they increase in number over time. Additionally, corrosion can appear and local steel mass losses can be observed, but these types of rope wear are not so frequent. Therefore, they will not be considered further here.

Looking at many wear processes in ropes of different constructions, it seems that a general model can be drawn. It is presented in Figure 5.8. The main point of consideration will be the phenomenon of the accumulation of rope wire cracks.

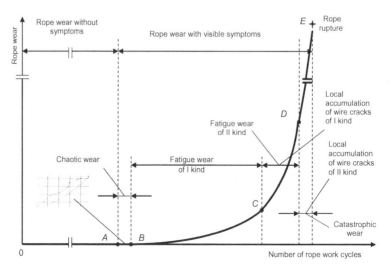

Figure 5.8.   Model of the hoist head rope wear process of the fatigue type.

As a preliminary statement it has to be recalled that, due to the construction of a rope, even if a single wire fracture appears this wire can carry its load after approximately five to six of spiral leads of the rope because the fractured wire will be seized by neighbouring wires as a result of friction forces acting between these wires. Thus a given single wire can have many cracks along its length.

When a rope starts its duties during a certain period of time no effects of wear are visible $(0, A)$. Finally, the first wire cracks are observed. Usually, these first few cracks (and sometimes more) are the result of both the fatigue process and local faults in the material used during the wire and rope production processes. Some faults may result from the steel used for production, some faults may have been generated during the production process of the whole rope. This period (A, B) is usually short and is chaotic in character. Sometimes it is almost not observed.

When the number of cracks in the wires starts to grow, their total number noted in time or their total number versus the number of work cycles executed (a better parameter) begins to lie approximately along a certain line. Usually it lies along an increasing curve (B, C); in some cases it lies along a straight line, especially during the inception period. Rarely does the speed of the increment of the total number of broken wires decrease.

After a certain period of time during which usage of the rope is continued, the accumulation of cracks in the wires on a certain rope section becomes significant. On one rope section (later sometimes there are two) the total number of cracks is distinctly higher than on other rope sections. In the previous period (A, B) cracks in the wires were uniformly distributed in a stochastic sense almost along the whole rope length. Now, this regularity is disturbed and is changed (see Figure 5.9). This means that apart from the regular fatigue process a new component has appeared. For this reason the previous period will be named: *Fatigue wear of the I kind*. Now is the period of *Fatigue wear of the II kind*—and the two processes running in time are important for the rope (C, D). The first process, the fatigue of the material for the whole rope, makes a random but uniform distribution of cracks in the wires and the second one results in the accumulation of cracks in a certain rope section or sections. This situation in the rope wear process becomes dangerous. It is well-known that rope rupture occurs when such a number of cracks appears on a certain rope section that the course of action changes its character—in a dramatically short period of time the rope will be torn off (D, E). Notice, that in period (C, E) the local accumulation of cracks in the wires is observed; however, in period (C, D) this process has one character, whereas in (D, E)

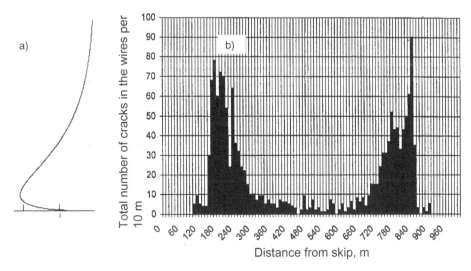

Figure 5.9.   a) Density function of the number of cracks accumulated on a given rope section; b) Distribution of cracks in the wires along the rope length (Tytko and Nowacki 2006).

it is completely changed. A different quality appears in a physical sense. It is worth noting that between points 0 and E a few different physical processes are running that create rope wear. One process is running continuously from the first day of rope operation—this is the fatigue process. A second process runs over period (A, B) but it dies out at the end of this phase. A third process appears at the beginning of stage (C, D) and it runs till the end. Similarly, a new component occurs in period (D, E) and it also runs till the end.

One conclusion seems irresistible here—it looks rational that different approximation functions should be applied to analytically describe the course of action at stages where a set of constituents changes its composition.

From a mathematical standpoint the rope wear process is a stochastic one with very complicated properties that change over time.[9] At the very beginning it can be presumed that it is stationary one. No effects of wear are visible. Next, local cracks in the wires start to occur and these grow in number over time. From this moment the process is a non-stationary one. At the beginning the probability of the occurrence of two or more cracks is negligible, so it can be assumed that the process is a singular one. Nevertheless, this probability increases over time and—if rope operation continues—it is very likely that the occurrence of two or more cracks will be visible. In addition, a memory in the process is noticed, which is expressed by autocorrelation. During this period the process is non-stationary, non-singular and with memory. If the rope's operation continues, the process changes its character dramatically at a certain moment. The accumulation of cracks in a certain rope section will be so great that a rope rupture will be inevitable.

Mining engineers are quite conscious that a hoisting rope is greatly important for the entire mine production and that it is responsible for safe transportation in a shaft. They are aware of the fact that the wear process of rope is extremely complicated. These are the reasons why particular attention is paid to rope exploitation as a rule. Diagnostics of ropes are done with great concentration, and the outcomes are usually analysed and noted. In the early days of mining only a visual inspection was applied. However, for the past fifty years the basic method has been electromagnetic particle crack detection. Other techniques are of less significance. This method is oriented to tracing the weakest section of a rope and almost nothing else. It neglects the fact that the wear process of a rope runs over time and its symptoms are visible. Making use of this information, more valuable inferences on the actual state of a rope can be made; better than those based exclusively on one rope section even when this section is the weakest one.

Let us look at the process of rope wear more closely. The first two periods of rope wear are of no interest. During the first stage no symptoms are visible; during the second stage symptoms are insignificant. The last stage is also of no interest because—fortunately—it is very rarely seen in practice. Therefore, the points of interest should be the third and fourth stages of the rope wear process.

5.3.2 *Applying approximation functions*

Look at some example courses of the rope wear process in Figure 5.10. Looking at this figure a conclusion can immediately be drawn—the rope wear process, understood as the process of the accumulation of cracks in the wires over time, can be approximated by a straight line or by a convex curved line.

Among convex curved lines a power function and an exponential function can be useful. Thus, we can write down:

$$\Theta_t = \alpha_1 t + \alpha_0 + \gamma_t \tag{5.33}$$

---

[9] Usually, it is presumed that a parameter of a stochastic process is a time. This approach is common in econometrics or demography for example, but in engineering practice time has a lower significance. An alternative process parameter should be the number of tons excavated, transported or dumped. In some cases, the number of cycles executed—a hoist rope—should be considered at first. If the intensity of usage of a hoisting installation is approximately constant during a rope's life, time can be taken into account instead of the number of hoist work cycles.

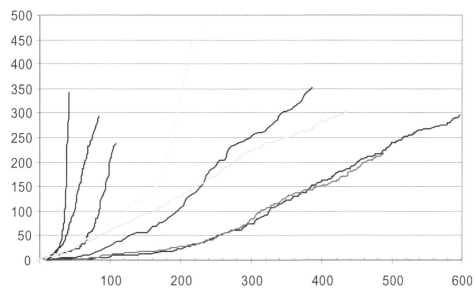

Figure 5.10. Diagram of the functions of the total number of cracks in the wires for different hoist head ropes 60 mm in diameter vs. the number of rope work cycles $q \times 10^3$ for one shaft.

$$\Theta_t = \beta e^{(\beta_0 t^2 + \beta_1 t + \xi_t)} \tag{5.34}$$

The power function is determined by formula 3.20:

$$\Theta_t = \delta t^\chi c^{\zeta_t}$$

where Θ_t is the total number of cracks in the wires of a rope, α_0, α_1, β, β_0, β_1, δ, χ are structural functions parameters and γ_t, ξ_t, ζ_t are random components.

The power function has been used in Poland since the late 1950s (see, for example, Kowalczyk 1957, Kowalczyk and Steininger 1963, Kowalczyk and Hankus 1965, 1966). The idea for its usage is as follows.

Usually ropes are worn in a fatigue manner only. In those days triangle strand ropes were used almost exclusively as hoisting elements. It was mandatory to note the total number of cracks in the wires for each rope at each mine every weekend. In the majority of cases, a graph of the total number of cracks in the wires of a rope plotted against time could be plotted as a dotted line creating a shape similar to a power function plot. Two theoretical functions could have been taken into consideration in those days; however, a power function was selected because it was assumed that a good measure of increasing rope fatigue was the speed of the increase in the number cracks in the wires. And this speed was presumed as a basic measure of rope weakening.[10] Because the first derivative of function 5.34 is complicated, a power function has been chosen in spite of the fact that an exponential function has three structural parameters compared to the two parameters of a power function. Obviously, functions with three parameters describe data better than functions with two parameters.

---

[10] This idea has been abandoned. Even today it is difficult to determine the critical value of speed, especially when there are several different definitions of the speed of the increase in cracks in ropes (Czaplicki 2009a).

The way of estimating the structural function parameters was quite simple. Linearisation was done to obtain a straight line function and the least squares method was applied in order to get the desired estimates. Approximation theory was at an early stage in its development and, for this reason, no random component was taken into account in those days. At that time electromagnetic particle crack detection was just coming into existence and the importance of investigations based on notations of cracks in the wires lost its position in a short time. Soon it was completely abandoned, wrongly in my view.

There is no doubt that the purpose of finding an increasing number of cracks in rope wires is to create a picture of the developing wear process of a rope, but the problem is how to investigate and 'read' this picture properly and comprehensively.

Consider now a way to estimate structural parameters. At first linearisation is done and the least squares method is applied. The best fit in the least-squares sense is that instance of the model for which the sum of the squared residuals has its lowest value, a residual being the difference between an observed value and the corresponding value given by the model. But for the case being considered there are no original values, only their logarithms (linearisation). The method assures the minimum for a linearised function but does not assure the minimum for the original one. Fortunately, any error made is not so great.

Moreover, in the classical model of linear regression three assumptions are important concerning the type of function being applied (Goldberger 1966, for example), namely:

a. $E(\zeta) = 0$
b. $E(\zeta^2) = \sigma^2$
c. $E(\zeta_s \zeta_t) = 0 \quad s \neq t$.

- The first assumption says that the random component of the model has an expected value of zero.
- Assumption (b) states that the variance of a random component is constant in time.
- The last assumption focuses on the non-correlation of the random component.

If these assumptions are fulfilled, estimators obtained from the application of least squares method will have good, desirable statistical properties. Indeed, the application of the least squares method assures the fulfillment of assumption (a) but not entirely because of linearisation.

Consider assumption (b). Let us recall from Chapter 3.3 that a random component can be defined in two ways:

- as a sequence of differences between the empirical values of θ_i and its theoretical counterpart; for formula 3.18 it is a sequence:

$$u_i = \theta_i = at_i^b \quad i = 1,2,\ldots,N$$

- as a sequence of residuals defined by pattern:

$$\hat{u}_i = \ln\frac{\theta_i}{at_i^b} \quad i = 1,2,\ldots,N$$

which is the result of an appropriate conversion of formula 3.18. Notice, that a residual here is the index of power $c = e$ and obviously N is a sample size.

To verify assumption (b) mine data was taken and samples were divided into two approximately equal parts and for each sub-sample the corresponding empirical standard deviation was calculated. A test employing the F-Snedecor distribution was applied to verify the hypothesis stating that both standard deviations are the same. Many samples of ropes of different constructions were considered and the result was:

'In many cases the test gave grounds to reject the verified hypothesis; empirical standard deviation of the second part was significantly greater than that of the first part'.

Moreover, this statistical way of reasoning was conducted for both means of residuals determination. The outcomes were identical. Some examples of these considerations were presented in Czaplicki's publications (1999, 2006c).

This statistical information was immediately converted into a physical, engineering model—it suggests that for many ropes their wear processes develop in such a way that generates an increase in the dispersion of the total number of cracks in the wires with regard to its mean value. The prediction of the future number of cracks in the wires becomes more difficult, with lower confidence when the number of rope work cycles executed increases. However, we increasingly desire good predictions when the number of cracks increases significantly.

Now consider assumption (c). The first statistical investigations on tracing autocorrelation in a time series of residuals connected with the application of power function 2.20 for a description of the accumulation of cracks in ropes was carried out in 1999 (Czaplicki). The Breusch-Godfrey test was applied supported by the test *J* derived by Pawłowski (1973) and autocorrelation up to the fourth order was investigated. In more than 70% of the cases investigated, autocorrelation was found by both tests.[11] Three conclusions were drawn:

- existence of autocorrelation has serious repercussions on the properties of estimations of structural model parameters
- autocorrelation can be used during prediction to increase the accuracy of the prognoses being constructed
- there must be a physical cause generating this regularity.

More comprehensive research was done by the same author some time later, resulting in a further publication (Czaplicki 2000).

Much data on rope wear processes for ropes of different constructions was investigated. Large samples were divided into three approximately equal sub-samples and autocorrelation up to eighth order was tested.[12]

Some interesting results were obtained and further statements were formulated:

- autocorrelation of residuals has been found for ropes of different construction
- rarely does the significant autocorrelation of residuals have a negative value
- sometimes autocorrelation increases over time
- there was no decrease in autocorrelation during the time observed.

5.3.3 *Memory in the wear process of ropes—further approximation functions*

A literature review was performed (see Ditlevse and Sobczyk 1986, Sobczyk and Spencer 1992) in an attempt to find physical grounds for the observed phenomenon. It was discovered that it was because of the nature of the fatigue accumulation process, *the rope after a certain period of its operation commences to 'remember' its degradation and later 'remember' it.* This supposition allows the suggestion that perhaps a new and different function-model should be applied.

Meanwhile the second function, the exponential equation 5.34, was taken into consideration and similar investigations were carried out. The outcomes were not astonishing—the residuals showed identical regularity. In a great number of cases autocorrelation was found.

It is worth noting that both of these functions have no physical sense. They only allow the total number of cracks in wires for the rope being investigated to be estimated for a given time (or for the number of work cycles executed)—and that is all. However, the accuracy of this estimate decreases over time as a rule. This means that when rope weakening increases and a more precise prognosis of the state of the rope is highly desirable, the mathematical tools applied become

---

[11] In Dłubała's (2009) investigations a higher result was obtained—approximately 80% of the cases.
[12] The author is aware of the fact that in autocorrelation of a high order there is a kind of memory associated with previous orders.

less accurate. These functions describe the developing tendency of the process. This corresponds approximately with the function of an expected value (a general tendency, according to the nomenclature used in econometrics). But in their construction, no 'mechanism' is included which shows that the state of a rope depends on the state of the rope just before. In mathematics a function that contains this type of information is an autoregression function. Its general form is:

$$X_t = \sum_{i=1}^{w} \alpha_i X_{t-i} + \alpha_0 + \zeta_t$$

where:

α_0, α_i, $i = 1,2,\ldots,w$ are structural function parameters
ζ_t is random component.

The author has carried out comprehensive research to find the best 'prescription' among this class of functions and the result is:

$$\theta_t = \varphi_1 \theta_{t-1} + \varphi_2 \theta_{t-2} + \psi t + \chi_t \tag{5.35}$$

which is a modified version of an autoregression function. The only reason for applying this function is to fit empirical data in the best way and to allow a prognosis on the future state of the rope to be constructed if the prediction horizon is not too distant.

Let us now add one more item taken from pure theory to our considerations. The term 'memory of a stochastic process' exists in the theory of random processes and this term has been known for many years (see, for example, Gnyedenko and Kovalenko 1966). In general, it can be stated that a given stochastic process has no memory if the future realisation of the process depends solely on the value of the current process and does not refer to any previous realisations. Actually, mathematicians distinguish a few different types of memories, such as short memory (see So 2002) and long memory (see Granger, and Zhuanxin 1996), in connection with different types of processes (continuous ones, time series, etc).

The wear process of a rope has a stochastic nature. Each random process can be either memoryless or have a memory. Actually, it can be clearly stated that this process has memory and this property—according to the best knowledge of the author of this book—has not yet been investigated comprehensively.

Before the continuation of further reasoning let us consider an example.

■ Example 5.1

Data concerning a hoist head rope that operated in a main output shaft was obtained. This data provided evidence of the wear process during the realisation of haulage tasks from 24×10^3 work

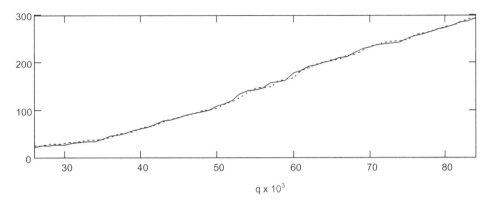

Figure 5.11.   Empirical data on the total number of cracks (dotted line) and a theoretical plot based on modified autoregression (continuous line).

cycles to 84×10^3 cycles. Information on the total number of cracks is shown in Figure 5.11. The total number of cracks reached almost 300.

A set of equations allowing the unknown structural parameters in function 5.35 to be estimated is as follows:

$$\varphi_1 \sum_{i=3}^{N} \theta_{i-1}^2 + \varphi_2 \sum_{i=3}^{N} \theta_{i-1}\theta_{i-2} + \psi \sum_{i=3}^{N} q_i\theta_{i-1} = \sum_{i=3}^{N} \theta_i\theta_{i-1}$$

$$\varphi_1 \sum_{i=3}^{N} \theta_{i-1}\theta_{i-2} + \varphi_2 \sum_{i=3}^{N} \theta_{i-2}^2 + \psi \sum_{i=3}^{N} q_i\theta_{i-2} = \sum_{i=3}^{N} \theta_i\theta_{i-2} \qquad (5.36)$$

$$\varphi_1 \sum_{i=3}^{N} q_i\theta_{i-1} + \varphi_2 \sum_{i=3}^{N} q_i\theta_{i-2} + \psi \sum_{i=3}^{N} q_i^2 = \sum_{i=3}^{N} \theta_i q_i$$

Solving this set of equations we obtain these estimates:

$$\hat{\varphi}_1 = 0.993 \quad \hat{\varphi}_2 = -0.013 \quad \hat{\psi} = 0.136$$

and so:

$$\Theta_i = 0.993\,\theta_{i-1} - 0.013\,\theta_{i-2} + 0.136\,q_i + \chi_i$$

Now an evaluation of the goodness of fit of the estimation can be made. The sequence of residuals defined as differences between the empirical and theoretical values is presented in Figure 5.12.

The average value of residuals is 0.08 whereas the corresponding standard deviation is 3.05. It can be proved that this average value differs non-significantly from zero.

Let us now divide the data into two sub-samples. For the first sub-sample the standard deviation is 2.65 and for the second it is 3.58. To verify the hypothesis that the difference between these standard deviations is non-significant, we calculate the ratio:

$$\left(\frac{3.58}{2.65} \right)^2 = 1.825$$

The corresponding critical value from the F-Snedecor statistics is $F_{\alpha=0.05}(26, 26) = 1.86$. There is no ground to reject the hypothesis that these values are non-significantly different from each other.

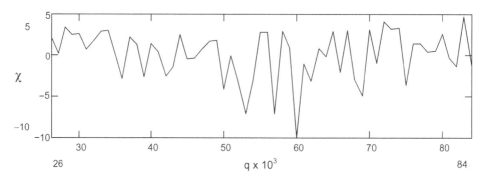

Figure 5.12.   Sequence of residuals—differences between the empirical and theoretical values obtained from an autoregression function.

Finally the autocorrelation in a sequence of residuals was traced. Autocorrelation of first and second order were investigated receiving Pearson's correlation coefficients −0.01 and −0.035. It can immediately be stated that there is no autocorrelation in the sequence of residuals.[13]

Additionally, an investigation of the supposition that the sequence of residuals is in fact a realisation of a Gaussian process was carried out. The Kolmogorov test applied gave no grounds to reject this hypothesis.

Therefore, it can be stated that in the case being considered the modified autoregression function is an excellent tool to describe the empirical data well.   ◄

Let us now look at this model more carefully. The resulting function has a physical sense. The left-hand side of the equation is the total number of cracks in the wires of a rope, which is obviously described by the right-hand side of this equation.

The first component is the total number of cracks in the previous data recording $(t − 1)$ multiplied by a non-dimensional coefficient whose value is near 1 (a property of the autoregression function).

The second component is the total number of cracks in an earlier recording $(t − 2)$ multiplied by a second non-dimensional coefficient, of a lower value than the first one in most cases. Its sign is positive in almost all instances; however, sometimes it can be negative.

The third component is the product of a certain constant and a time t counted from 0 up to the moment for which the total number of cracks is being considered. Because the whole function has a physical sense it is easy to recognise that γ has the dimension of the speed of the accumulation of cracks in the wires.

This function has:

– three structural parameters, two non-dimensional constants φ_1 and φ_2, and one constant ψ, being the speed of the accumulation of cracks in the wires; these constants are determined from the whole period of observation up to the last moment of the observation of the rope and it is the so-called 'correction speed' (however, it has no special physical meaning)
– variable t
– two empirical values θ_{t-1} and θ_{t-2}, which are local characteristics allowing the use of the latest data recordings of the total number of cracks.

It is worth adding that there is no need to linearise equation 5.35. Applying the least squares method assures a true minimum in this regard.

---

[13] In the case analysed no autocorrelation was traced in the modified autoregression function residuals. However, in many cases autocorrelation is still observed in residuals.

After the development of a modified autoregression model, further investigation was focused on verifying whether previously stated regularities can be observed in connection with this new function.

First, a test was employed to check whether the variance of the wear process is constant in time. In the majority of cases there was no ground to reject this hypothesis. This means that model described by equation 5.35 follows future changes and takes their intensity into account.

Second, autocorrelation was investigated in residuals, where the residuals represent a series of differences between real values and theoretical ones. In many cases autocorrelation was stated.

Because the speed of the accumulation of cracks in the wires is a measure of rope weakening, let us determine this speed based on formula 5.35. Here we have:

$$\frac{\theta_t - \theta_{t-1}}{\Delta t} = \varphi_1 \frac{\Delta \theta_{t-1,t-2}}{\Delta t} + \varphi_2 \frac{\theta_{t-2,t-3}}{\Delta t} + \psi = \overline{V}_{t,t-1} \tag{5.37}$$

where:

Δt is the series range
$\Delta \theta_{t-1,t-2}$ is the empirical increment of the total number of cracks in the wires in a time interval $(t-2, t-1)$.

Because a sequence of points of cracks notation has a discrete type, formula 5.37 determines the average speed $\overline{V}_{t,t-1}$ in a time interval $(t, t-1)$ of length Δt.

Notice that formulas 3.20, 5.33 and 5.34 take into account all the data and this information is treated uniformly. Information obtained a long time ago has the same weight as information noted just recently. But this does not hold true in practice. The model described by formula 5.35 is in fact a combined one in which the latest data recordings play the main part. In addition, it has a physical sense. Nevertheless, readers should realise that formulas 3.20, 5.33 and 5.34 describe in a better or worse way the general tendency in the wear process of a rope; the tendency that is associated with the expected value of the function. A modified autoregression model is excellent for prediction—to answer the question: what will the state of the rope be one or two steps ahead? Additionally, if autocorrelation exists in a model of residuals for a certain rope, it can be used to improve prediction and to get a more accurate prognosis.

There are many merits of a modified autoregression function as have just been stated, however, it has one important demerit. It is heterogeneous. The wear process of a rope as a random process has a parameter—time or the number of work cycles executed by the rope. Depending on which parameter is taken into consideration, significantly different estimations of its structural parameters will be obtained. This fact is the source of the heterogeneity.

Let us return to the concept of a memory existing in the wear process of ropes. If we are aware of the fact that a memory exists and that the wear process has a developing character, the question can be formulated as to whether this memory is constant in time. To get the answer to this question we should investigate a function that describes this memory and this function must be homogeneous. This requirement means that the modified autoregression function cannot be used. Its construction must change to release it from heterogeneity. This can be done simply by rejecting the third component of the sum in equation 5.35. If so, we have new approximation function, one that is purely autoregressive:

$$\theta_t = \varphi_1 \theta_{t-1} + \varphi_2 \theta_{t-2} + \varepsilon_t. \tag{5.38}$$

As investigations have shown this function fits empirical data well, only a bit worse when compared to the modified autoregression function.

Now we commence investigations allowing information on changes in the power of memory in the wear process of rope to be obtained.

From the set of data that was illustrated in Figure 5.10 five ropes were selected giving a wide variety of realisations of wear processes. They are presented in Figure 5.13. The numbers ascribed to each rope were those originally given during the investigations.

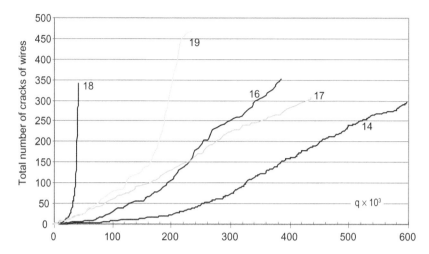

Figure 5.13.   Diagram of selected functions of the total number of cracks in the wires for different hoist head ropes 60 mm in diameter vs. number of rope work cycles $q \times 10^3$ for one shaft.

Estimations of the structural parameters of function 5.38 were calculated obtaining:

$$
\begin{array}{lll}
\text{Rope 14:} & \varphi_1 = 1.387 & \varphi_2 = -0.381 \\
\text{Rope 16:} & \varphi_1 = 1.499 & \varphi_2 = -0.485 \\
\text{Rope 17:} & \varphi_1 = 1.611 & \varphi_2 = -0.604 \\
\text{Rope 18:} & \varphi_1 = 1.125 & \varphi_2 = -0.037 \\
\text{Rope 19:} & \varphi_1 = 1.889 & \varphi_2 = -0.886.
\end{array}
$$

All approximation functions fit the empirical data satisfactorily.

Looking at these estimations of the parameters for formula 5.38 some remarks can be made. These are formulated taking into account the results of investigations on a larger sample.

– Values of the first proportional coefficient standing at the total number of wire cracks noted just before are high. They are near one and they are always positive.
– Values of the second proportional coefficient standing at the total number of wire cracks noted at an earlier time are much lower than the first proportional coefficient. They can be either positive or negative.

These observations strengthen the argument that memory exists in the wear processes of ropes.

To verify the supposition as to whether memory is stable or not in a given wear process the following expression was calculated:

$$
\frac{100\varphi_1(q)n_{q-1}}{\varphi_1(q)n_{q-1} + |\varphi_2(q)|n_{q-2}}\% \tag{5.39}
$$

for a sequence of the total number of cracks for each rope. This quotient gives information on the percentage of participation of the first parameter in the total number of cracks at the moment of investigation. The outcomes of the computations are shown in Figures 5.14–5.18.

By analysing these figures some comments can be made:

– during the inception stage, in spite of the fact that a process runs over a certain period and is characterised by high dispersion, it looks as if memory is being stabilised
– after this initiation the power of memory looks stable

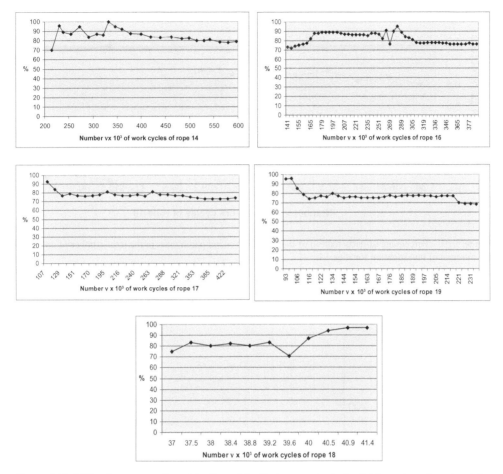

Figure 5.14–5.18. Percentage of participation of the first parameter in the total number of cracks at the moment of investigation for ropes being investigated.

– in all of the cases analysed the state of the rope at a given moment depends strongly on the state before, and this influence is counted at approximately 80%
– the remaining 20% belongs to second component.

After a careful study of the plot shown in the last of these five charts (Figure 5.18), some attractive information can be obtained. This diagram concerns the rope with the most rapidly increasing number of cracks. The statistical test applied gave grounds to reject the hypothesis stating that the power of memory is constant in time versus the alternative supposition that it is increasing. Let us make a more penetrative in-depth examination.

Estimations of the structural function parameters φ_1 and φ_2 were completed for the sequence of the total number of cracks. Results of computations are shown in Figure 5.19.

This figure holds some interesting information. The power of memory seems stable up to the moment of the execution of slightly fewer than 40×10^3 work cycles. Then a significant change occurred—the importance of the first component drastically increased and the value of the second changed to negative. This type of modification must have a physical background. A supposition can be formulated that perhaps a new component in the wear process of the rope has just appeared.

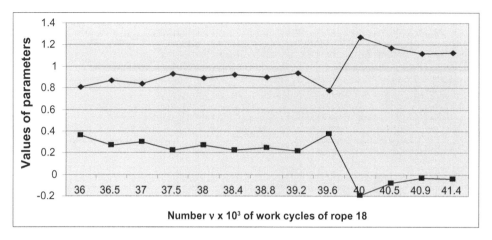

Figure 5.19.   Percentage of participation of the first and second parameters in the total number of cracks at the moment of investigation for rope 18.

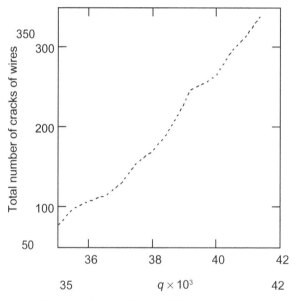

Figure 5.20.   Total number of cracks in the wires for rope #18 vs. number of rope work cycles $q \times 10^3$.

Let us look more closely at the degradation of the rope as expressed by the number of cracks noted. Figure 5.20 illustrates the course of the function: the total number of wire cracks versus the number of work cycles executed in a more precise form than was shown in Figure 5.13.

It is worth noticing that there is no indication in this figure that something happened in the wear process. This means that it is not enough to have records on the cumulative number of cracks in rope wires. A more subtle analysis has to be done to get information on the phenomena that may be occurring. The author is not in possession of more information in regard to this example. However, the management of the underground mine decided to remove this rope from further operation a short time later. The rope executed 41.4×10^3 work cycles. The possibility cannot be excluded that the wear process of the rope had entered the fourth stage of its course.

This example confirms a statement that was made at the end of Chapter 2—statistical diagnostics can give vital information; information that is sometimes difficult to obtain in a different way.

5.3.4 *Rope reliability*

According to mining regulations a hoist head rope must be withdrawn from further operation if the loss in a load carrying section is greater than a determined value, say 20%, provided that this loss concerns a certain determined rope length counted in the number of rope spiral leads. Denote this number by k_r.

Presume that a rope of metallic area S_n is given. Presume further that the rope is constructed of wires of an identical diameter d_d, in order to get more communicative results. Consider two extreme cases—first, the minimum lifetime of the rope. This case is noted if all the first cracks in the wires are located on a certain rope section of k_r spiral leads. In this case the rope length is non-significant; however, the longer the rope, the lower the probability of the occurrence of such an event. Calculate the least number φ_{min} of cracks in the wires. The following equation holds:

$$\varphi_{min} = \frac{0.2 S_n}{\pi \frac{d_d^2}{4}} = 0.255 \frac{S_n}{d_d^2}. \tag{5.40}$$

Note that the that appearance of such an event can be assessed as having a very low probability. If, before the occurrence of this number of cracks on a rope section of k_r spiral leads, some cracks are noted on other sections of the rope, this event will be evaluated as more probable. Following this line of reasoning, a conclusion can be drawn that events when $\varphi_{min} + 1$, $\varphi_{min} + 2$, $\varphi_{min} + 3 \ldots$ number of cracks force the withdrawal of the rope will be more and more probable. But this holds only up to a certain point.

Consider a second extreme case—the maximum rope life. The maximum number of cracks in the wires φ_{max} may be observed if on all sections of k_r spiral leads of rope $\varphi_{min} - 1$ cracks are noted. The next crack—it does not matter where it will happen—makes withdrawal of the rope compulsory. Thus:

$$\varphi_{max} = \frac{i_s}{k_r}(\varphi_{min} + 1) + 1 \tag{5.41}$$

where i_s is the total number of the spiral leads of the rope.

Again, such an event will be assessed as extremely rare. All cases where a rope is removed from the hoist with a smaller number of cracks in the wires $\varphi_{max} - 1$, $\varphi_{max} - 2$, $\varphi_{max} - 3 \ldots$ can be evaluated as more probable. Again, only up to a certain point.

Therefore, when considering the probability of the withdrawal of a rope as a function of the total number of cracks in the wires, we can state that:

– this probability function possesses a support $[\varphi_{min}, \varphi_{max}]$
– this probability function increases from the left-hand side boundary up to a certain point and then decreases attaining at point φ_{max} its value near zero.

Assume now that our considerations are associated with a rope wear process of the I kind exclusively.

Because there is no reason to distinguish any rope section or to consider any stochastic 'mechanism' disturbing the process of the accumulation of cracks in the wires, this probability distribution is a symmetric one of expected value:

$$E(\vartheta) = \frac{1}{2}(\varphi_{min} + \varphi_{max}) \tag{5.42}$$

where ϑ is a random variable, denoting the total number of cracks in the wires at the moment of the withdrawal of the rope.

As was proved by Kopociński (Kopociński and Czaplicki 2007) a probability distribution of the random variable ϑ can be approximated by the normal distribution. If so, the corresponding probability density function is determined by the pattern:

$$f_\vartheta(\varphi) = \frac{1}{G\sigma_\varphi\sqrt{2\pi}} e^{-\frac{(\varphi-\bar\varphi)^2}{2\sigma_\varphi^2}} \qquad \varphi_{min} \leq \varphi \leq \varphi_{max} \tag{5.43}$$

where G is a normalisation constant because this function has limits on its existence. However, it was proved (Czaplicki 2007) that for practical purposes $G = 1$.

Because it was assumed that the wear process of a rope is of the I kind, cracks in the wires are uniformly distributed in a stochastic sense over the whole rope length. If there are some sections with a significantly higher number of cracks, rope removal occurs earlier on average. Thus, a conclusion can be formulated that if cracks in the wires are regularly distributed over the whole rope length, the number of cracks reaches a stochastic maximum. If this regularity is disturbed, it will cause a shorter rope life in stochastic logic. This disturbance also increases the dispersion of the random variable ϑ. Furthermore, for a uniform distribution of cracks the density function shown in Figure 5.9a will have a normal character. For a more irregular distribution of cracks this distribution becomes asymmetric, as is shown in Figure 5.9a.

But let us keep the assumption of a rope wear process of the I kind. Having defined the probability distribution of the total number of cracks at the moment of the withdrawal of the rope (because of the loss of a load carrying area on a certain rope section), a survival function can be defined based on a well-known formula in reliability theory:

$$R_\vartheta(\varphi) = P\{\vartheta \geq \varphi\} = 1 - \int f_\vartheta(\varphi)d\varphi$$

which yields:

$$R_\vartheta(\varphi) = \frac{1}{\sigma_\varphi\sqrt{2\pi}} \int\limits_\varphi^\infty e^{-\frac{(\varphi-\bar\varphi)^2}{2\sigma_\varphi^2}} d\varphi \tag{5.44}$$

An example plot of this function for a given rope is shown in Figure 5.21. The data allowing the plot to be constructed was taken from mining practice.

A hazard function is obviously determined by the formula:

$$\lambda(\varphi) = f_\vartheta(\varphi)/R_\vartheta(\varphi). \tag{5.45}$$

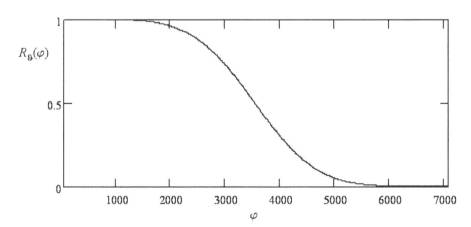

Figure 5.21.   An example of a rope reliability function vs. the total number φ of cracks in the wires.

Pay close attention to Figure 5.21. Keeping in mind that there should be a high level of safety in the operation of a hoist, we are inclined to accept only high values of the reliability function. If so, 2000 cracks in the wires looks acceptable provided that no sections with a high accumulation of cracks are observed. But it seems that this conclusion should be confirmed by some results of further investigations (neglecting here other forms of rope diagnostics).

Notice, that the current variable is here φ—the total number of cracks. For practical purposes this is inconvenient. The best solution is to convert this variable into the number of work cycles executed, q, or time.[14] And here we have the problem of which approximation function should be applied from among functions 3.20, 5.33 and 5.34.

We can conduct our considerations after selecting the function for which the measures of goodness of approximation have most convenient values.

Let us presume that the best function is an exponential one, formula 5.34, and the process parameter is q. The introduction of this function requires the determination of new parameters and a new probability function. The support of this function is determined by limited values:

$$q_{\min} = \frac{1}{2\beta_0}\left(-\beta_1 + \sqrt{\beta_1^2 + 4\beta_0 \ln\frac{\varphi_{\min}}{\beta}}\right) \tag{5.46}$$

$$q_{\max} = \frac{1}{2\beta_0}\left(-\beta_1 + \sqrt{\beta_1^2 + 4\beta_0 \ln\frac{\varphi_{\max}}{\beta}}\right) \tag{5.47}$$

Now, the construction of the probability density function of the random variable q is possible. Applying well-known principles of probability theory (see for instance Papoulis 1965 chapter 5–3) we have:

$$h(q) = \frac{\beta\sqrt{\beta_1^2 + 4\beta_0 q(\beta_0 q + \beta_1)}}{\sigma_\varphi \sqrt{2\pi}}\exp\left(\beta_0 q^2 + \beta_1 q - \frac{\left[\Theta(q) - \bar{\Theta}\right]^2}{2\sigma_\varphi^2}\right) \tag{5.48}$$

The reliability function is obviously determined by pattern:

$$R(q) = 1 - \int_{q_{\min}}^{q} h(q)\,dq \tag{5.49}$$

These functions are presented in Figures 5.22 and 5.23 for the same example. Looking at these plots one piece of information seems clear—in order to assure a high level of safety in the operation of the rope, the rope should be withdrawn after the execution of 60×10^3 work cycles or more precise and penetrative rope diagnostics should be deployed.

Perhaps, the diagnostics should be done more frequently. From both figures the same fact can be stated—after 60,000 cycles the reliability function decreases considerably and the hazard function begins to grow intensively. At this number of work cycles there are 1.34×10^3 cracks in the

---

[14] There is also a different way to evaluate the service life counted in the number of work cycles of a rope (see, for example, Feyrer 1994, Beck and Briem 1995, Briem 2000) but this method is typically for engineers and takes into account the technical parameters of the whole hoist in which the rope operates and the durability depends on many constants ('factors') determined for each case being considered. There is also the term 'correlation' used in connection with this approach, but do not expect any correlation analysis *sensu stricto*. This term is not defined and has a rather more colloquial than scientific meaning.

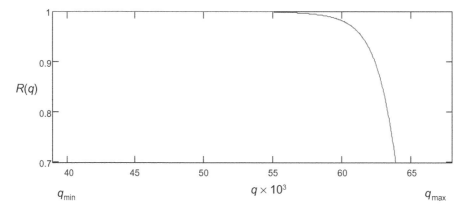

Figure 5.22.   An example of the rope reliability function vs. the total number of rope work cycles q.

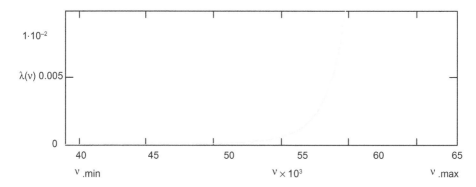

Figure 5.23.   An example of the rope hazard function vs. the total number of rope work cycles q.

wires. This number is significantly lower than that obtained from an analysis of the function $R(\varphi)$. In the case being considered a 30% coefficient of variation was assumed, hence this can be evaluated as a cautious inference.

These results are in accordance with information taken from reliability theory. In the Polish *Reliability Handbook* (Migdalski 1982 p. 86) Bobrowski states: '*Hazard function in a case of normal distribution increases slowly for small values* [here, a small number of work cycles] *and a near expected value commences to grow intensively and runs towards an oblique asymptote. The slope angle of this asymptote is greater for smaller standard deviation ... therefore a normal distribution is a proper tool to describe the time of the object's ability to fulfil its function in a case when failures* [here, cracks] *are caused by gradually occurring irreversible changes of an aging character.*' One can find a similar statement in Gnyedenko's monograph (1965).

To improve this way of reasoning, consideration can be directed towards the approximation function selected and a comprehensive analysis of residuals can be done. Information obtained from such an analysis permits, as a rule, better results to be obtained.

At the end of this part of our considerations, a general remark has to be made. All the reliability measures that have been obtained are rational ones provided that the rope wear process is of the I kind only. For this reason the above analysis can be treated as conditional and the reliability measures should also be considered as conditional functions.

It should be clearly stated here that the formulated recommendation—after inclusive reliability analysis—on the withdrawal of the rope under investigation or on changes in rope diagnostics has no absolute character. It may even happen that the rope will be removed from operation earlier than necessary; however, this event can be evaluated as having a low probability. It is necessary to remember that the phenomena considered here have a stochastic character. The best solution is to combine information on the state of a rope based on evaluations from different sources, that is from electro-magnetic investigations, visual ones and empirical-theoretical ones.

Consider now a case where the rope wear process enters the next stage and becomes a process of the II kind.

There are two ways to recognise that the rope wear process is at the type II stage. The first method relies on a careful analysis of a chart showing the distribution of cracks along the length of the rope. If the data is trustworthy, it will be easy to find a rope section where the accumulation of cracks is significantly higher than that for the rest of the rope.

If the data is incomplete and information is in the form of the total number of cracks against time, it is sometimes possible to recognise changes in the speed of the increase of the number of cracks. Unfortunately, in some cases this change in speed is smooth not radical. An analysis of changes of this speed determined by formula 5.37 would be advantageous here.

Let us now analyse data concerning the wear process of a rope 54 mm in diameter and 330 m in length (L). The data presents the total number of cracks against the number of work cycles executed by the hoist (see Figure 5.24). In this figure two approximation functions are presented. It is easy to see that both functions are very poor in application and should be rejected.

Looking more carefully at the course of the empirical values we can detect that after the execution of approximately 160×10^3 work cycles the speed of the increase in the number of cracks increased significantly. A hypothesis can be formulated stating that a physical source is generating this phenomenon. It is very likely that the stage of the I kind rope wear has ended and the II stage has commenced.

If this supposition is formulated then:

– a careful visual analysis of the rope is recommended in order to trace a possibly weaker section
– if such a section is identified, frequent watchful observation is highly recommended.

Before further analysis can be conducted it is worth realising that if recordings of cracks are made for every 10 m of rope length, it is difficult to get a clear picture of the situation. Mining regulations state, for instance, that a rope must be removed from further exploitation if the loss of the total load carrying area of the rope is 20% or more on 5–6 rope spiral leads, and a 10 m rope section contains several such *criterion sections*. It is a separate problem to construct a criterion

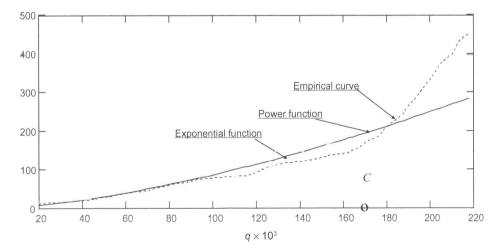

Figure 5.24.   Empirical curve of the total number of cracks in the wires vs. the number q of executed work cycles and approximation functions.

function to assess what number of cracks in a 10 m rope section indicates that we have a significant accumulation of cracks.

As has been stated, if a new stage in the rope wear process commences, a new approximation function should be applied. Therefore, one function should be applied up to the moment of the occurrence of point C in Figure 5.24 and a second approximation function from this point. In both cases the exponential function 5.34 was applied.

Making the necessary calculations one obtains:

– for the first 31 pairs (q_i, θ_i) estimates of the structural function parameters are:
$\hat{\beta} = 5.22$; $\hat{\beta}_0 = -1.1 \times 10^{-4}$; $\hat{\beta}_1 = 0.038$ and the function is denoted by M_1; see Figure 5.25.
– for a further 14 pairs that were observed the estimates are:
$\hat{\beta} = 0.34$; $\hat{\beta}_0 = -6.85 \times 10^{-5}$; $\hat{\beta}_1 = 0.048$; for the function denoted by M_2; see Figure 5.26.

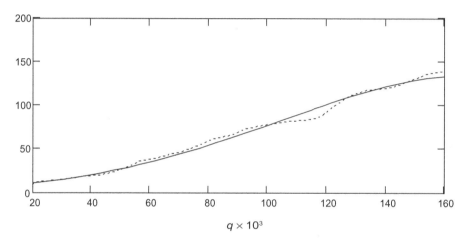

Figure 5.25.   Total number of cracks vs. the number q of work cycles up to point C: dotted line is the empirical values, and the continuous line is the theoretical exponential function M_1.

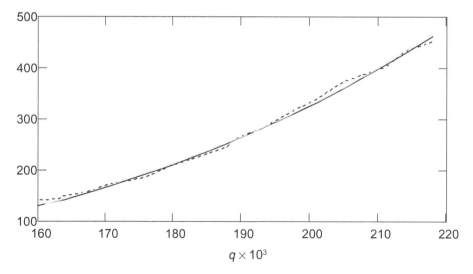

Figure 5.26.   Total number of cracks vs. the number q of work cycles from point C: dotted line is the empirical values, continuous line is the theoretical exponential function M_2 and the dashed line is theoretical power function.

Looking at these figures we can evaluate that the goodness of fit between the theoretical function and the description of the empirical values is very high.

Let us now consider a sequence of residuals for the first function. This is illustrated in Figure 5.27.

Analysing this plot it is easy to see that when the number of work cycles executed by the rope increases, the discordance between the empirical values and the theoretical values increases. One can suppose that there must be a certain physical process generating this dispersion. This process precedes the quality change that happens at point C.

Let us now consider function M_2. Because this concerns the II stage of rope wear, limited values for the numbers of work cycles executed should be found. Unfortunately, the exponential function is useless—there is no maximum value. Therefore, the approximation function must be replaced. Hence, a power function was introduced. Estimates of the structural parameters were as follows:

$$\hat{\delta} = 9.36 \times 10^{-10} \quad \text{and} \quad \hat{\chi} = 4.15$$

and a sequence of residuals is shown in Figure 5.28.

An investigation of these residuals gives no grounds to reject the hypothesis that their sequence is constant with regard to q, although the dispersion between the empirical and theoretical values is high. But remember that during this stage significantly higher values for the total number of cracks are observed. Additionally, autocorrelation in the sequence was investigated and no significant autocorrelation was found. The approximation function fulfils the requirements.

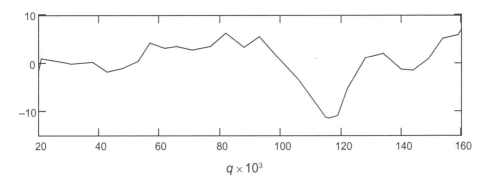

Figure 5.27.   Residuals for the first period considered—up to point C.

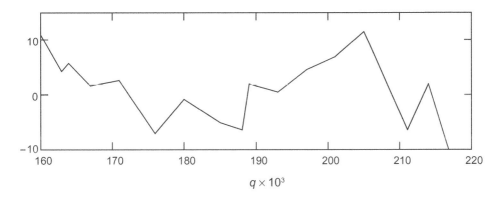

Figure 5.28.   Residuals for the second period considered—from point C.

Now we will produce a reliability evaluation for the rope being investigated. We know that the probability distribution of the total number of cracks in the wires at the moment of the withdrawal of the rope from service can be approximated by the Gaussian distribution provided that the rope wear process is of the I kind. In the case being considered here, the next stage of the rope wear process is taken into account.

From a lecture given by Kopociński (2007) a conclusion can be drawn that the distribution of the random variable of interest can be approximated here by the distribution of the sum of random variables normally distributed. Thus, we still have the Gaussian distribution of appropriately modified parameters. So, the previous way of reasoning can be repeated.

For this method it can be stated that the formulas for determining the limited values θ_{min} and θ_{max} remain unchanged. Assume additionally that the average value is still the arithmetic mean of limited values. The principal change concerns the speed of the increase in the number of cracks. At the first stage of the rope wear process this speed is slow; at the second stage, the speed is high.

Let us estimate the mass of probability of the density function of a random variable: the total number of cracks in the wires up to the moment of the appearance of point C.

Limited values are as follows:

– Limited total number of cracks: $\theta_{min} = 52$ $\theta_m = 136$
– Corresponding numbers of work cycles with these values, from formulas 5.46 and 5.47 are:

$$q_{min} = 78.2 \times 10^3 \quad q_m = 159 \times 10^3$$

Thus, the mass investigated is determined by the formula:

$$\int_{q_{min}}^{q_m} \frac{1}{\sigma\sqrt{2\pi}} \exp\left(-\frac{\left[\theta(q) - \bar{\theta}\right]^2}{2\sigma^2}\right) dq \tag{5.50}$$

and is approximately equal 2×10^{-5}.

This mass can be neglected.

Conclusion: the main phase of rope wear is of the II kind of fatigue wear and this phase should be the main point of further interest.

Now consider the rope reliability function. The limited values for the total number of cracks are:

$$q_m = \left(\frac{\theta_m}{\hat{\delta}}\right)^{1/\hat{\chi}} = 162 \times 10^3 \quad q_{max} = \left(\frac{\theta_{max}}{\hat{\delta}}\right)^{1/\hat{\chi}} = 424 \times 10^3 \tag{5.51}$$

Now construct a probability density function of the random variable for the number of work cycles for which the process of the accumulation of the number of cracks is determined by a power function with the parameters that were found, that is:

$$\hat{\delta} = 9.36 \times 10^{-9} \quad \text{and} \quad \hat{\chi} = 4.15$$

Making all the necessary mathematical transformations one obtains:

$$h(q) = \frac{\delta\chi}{\sigma\sqrt{2\pi}} \frac{1}{q^{1-\chi}} \exp\left(-\frac{\left[\delta q^\chi - \bar{\theta}\right]^2}{2\sigma^2}\right) \tag{5.52}$$

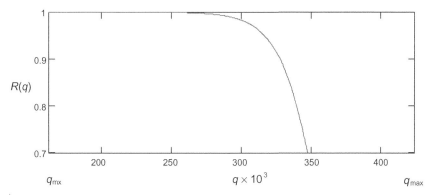

Figure 5.29.   Reliability function R of the rope being considered vs. the number q of work cycles.

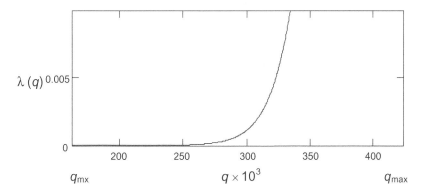

Figure 5.30.   Hazard function λ of the rope being considered vs. the number q of work cycles.

Figure 5.29 illustrates the reliability function of formula 5.49 and the corresponding hazard function is shown in Figure 5.30.

By analysing both plots it is easy to notice that the rope can execute approximately 270×10^3 cycles. After about 300×10^3 cycles the situation changes—the survival function begins to go down drastically and the hazard function increases considerably. Depending on the state of the rope being investigated a decision should be made about either withdrawing this rope from further operation or changing its diagnostics qualitatively. This number of cycles corresponds with a total of 1750 cracks in the wires.

One important statement must be repeated one more time: all these considerations are correct provided that during visual and electromagnetic diagnostics no rope section of 5–6 spiral rope leads with a number of cracks reducing load carrying area by 20% or more has been found.

An interesting characteristic from both the theoretical and empirical points of view is the average remaining lifetime of the rope. This is a local characteristic, a conditional one. Its meaning is as follows. The rope being investigated has executed some number of cycles and is good for further operation. Denote this number of cycles by q_h. The point of interest is the expected remaining time of the rope's operation. Let us denote this average by $E\{M(q_h)\}$. Making use of well-known reliability function transformations (see, for example, Gertzbah and Kordonsky 1966) we get:

$$E\left\{M\left(q_h\right)\right\} = \int\limits_{q_{\max}}^{q_{\max}} \frac{R\left(q_h + q\right)}{R(q)}\,dq \qquad (5.53)$$

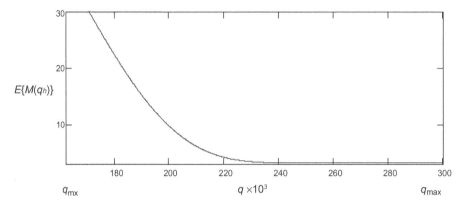

Figure 5.31.   Expected remaining lifetime $E\{M(q_h)\}$ of the rope being considered vs. the number q of work cycles.

Here again—similar to the reliability and hazard functions—there is no possibility of obtaining an explicit mathematical form of this function. However, by using some computer programs a diagram of this expected value can be drawn. It is shown in Figure 5.31.

Analyse this plot carefully. The expected remaining lifetime of the rope decreases exponentially, which should have been expected. This means that a point exists from which, it can be presumed, that the remaining lifetime is constant. The relationship illustrated in Figure 5.31 is a characteristic one. Therefore, the information contained in this picture is that after the execution of approximately 240×10^3 cycles the remaining lifetime is a little above 3,000 cycles.[15]

It looks as though from this moment in time the rope diagnostics should be changed. Moreover, based on the Figure 5.31 a suggestion can be formulated that if a decision is made to continue to use this rope, then perhaps the most convenient course of action will be to diagnose this rope for fewer than 3,000 cycles and, perhaps, more frequently. This figure (of 3,000) can be treated as an assessment based on above analysis.

Because a condition was created where these considerations are correct, keeping in mind the stipulation that there is not a rope section (5–6 rope spiral leads) with a loss of a load carrying area not less than 20%, one can state that these reliability characteristics are conditional ones. For this reason, a careful observation must first be made to see whether such a rope section has occurred and, if not, whether these considerations can be utilised.

5.3.4.1   Lets make a short recapitulation

For the first time empirical and theoretical investigations have allowed the construction of the reliability function for a rope that must be withdrawn from further operation because the degree of its degradation has attained the limit defined by mining regulations. The moment of the rope's removal is a stochastic one, but the probability distribution of the rope's lifetime has been found which allows for the construction of further reliability characteristics such as a hazard function and a function for the average remaining lifetime of the rope. This allows recommendations to be formulated for when a rope should be withdrawn from further operation or when the method of rope diagnostics ought to be changed. What is more, using in-depth empirical-theoretical reasoning it is possible to formulate a suggestion as to what the time interval should be between neighbouring diagnostic actions.

Generally, the results of these considerations should be treated as a supplementary tool to accompany the outcomes of visual and electromagnetic investigations.

---

[15] *Nota bene*, when the rope had executed 218×10^3 cycles and 452 cracks in the wires were noted, the mine management decided to remove the rope from further operation.

It also has to be emphasised that in some cases the course of the rope wear process can be different. For some rope constructions no effects of a wear process are visible for a relatively long period of time but when cracks in the wires start to occur their intensity is high right from the beginning and the rope must be diagnosed more frequently than before. Very often in a short time after the first cracks appear the rope has to be removed.

Additionally, in the regular course of the rope wear process some periods may be noted when by examining the cracks in the wires exclusively, it looks as if the wear process is retarding. It seems like the rope is *resisting* wear—the number of cracks increases slowly, but after a short period of time it looks as if it is *giving up* resulting in a sudden increase in the number of cracks. The reason for this phenomenon is still unknown.[16]

However, it is worth finally remembering that rope transportation is one of the safest displacement methods.

---

[16] The author is not going to say frankly that this is sometimes connected with the inaccuracy of recordings made by mine personnel.

CHAPTER 6

Systems of machines operating in parallel

In mining there are several systems of machines operating in parallel in a reliability sense—that is, machines operate independently of each other. In the above photograph there are two types of machinery systems operating in parallel clearly visible. They are a system of drillers and a system of power shovels.[1]

For a system of this kind, two factors have great significance for the exploitation process and the manner in which it is analysed and calculated. They are:

1. The homogeneity of the system
2. Existence of spare machines that can replace failing units.

Homogeneity is usually understood as the quality of being similar or comparable in kind or nature. The homogeneity of a system is here understood as the homogeneity of machines—that is, their functioning is the same, they execute identical duties and their basic parameters differ in a random way only.

Let us presume that the system which will be considered is:

1. Homogeneous
2. One for which no spare units are available.

---

[1] Actually there is a third machinery system in this photo—a system of auxiliary machines—but it is difficult to see them.

If so, by applying basic reliability principles we can determine the probability distribution of number D of machines in a work state. This number is obviously a random variable. This distribution is determined by the formula:

$$P\{D = d\} = P_d^{(p)} = \binom{n}{d} A^d (1 - A)^{n-d} \quad d \le n \quad n \ge 2 \tag{6.1}$$

where:

$P_d^{(p)}$ is the probability that d machines are in a work state (superscript (p))
A is the steady-state availability of machine
n is the number of machines in the system.

It is easy to identify that this is a binomial distribution describing the number of successes in n independent trials. Two basic statistical parameters—the mean and the standard deviation—are determined by the formulas:

$$E\{D\} = n A \tag{6.2}$$

$$\sigma\{D\} = \sqrt{nA(1 - A)} \tag{6.3}$$

Let us move our consideration to the engineering field. Formula 6.2 determines the average number of machines in a work state in the system, and formula 6.3 gives the corresponding mean deviation of the random variable D with regard to the average value.

The probability of an event that all machines are in an up state is given by:

$$P_{d=n}^{(p)} = A^n$$

and this means full output of the system because of full system availability.

If succeeding machines are in a failure state, the degradation of the availability of the system is visible. Note that the availability of this system is not a zero-one type but it is a function of the number of machines in an up state. Figure 6.1 illustrates the probability distributions of the number of machines in a work state for different numbers of machines in the system ($n = 5$, 7 and 9) where the steady-state availability of machines is $A = 0.700$.

Notice, there is a certain stochastic subtleness. Usually, the binomial probability distribution is connected with the so-called Bernoulli experiment and it determines the probability of the occurrence of successes in a sequence of n independent 0/1 experiments, each of which yields

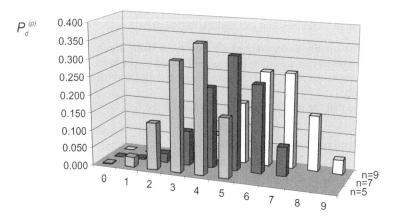

Figure 6.1.   Probability distributions $P_d^{(p)}$ of numbers of machines in a work state for different numbers n of machines in the system with steady-state availability $A_k = 0.700$.

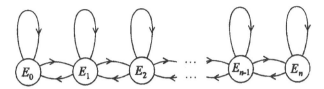

Figure 6.2. Process of changes of states of a system of *n* machines operating in parallel.

success with a certain constant probability. In the case of a system of *n* machines operating independently, we observe an exploitation process similar to the birth-and-death process of a Markov type in the sense that possible transitions happen only between neighbouring states. In mining practice in a general case, probability distributions are not exponential as in the Markov model. If the times of a given state are described by a certain probability distribution (general case), and random variables associated with particular states are independent of each other (which usually holds), we can state that here we have a birth-and-death semi-Markov process limited from above. If *k* out of *n* power shovels are in failure the only possibility to pass to the next state is either when one out of *k* power shovels resumes its duties (repair is finished and renewal occurs) or when—before the occurrence of this event—another power shovel (one out of the *n–k* remaining good units) goes down. This process is illustrated in Figure 6.2, where E_k denotes a state in which *k* machines are down.

The process starts in E_0 and from time to time jumps through some states to the right and later—after some time—returns to the left. In some cases, when moving considerably away from E_0 a return to the left can be done with a small jump of one or two states to the right. For more reliable machines the process spends a lot of time in E_0. The greater the frequency of the appearance of the state E_0, the shorter the periods of the process movement to the right. Let us now make a short translation of this probable scheme into engineering language. Truck dispatchers usually make their decisions by predicting the next state of the exploitation process—some shovels are down but the dispatcher has got information that soon they will be repaired and for this reason he commences to direct more trucks to the pit.

Let us return to the main consideration of this chapter. Two additional problems should be taken into consideration here.

First, even if a machine is very expensive and should operate round-the-clock, this does not mean that the machine does so. Some breaks in its operation are inevitable. A shovel, for instance, after loading up the total amount of blasted rock at a given place must move to a new loading point. A driller after making a hole in one place must move to a new point marked by a peg. Moreover, operators of machines must be changed. Some breaks during longer periods of work are compulsory, and so on. All these mean that apart from reliability states such as work and repair, a further state must be taken into account—the state of the inaccessibility of a machine for the execution of its duties. This is not a reliability state but an exploitation one, as from a reliability standpoint such a machine is still in a work state.

A basic measure of this state is an accessibility coefficient, *B*. Taking into account that:

a. Machines' states of repair and accessibility to execute duties are independent of each other.
b. Machines operate independently of each other.

we can construct the probability distribution of the number of machines in a state of accessibility to execute their duties. This distribution is given by:

$$P_d^{(zd)} = \binom{n}{d} G_k^d (1 - G_k)^{n-d} \quad d = 1, 2, \ldots, n \quad G_k = AB \tag{6.4}$$

where $P_d^{(zd)}$ is the probability that *d* machines are in a state of accessibility to execute their duties, that is *d* machines are able to accomplish their tasks.

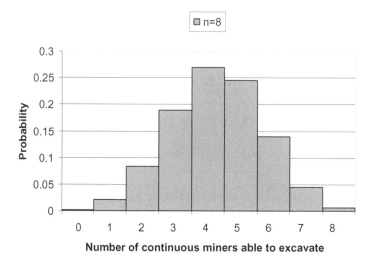

Figure 6.3.   Probability distribution of the number of continuous miners able to excavate.

Figure 6.3 shows the probability distribution of a number of continuous miners able to excavate operating at a certain level of a coal mine. The system consists of 8 units, $G = 0.47$ and their exploitation processes run in parallel.

Let us look more carefully at the exploitation process of n machines working in parallel. One event—a state occurrence—is a peculiar one. This is when all machines are down. The system has stopped performing its task entirely. Following the previous line of reasoning, the area of consideration can be developed. A system stops performing its task completely because either all machines are in repair, or some machines are in repair but some are in a state of inaccessibility for loading, or all machines are in a work state but in a state of inaccessibility for loading.

Neglecting here how rare such an event is for a given type of system and the reliability of its machines, let us look at this state and consider the consequences. Generally, there are two types of effects:

a. If all machines in this system do not cooperate directly with other pieces of equipment, there is no repercussion apart from the fact that the output of the system is nil. This might be the case, for example, for a system of drillers making blasting holes for excavation.
b. If machines in this system cooperate with other pieces of equipment, the repercussion can be significant for the exploitation process of the general system.

If, for example, the time to finish the nearest repair of the shovel (the earliest renewal point in the exploitation process of shovels) or the time to terminate the inaccessibility shovel state is longer than the truck time work cycle, the whole process of exploitation of the system regenerates itself. In that time all trucks will reach their points of destination, some trucks will be in queues at shovels waiting for a restart of loading, some trucks will be in reserve, some trucks will be in a state of inaccessibility for loading—refilling their tanks, or through drivers having a coffee break, etc. Truck movement deteriorates. When the process of loading is recommenced and one after another shovels begin to load, the exploitation process characteristics in this period of time are different from those when the process is stabilised and runs regularly. After loading restarts the shovels operate continuously because there are many truck boxes to fill. The system of hauling units is not intensively used, and some dumpers are still in queues. Some time is needed to stabilise the general system characteristics.

It is easy to notice that the fewer the shovels in the system, the lower their reliability and accessibility, and the higher the probability of the occurrence of such a singular state.[2] In spite of the fact that this peculiar event is rare, the probability of its occurrence is important. Making calculations of the whole machinery system in all possible states must be considered and the sum of all probabilities must be close to unity. Therefore, in some formulas a normalisation constant must be included to obtain the correct outcomes for the calculations. This constant is a function of the probability of the occurrence of this singular state.

It is worth pointing out that a second event of this type occurs when all trucks do not transport. The probability of the occurrence of this state decreases with an increase in the number of trucks being used in the system and increases when the reliability of hauling machines is reduced. For the majority of large mine machinery systems, the probability of the occurrence that all hauling units are down is so small that it can be neglected.

Similar considerations to a certain degree can be formulated for the exploitation process of continuous miners cooperating with shuttle cars, for example. However, for some machinery systems the importance that should be attached to a singular state is very small and for some systems the problem does not exist.

If a given machinery system has a reserve of units, the probability of the occurrence of a singular state is diminished. If the reserve size is sufficient, it can be assumed that when any machine in the system is down there will always be a spare unit to make up the system loss. In such a case the probability of the occurrence of a singular state is practically zero and the normalisation constant is 1 and can be removed from formulas. An example of such a system is a shovel system with a system of wheel loaders, which are usually plentiful in large open pits.

One further important point should be taken into account in connection with this analysis. If for a given machinery system spare units are identical to the regular units, no special additional consideration is necessary. Conversely, however, if the spare machines are different, the homogeneity of the system is not preserved. The method of system calculation must be modified (see Chapter 10).

---

[2] Appearance of this state in shovel-truck system was considered more comprehensively in Czaplicki's monograph (2009, pp. 92–93).

CHAPTER 7

Systems of machines in continuous operation

7.1 INTRODUCTION

Some basic types of mine machinery systems (Chapter 2.2) are involved in so-called 'continuous mining'. This concerns both underground and surface mining operations. A general principle for machinery systems of this kind is that excavation, hauling and dumping is accomplished continuously. There are also some particular pieces of mine equipment that can be encountered in this class but this is hard to detect at first glance (see Chapter 7.6).

In underground mining a continuous type of operation concerns primarily coal mining but also mining of mineral deposits that are easy to excavate, such as salt, trona or potash. For surface mining, this term is associated mainly with opencast operations, including mining of bedded type deposits such as lignite, iron ore, copper ore—and all of those mineral deposits from an early stage of a geological age. Continuous mining concerns first of all young rocks that are easy to extract. Mining under water (dredging for instance, see Golosinski and Boehm 1987) is also done by applying continuous mining in the majority of cases.

In surface mining the largest machines are used for this type of mining. Bucket wheel excavators, bucket chain excavators, surface continuous miners, many different types of conveyors with several belt units, belt wagons, stackers and reclaimers are machines involved in this type of mining.

In underground operations involving continuous mining, first and foremost are the machines that extract coal: primarily continuous miners and longwall units, shearers and ploughs. Furthermore, a large number of chain conveyors and belt conveyors operate underground. Additionally, pneumatic and hydro-transportation can be added to the list with some pieces of equipment operating in dressing plants.

Let us now look at these pieces of equipment that comprise large machinery systems, keeping in mind their exploitation process. Luckily, looking at the processes of changes of states of these machines, it can be recognised that the courses of action can be successfully described by Markov processes. The vast number of reliability investigations done in Poland (mainly in the 1970s) into shearers, ploughs, AFCs, and underground belt conveyors as well as surface units, stackers and reclaimers allow the conclusion that their processes of changes of states of the work-repair type are of a Markov[1] type in a great number of cases, that is probability distributions of work and repair times can be satisfactorily described by exponential distributions and times of states are mutually independent. Therefore, a method for calculations of such systems should be placed in the part of the theory of probability that concerns Markov processes. This fact is extremely practical because almost all of the basic functions obtained are easy to calculate. In the 1970s after several years of both empirical and theoretical investigations in Poland, two procedures were constructed allowing for the comprehensive analysis of the exploitation processes of these systems. One procedure was developed by the Wroclaw University of Technology. A second one, a bit later, at the Silesian University of Technology. The latter procedure is simpler and it will be presented here.[2]

This method is a precise one that takes into account all possibilities. It is communicative and easy to understand. However, for large and complicated systems, it is too detailed. The most painful and tedious point is the identification of a set of technically unfeasible states for the exploitation process of the system being analysed.

Let us discuss this modus operandi. It can be presented as a sequence of steps designed to get a final result. These points are as follows:

1. Determination of a system $\mathfrak{S}$ to be considered

   a. Determination of the elements of the system
   b. Determination of the reliability of the elements of the system
   c. Determination of the reliability structure of the system
   d. Determination of the method of utilisation of the system
   e. Determination of the method of maintenance of the system

---

[1] To be precise, a stochastic process whose state at time t is $X(t)$, for $t > 0$, and whose history of states is given by $x(s)$ for times $s < t$ is a Markov process if

$$P\{X(t+h)=y|X(s)=x(s),\forall s\leq t\}=P\{X(t+h)=y|X(t)=x(t)\} \quad \forall h>0$$

That is, the probability of its having state y at time $t + h$, conditioned on having the particular state $x(t)$ at time t, is equal to the conditional probability of its having that same state y but conditioned on its value for *all* previous times before t. It has been proved that the Markov process is a mathematical model for the random evolution of a memoryless system.

[2] By the way, in 1989 the Pavlović publication where theoretical problems of the reliability of continuous systems are considered was published. Reading this book two impressions are inevitable—it is the only book on this theme in the world (there is no reference at all) and no connection with mine practice is visible.

2. Reduction of the elements in series
3. Determination of an exploitation repertoire ℭ

    a. Determination of the number of theoretically possible states
    b. Determination of the states

4. Exploitation graph construction

    a. Graphical project of the transition between states
    b. Verification of states that are technically unfeasible

5. Construction of the matrix of transition intensities between states
6. Construction of equations determining the limited probability of states
7. Solution of the equations
8. Verification of the results obtained, conclusions.

The presented methodology comprises many points and contains some terms that should be defined. Let us discuss them in a sequence.

The first point is the determination of the system for which an exploitation process will be analysed. An adequate set of information is required. At the beginning we should know the number and types of elements in the system (1a). Obviously, information is also needed on the reliability of these elements (1b). Information on the reliability structure[3] of the system ℨ is also needed (1c). If the system being considered has a hierarchical structure[4] (some elements are more important than others), further comprehensive information is compulsory:

- the relationship between the elements of the system must be specified
- if there is a reserve in the system, information on the elements in the reserve must be given together with the sequence of how particular elements will be switched in order to work; likewise information on the type of reserve applied is necessary.

Determination of the method of system utilisation (1d) should include a description of how particular elements will operate, sometimes in relation to some other system elements.

Determination of the method of system maintenance (1e) will require information on the number of repair teams available together with information on the order of repairs in relation to the occurrence of failures. If the number of repair teams is considered to be a subsystem of maintenance, it is assessed as:

- sufficient, if the number of repair teams is identical to the maximum number of repairs that can occur in the system simultaneously—this means that all failures will be cleared out without delay; notation $\mathfrak{Q}_1$
- insufficient, if otherwise; notation $\mathfrak{Q}_2$.

In the second case further information is required describing in detail the number of repair teams and the order in which failures will be cleared out.

A reduction of the elements connected in a series (2) to one conventional element is possible and allows, sometimes significantly, the number of system states it is necessary to consider to be reduced.

---

[3] A definition of the reliability structure of a system does not exist in the most books on the theory of reliability but many authors use this term. It looks as though the definition given by Bobrowski (1985) is just right for our considerations: 'This is a function assigned to the states of the elements state of the system created by these elements'. According to some authors 'this is a system description that takes into account reliability events that occur in the system generated by its elements'.

[4] Again, no definition of a hierarchical reliability structure can be found in books on the theory of reliability. Perhaps, the most communicative is a literary definition: 'All … are equal, but some … are more equal'—Orwell.

The next point (3) relies on the determination of the exploitation repertoire $\mathfrak{E}$. Recall, an exploitation repertoire is a defined set of all possible states of the system. Calculation of all theoretically possible states of the system is based on the following principle.

If the number of elements of the system is N, where n_1 elements can be in k_1 states, n_2 elements can be in k_2 states and so on up to m, that is:

$$\sum_{i=1}^{m} n_i = N$$

then the number of theoretically possible states is determined by formula:

$$L_t = \prod_{i=1}^{m} k_i^{n_i} \tag{7.1}$$

Determination of states (3b) relies on the enumeration of all possible states of the system.

Recall, an exploitation graph (4) is a set of vertices that illustrate states connected by arcs in such a way that each arc begins in one vertex and ends in another vertex. Vertexes represent states and are numerated, and arcs represent possible connections, transitions between states. Arcs have directions showing that passage from one state to another is possible. The intensity of this passage is assigned to each arc if needed and therefore an exploitation graph is a picture of the principles of transition between the states of system, or to be more precise, the graph illustrates transition between states that are components of an exploitation process of the system.

The procedure of the construction of an exploitation graph is as follows. All states contained in the exploitation repertoire $\mathfrak{E}$ are represented by vertices. Knowing what each state means, consideration is focused on identifying all feasible one-step changes of a given state. These changes are next shown in the graph by appropriate arcs. If a state exists that is not connected by an arc to any other state, this means that this state is technically unfeasible and it should be rejected from further considerations.

The construction of an exploitation graph is not indispensable. This point can be neglected and you can move on to consider point 5. However, the construction of an exploitation graph—if it makes sense—possesses two crucial merits:

- it is easier to determine a set of system states that are technically feasible[5]
- the graph illustrates the principle of transition between states well and—speculatively—the exploitation process of the system; it has great didactic merit.

For an exploitation process with a large number of states the construction of a graph is useless as a rule, as such a graph is unreadable.

The next step (5) is the construction of a matrix of transitions between states. This matrix is an equivalent of an exploitation graph. The matrix is square, with the number of rows and columns corresponding with the number of states. This principle holds: 'From state of number … (number of row) to state of number … (number of column)'. Elements lying on the main diagonal are calculated knowing the principle from the Markov theory of states that says that the sum of all intensities in a given row must be zero (see, for example, Kopociński 1973 p. 38). An element of the main diagonal is the intensity of the remaining process at this state. Usually, in the matrix only non-zero intensities are shown presuming that all intensities not specified are nought.

---

[5] There are three methods for determining whether a set of states is technically feasible, namely by means of an analysis of the schemes of passages in an exploitation graph, by an analysis of the matrix of transitions between states and by using some computer programs.

This matrix allows all unfeasible states to be traced. If any column (row) has no intensity assigned, it means that this state is technically unfeasible.

Construction of equations to determine the limited probability of states (6) is made using information contained in the matrix. Equations are based on the principle that a sum of products: the probability by intensity for a given column must be zero. The number of the row indicates which limited probability is being considered but intensities are taken one by one from a given column. In such a way the number of equations equals the number of unknown limited probabilities. It looks like a solution for the set of equations of the probabilities being searched can be obtained. This is not true. This set of equations is indeterminate. It is necessary to reject one equation and to add an equation that the sum of all probabilities must be closed to unity.

Verification of the results obtained and conclusions (8) are made separately for each case being considered.

7.2   PRINCIPLES OF REDUCTION OF SERIES SYSTEMS

There were two independent developments in this part of the theory of Markov processes. Both roughly occurred at the same time—in the early 1960s. The first was connected with the development of mining problems in Poland. Possibly as a result of the need to export greater amounts of coal, there was a distinct call addressed to research centres to develop methods for the precise calculation of belt conveyor systems. In those days two approaches were being used in this field—analytic and simulation ones.

The application of computer techniques to analyse belt conveyor systems and later systems of continuous operation was undergoing a great evolution throughout the world during this decade (see, for example, Rist 1961, Teicholtz 1963, Bishele et al. 1964, Harvey 1964, Aurignac et al. 1968, Bucklen et al. 1968, Eichler 1968, Juckett 1969). In Poland in addition to simulation, which progressed at a relatively leisurely pace, significant progress was observed in analytical methods of calculating these systems (witness Gładysz 1964, Battek 1965, Battek et al. 1969 or Sajkiewicz 1973–75).

Generally, the theory of mine machinery systems of continuous operation which developed in a mature form in the 1970s has remained in an almost unchanged form since then. There was no special need to conduct further theoretical investigations, especially because of continuous progress in the field of simulation. Regularly every few years a new enhanced method of simulation, just for mining purposes, came into being and created some new possibilities in this regard, see, for example, Mutmansky and Mwasinga (1988), Sturgul (1989), and Tan and Ramani (1988). Both ways of modelling, analysis and calculation, have their own merits and demerits but—what is important—they complement each other. Nevertheless, it is more appropriate to discuss some problems analytically, and analyse other problems using simulation.

It is more practical to consider problems related to servicing and making repairs methodically by considering the transitions between the states of a given system, if the system is a basic or simple one and not too complicated. For complex systems a matrix of transitions between states and the corresponding exploitation graph is complicated, expanded and usually difficult to decipher. The application of simulation is a more practical tool for analysis in such cases.

The second general development in the theory of Markov processes was not connected with mining but with the considerable progress being made in reliability theory (Gnyedenko et al. 1965, for example). Equations appeared from these two independent developments that became the principles for the reduction of systems of elements connected in a series.

Before presentation of these principles, recall the definition of a series system in a reliability sense.

One definition says that a system has elements connected in a series if, and only if, any failure of any element means the failure of the whole system. This kind of reliability structure is the worst of any possible system structure.

7.2.1   *The first principle of reduction of elements in series*

The intensity of failures λ_s in a system of n elements connected in a series is the sum of the intensities of its elements λ_i provided that the process of changes of states of the system is a Markov one, that is:

$$\lambda_s = \sum_{i=1}^{n} \lambda_i \tag{7.2}$$

It worth noting that if the probability distributions of work and repair times do not belong to an exponential family of distributions, this equation holds but is limited, that is it is correct if the system operates over a long period of time (Gnyedenko et al. 1965).

7.2.2   *The second principle of reduction of elements in series*

The repair rate κ_s of a system of n elements connected in a series is the sum of the repair rates of its elements κ_i provided that the process of changes of states of the system is a Markov one, i.e.

$$\kappa_s = \sum_{i=1}^{n} \kappa_i \tag{7.3}$$

Again it is true that if the probability distributions of work and repair times are not an exponential one, the above equation holds but is limited, that is it is correct if the system operates over a long period of time (Gnyedenko et al. 1965).

If all elements are the same:

$$\lambda_s = n\lambda_i \text{ and } \kappa_s = n\kappa_i$$

which is obvious.

Recalling equation 5.19:

$$\lambda = (T_w)^{-1}$$

we are able to calculate the average work time T_{ws} of the system according to the formula:

$$T_{ws} = \left\{ \sum_{i=1}^{n} \frac{1}{T_{wi}} \right\}^{-1} \tag{7.4}$$

where T_{wi} is the average work time of the i-th element.

Similarly, the average of repair time T_{ns} of the system can be calculated from the formula:

$$T_{ns} = T_{ws} \sum_{i=1}^{n} \kappa_i \tag{7.5}$$

If all elements of the system are identical:

$$T_{ns} = T_{ni}$$

Using formulas 5.19 and 7.3, the following relationship can be obtained:

$$\mu_s = \frac{\lambda}{\sum_{i=1}^{n} \kappa_i} \tag{7.6}$$

which determines the intensity of system repairs μ_s.

Remembering formulas 5.23 or 5.25, we can state that:

$$A_s = \frac{1}{1 + \sum_{i=1}^{n} \kappa} \tag{7.7}$$

which means that the steady-state availability (that is the probability that at any moment in time the system is in a work state) of a system of this kind is determined by formula 7.7. This formula illustrates how the reliability of a system decreases if the number of elements applied increases.

Notice, that principles of reduction—formulas 7.2 and 7.3—permit the whole series system to be substituted by single conventional element λ_s and κ_s reliability indices.

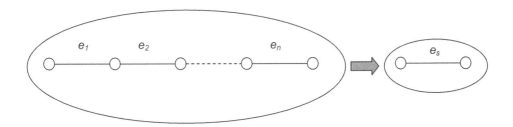

7.3 CASE STUDIES

Consider now a few examples of machinery systems that are in frequent use in mines. Let us do an analysis of these systems according to the methodology we have outlined.

Let us start from a basic system of a chute type, that is two elements that deliver masses I_1 and I_2 onto a third element.

- **Example 7.1**

1. There is a system

$$\text{❧} : < e_i, i = 1, 2, 3; \lambda_i, \mu_i; \text{Ω}_1 >$$

In this notation, the entry '❧:' means 'system ❧ is determined by'. The system consists of three elements of intensities of failures λ_i and intensities of repair μ_i. The maintenance subsystem is sufficient—notation Ω_1, which means that all repairs will be cleared out without delay.

The reliability structure of the system is shown in Figure 7.1. Let us solve this system.

2. Reduction of the elements in a series is obviously not needed.

It could be suggested that perhaps all these elements are just what remains after reduction. We may guess that element e_1 and e_2 before reduction comprised a winning machine of continuous operation and several hauling units, e.g. a shearer, two AFCs and some belt conveyors if underground mining is being investigated. For opencast mining it may be, for example, a BWE and some belt conveyors and inclined chutes. The third element may consist of a certain number of belt conveyors, or a combination of belt conveyors and a stacker if opencast mining is being investigated.

3. An exploitation repertoire ☞

The number of theoretically possible states is:

$$L_i = 2^3 = 8$$

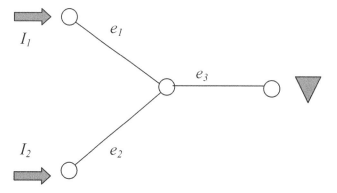

Figure 7.1.   Scheme of three-element chute system.

Table 7.1.   Exploitation repertoire $\mathfrak{E}$ for a three-element chute system.

State of system	Element			Effect
	e_1	e_2	e_3	
$\mathfrak{s}_1$	W	W	W	$I_1 + I_2$
$\mathfrak{s}_2$	R	W	W	I_1
$\mathfrak{s}_3$	W	R	W	I_2
$\mathfrak{s}_4$	S	S	R	0
$\mathfrak{s}_5$	R	R	S	0
$\mathfrak{s}_6$	S	R	R	0
$\mathfrak{s}_7$	R	S	R	0
$\mathfrak{s}_8$	R	R	R	(–)

Notation:
$\mathfrak{s}$ – state of the system
W – work
R – repair
S – standstill[6]
(–) – state technically unfeasible

Let us now construct a table in which all states will be shown. Additionally, the effect of whether mass is hauled or not in relation to the configuration of given states is also included. Results of this construction are shown in Table 7.1.

By analysing the information contained in Table 7.1 it is easy to recognise that:

- three states assure mass flow, from different sources, depending on the state of the system
- there are four states of the system repair: $\mathfrak{s}_4$, $\mathfrak{s}_5$, $\mathfrak{s}_6$ and $\mathfrak{s}_7$
- state $\mathfrak{s}_8$ is technically unfeasible.

A stochastic 'mechanism' that blocks passage into state $\mathfrak{s}_8$ is an assumption that an element in a standstill state will not fail.

---

[6] A standstill state is not the 'own' state of any element. It is a 'by-product' of the organisation of elements in the system.

4. Exploitation graph construction

An exploitation graph for this system is shown in Figure 7.2. It is clear that there is no arc directed to state ϑ_8 and for this reason this state has been excluded.

5. Construction of a matrix of intensities of transitions between states

By looking at Table 7.1 or at Figure 7.2 the matrix shown below can be constructed. Only non-zero matrix elements are incorporated. Elements on the main diagonal are obviously determined as follow:

$$\Delta_1 = -(\lambda_1 + \lambda_2 + \lambda_3)$$
$$\Delta_2 = -(\mu_1 + \lambda_2 + \lambda_3)$$
$$\Delta_3 = -(\mu_2 + \lambda_1 + \lambda_3)$$

and so on.

$$
\begin{pmatrix}
\Delta_1 & \lambda_1 & \lambda_2 & \lambda_3 & & & \\
\mu_1 & \Delta_2 & & & \lambda_2 & & \lambda_3 \\
\mu_2 & & \Delta_3 & & \lambda_1 & \lambda_3 & \\
\mu_3 & & & \Delta_4 & & & \\
& \mu_2 & \mu_1 & & \Delta_5 & & \\
& & \mu_3 & \mu_2 & & \Delta_6 & \\
& & \mu_3 & & \mu_1 & & \Delta_7
\end{pmatrix}
$$

6. Construction of equations determining the limited probability of states

$$P_i, i = 1, 2, \ldots, 7.$$

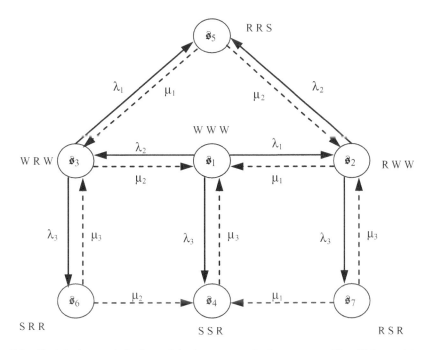

Figure 7.2.   Exploitation graph for the exploitation process of a chute system with sufficient maintenance.

Applying the principle defined in the methodology (see Chapter 7.1) these equations will be:

$$\Delta_1 P_1 + \mu_1 P_2 + \mu_2 P_3 + \mu_3 P_4 = 0$$
$$\lambda_1 P_1 + \Delta_2 P_2 + \mu_2 P_5 + \mu_3 P_7 = 0$$
$$\lambda_2 P_1 + \Delta_3 P_3 + \mu_1 P_5 + \mu_3 P_6 = 0$$

and so on.

Recall, it is necessary to reject one equation and to add:

$$\sum_{i=1}^{7} P_i = 1$$

The limited probability distribution will be obtained by solving these equations.

Having all the estimated probabilities, it is easy to evaluate the system's effective output, W_{ef}; bear in mind, that an effective output is such an output that can be expected over a long period of a system's operation.

Here we have:

$$W_{ef} = (l_1 + l_2)P_1 + l_1 P_2 + l_2 P_3$$

The probability of a system failure that means 'no mass delivery' is given by:

$$P_{sf} = \sum_{i=4}^{7} P_i$$

Let us now calculate this system based on some data taken from underground coal mining. Consider the system shown in Figure 7.3.

Elements e_1 and e_6 are coal winning machines (shearers), elements e_2 and e_6 are AFCs working at long walls, elements e_3 and e_8 are AFCs (stage loaders), and elements e_4, e_5, e_9, $e_{10}-e_{13}$ are belt conveyors hauling streams of coal.

The reliability parameters of these pieces of equipment were as follows:

$$\lambda_1 = 55 \times 10^{-3} \quad \lambda_2 = 38 \times 10^{-3} \quad \lambda_3 = 28 \times 10^{-3} \quad \lambda_4 = \lambda_5 = 9 \times 10^{-3}$$
$$\lambda_6 = 65 \times 10^{-3} \quad \lambda_7 = 30 \times 10^{-3} \quad \lambda_8 = 19 \times 10^{-3} \quad \lambda_9 = 8 \times 10^{-3} \quad \lambda_{10} = \cdots = \lambda_{13} = 4 \times 10^{-3}$$
$$\mu_1 = 0.92 \qquad \mu_2 = 0.83 \qquad \mu_3 = 1.02 \qquad \mu_4 = \mu_5 = 1.10$$
$$\mu_6 = 0.85 \qquad \mu_7 = 0.75 \qquad \mu_8 = 1.06 \qquad \mu_9 = 0.96 \qquad \mu_{10} = \cdots = \mu_{13} = 0.92$$

Remember, all these reliability indices have dimensions: here they are h^{-1}.

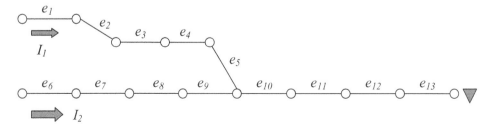

Figure 7.3.   Scheme of a winning-hauling system in an underground coal mine.

By making a reduction of all elements connected in a series, the system is reduced to the one shown in Figure 7.1. The reliability indices of new conventional elements are as follows:

$$\lambda_I = \lambda_1 + \lambda_2 + \lambda_3 + \lambda_4 + \lambda_5 = 139 \times 10^{-3}$$

$$\lambda_{II} = \lambda_6 + \lambda_7 + \lambda_8 + \lambda_9 = 122 \times 10^{-3}$$

$$\lambda_{III} = \sum_{i=10}^{13} \lambda_i = 16 \times 10^{-3}$$

$$\kappa_I = \frac{\lambda_1}{\mu_1} + \frac{\lambda_2}{\mu_2} + \frac{\lambda_3}{\mu_3} + \frac{\lambda_4}{\mu_4} + \frac{\lambda_5}{\mu_5} = 149.38 \times 10^{-3}$$

$$\kappa_{II} = \frac{\lambda_6}{\mu_6} + \frac{\lambda_7}{\mu_7} + \frac{\lambda_8}{\mu_8} + \frac{\lambda_9}{\mu_9} = 142.73 \times 10^{-3}$$

$$\kappa_{III} = \sum_{i=10}^{13} \left(\frac{\lambda}{\mu} \right)_i = 17.39 \times 10^{-3}$$

These parameters can now be included in the set of equations to get the limited probability distribution of the states of the system. These probabilities are:

$$P\{W,W,W\} = P_1 = 0.750 \quad P\{R,W,W\} = P_2 = 0.111 \quad P\{W,R,W\} = P_3 = 0.107$$
$$P\{S,S,W\} = P_4 = 0.015 \quad P\{R,R,S\} = P_5 = 0.016$$
$$P\{S,R,R\} = P_6 = P\{R,S,R\} = P_7 = 0.000$$

Let us go a little further in our considerations. Presume that the hauling capacity of all conveyors carrying the first stream of coal is $Q_I = 1200$ t/h and the hauling capacity of all the conveyors hauling the second stream is $Q_{II} = 1600$ t/h. The main line has 2800 t/h capacity. The question is what is the transporting capacity Q_{ef} of the system if its reliability is taken into consideration?

Keeping in mind the above probability distribution one can write down:

$$Q_{ef} = (Q_I + Q_{II})P_1 + Q_{II} P_2 + Q_I P_3 = 2405 \text{ t/h} \qquad \blacktriangleleft$$

This last equation describes the system's hauling capacity taking into account the reliability elements of the system. But there is one sensitive and extremely vital problem: whether the system has been selected properly. Analysing this issue in a more in-depth way one can see that at the very beginning there is the problem of a certain contradiction between the character of the stream of mineral generated by the winning machine and the hauling capacity of the conveyor. As mine investigations have shown, the stream of mineral generated by a winning machine like a shearer, plough or continuous miner can be satisfactorily described by a Gaussian process (see, for example, Antoniak 1990, Wianecki 1974) but the normal probability distribution applied has to be truncated at least to zero (Czaplicki 1994). Let us look at an example plot of a stream of mineral generated by a winning machine. It is shown in Figure 7.4. Variable *x* means the output.

Choosing a hauling device (say a conveyor) of a given capacity we make a cut by selecting a point on the x-axis x_{sel}. A stream of mineral of greater output than the conveyor's capacity will not be taken by the selected hauling unit. A portion of recently extracted rock will remain at the working face creating a slowdown in the cutting action of the winning machine—sometimes it will block this action for some time and sometimes this portion will be taken by the conveyor in the coming minutes. But the real output of the winning machine is reduced slightly. It looks as though this problem can be solved by designing a conveyor with a higher transporting capacity. In surface mining such an approach can sometimes be applied but in many cases it is uneconomical. The problem is not that the hauling unit's dimensions will be too large, but in the associated repercussions of changing the conveyor. In underground mining a larger size conveyor

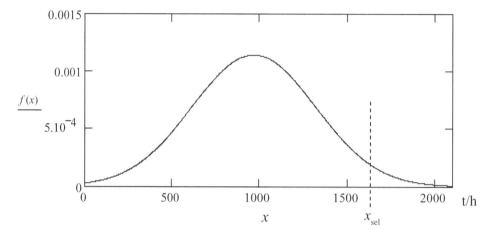

Figure 7.4.   Probability density function of a stream of rock extracted by winning machine.

needs a larger intersection of transporting drive, which is extremely expensive. In surface mines many winning machines have one or two conveyors that are constructional elements within these machines. If a greater size of conveyor is to be used, this may mean that a larger winning machine will be needed. Again, there are additional expenses, perhaps irrational ones.

Conceivably, this problem should be considered taking into account a different problem. How often will a given output occur in a given case? Such a question will lead to selecting level X_{sel} by applying the equation:

$$X_{sel} = \bar{X} + \vartheta \sigma_x$$

where:

$\bar{X}$ is the mean of the stream X of mineral
σ_x is the standard deviation of X
ϑ is the quintile of the standardised normal distribution.

Such an approach has been presented in some publications (see Franasik and Żur 1983, Antoniak 1990, Czaplicki 1994 and some Russian papers cf. Antoniak 1990). The problem that should be solved here is the proper selection of the quintile which determines the mass of probability to be taken into account.

Nevertheless, this attitude to the problem of the selection of conveyor output neglects the economic side of it entirely. If the value of the rock extracted is high we are inclined to agree to selecting a greater value of X_{sel}. Keeping this aspect in mind the theory of decision functions should be applied (Wald 1973, Matusita 1951) which allows this economic aspect to be included. This problem can be formulated in such a way. We are looking for an output of a conveyor for which the total economic losses are minimised. Two aspects of these losses are usually considered: losses due to the lower production of a winning machine and losses due to the application of a larger conveyor and the further pieces of equipment that are needed to use the larger conveyor effectively. Figure 7.5 shows an example plot for these functions presuming that they are linear ones. Production losses are more acute and for this reason the function increases rapidly when the output of the conveyor is reduced. Oversized equipment losses are not severe and the corresponding function increases slowly compared to production losses. In considering this problem one factor should be kept in mind. The stream of mineral flowing at a working face has a great dispersion. When the hauling distance increases, this stream is 'cooling down' and the dispersion becomes reduced. A greater dispersion of a stream of mineral calls for a greater hauling output but this requires money, work and time. In underground conditions there is a shortage of these items.

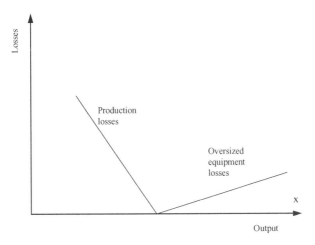

Figure 7.5.   Losses vs. selected output of conveyor.

Quite similar problems can be found in some other areas of mining engineering. For example, consider a fleet of trucks. When a hauler is down it is directed to the repair shop. The number of failed trucks in the system at a given moment is a random variable but the number of repair stands is a deterministic value. The problem is how to properly select the number of repair stands. The criterion proposed is to choose the number of stands such that the average queue of failed trucks waiting for repair will be negligible (Czaplicki's monograph 2009).

These considerations may be taken further but they are not entirely within the scope of this book.

Let us return to the analysis of sequent examples. Analysing the system we have been considering in Example 7.1 more carefully we might come to the conclusion that if the system is small and the occurrence of failures is not so frequent, perhaps there is no need to keep two repair teams available continuously. Let us discuss a case where only one team is available. Assumption $\mathfrak{A}_1$ must therefore be rejected.

If so, first a method of maintenance must be specified concerning the sequence of repairs to be made if two elements are in a failure state at the same time. Therefore, we have to carefully consider all states with two elements in failure, that is states $\mathfrak{s}_5$, $\mathfrak{s}_6$ and $\mathfrak{s}_7$. Notice that the methodology up to point 3 remains intact.

We can presume, for instance, that elements e_1 and e_2 are equally important but element e_3 should have priority because if it is in failure there can be no production. If so, we may presume that:

a. if failure occurs in the third element while element e_1 or element e_2 is being repaired, the current repair is stopped and the repair team goes to the third element and clears out the failure that has just occurred and then the team returns to finish the stopped repair
b. if failure occurs in element e_1 while element e_2 is being repaired or *vice versa*, the repair team does not halt its action.

Observe that one item in this consideration may be different in practice. If one stream of mineral is significantly greater than another a new priority can be formulated:

c. element e_3 is the most important
d. the element transporting greater mass (e_1 or e_2) is more important that the other one (e_2 or e_1).

One additional assumption is still compulsory. An answer to the following question must be found. What kinds of repercussions are connected with the fact that a repair might be stopped? This situation is similar to that in queue theory (considered in Chapter 8.1), which client should be serviced first?

Here a few different answers can be formulated, namely:

e. no repercussions: a repair after stoppage is continued, and there is no change to the time of repair
f. there is a repercussion: the repair must start again from the beginning
g. there is a repercussion: the repair is continued but it lasts a bit longer than without the stoppage.

It looks as if case (f) does not hold for the elements of continuous mining systems. Case (g) is more likely, however, as the extension in the time of repair is usually small. For calculation purposes, it is more practical if this increment is neglected. Therefore, we can presume that case (e) is the most frequent.

Notice that by selecting a given set of assumptions we obtain a different course of the exploitation process of the system. Based on ordinary logic we can construct an exploitation graph of interest.

Let us now look at effects of some assumptions. Figure 7.6 illustrates an exploitation graph for a chute system presuming assumption set: $\mathfrak{Y} = \{(a), (b), (e)\}$.

Let us construct matrix of transitions between states. It is given below.

$$
\begin{pmatrix}
\Delta_1 & \lambda_1 & \lambda_2 & \lambda_3 & 0 & 0 & 0 \\
\mu_1 & \Delta_2 & 0 & 0 & \lambda_2 & 0 & \lambda_3 \\
\mu_2 & 0 & \Delta_3 & 0 & \lambda_1 & \lambda_3 & 0 \\
\mu_3 & 0 & 0 & \Delta_4 & 0 & 0 & 0 \\
0 & \mu_2 & \mu_1 & 0 & \Delta_5 & 0 & 0 \\
0 & 0 & \mu_3 & 0 & 0 & \Delta_6 & 0 \\
0 & \mu_3 & 0 & 0 & 0 & 0 & \Delta_7
\end{pmatrix}
$$

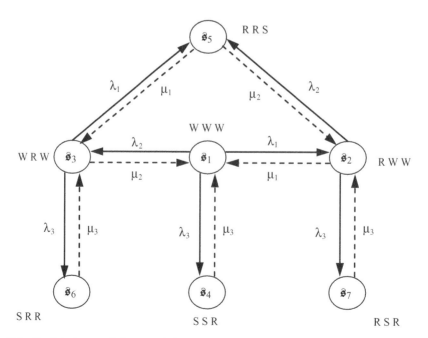

Figure 7.6. Exploitation graph for an exploitation process of three-element chute system with insufficient maintenance of a given type.

The further part of the analysis remains unchanged.

Essential remark: Note that different estimations for the limited probability distribution will be obtained.　　　　　　　　　　　　　　　　　　　　　　　　　　　　　　　　◀

■ **Example 7.2**

Let us consider a case in which three elements deliver mineral onto one trunk unit instead of two. Such an organisation is called a four-element chute system. Its graphical representation is shown in Figure 7.7.

　To make the consideration simpler it will be presumed that there is no problem with maintenance—all failures will be cleared out at once.

1. Therefore, the system is

$$\mathbf{\mathfrak{E}} : <e_i, i = 1, ...,4; \lambda_i, \mu_i; \mathbf{\mathfrak{D}}_1>$$

2. No reduction in the series elements is necessary. This is likely—it has been done before.
3. Calculate the number of theoretically possible states of the system:

$$L_t = 2^4 = 16$$

These states are enumerated in Table 7.2. Notice that all states are technically feasible.

5. The matrix of transition intensities between states is given below.

	1	2	3	4	5	6	7	8	9	10	11	12	13	14	15	16
1	Δ_1	λ_1	λ_2	λ_3	λ_4											
2	μ_1	Δ_2				λ_2			λ_3	λ_4						
3	μ_2		Δ_3			λ_1	λ_3				λ_4					
4	μ_3			Δ_4			λ_2	λ_4	λ_1							
5	μ_4				Δ_5											
6		μ_2	μ_1			Δ_6						λ_3	λ_4			
7			μ_3	μ_2			Δ_7					λ_1			λ_4	
8				μ_4				Δ_8								μ_3
9		μ_3	μ_1						Δ_9			λ_2		λ_4		
10		μ_4								Δ_{10}						μ_1
11			μ_4								Δ_{11}					
12						μ_3	μ_1		μ_2			Δ_{12}				
13						μ_4				μ_2	μ_1		Δ_{13}			
14								μ_1	μ_4	μ_3				Δ_{14}		
15							μ_4	μ_2			μ_3				Δ_{15}	
16		μ_4														Δ_{16}

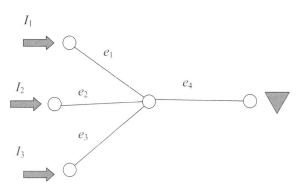

Figure 7.7.　Scheme of a four-element chute system.

Table 7.2.   Exploitation repertoire $\mathfrak{E}$ for a four-element chute system.

State of system	Element				Effect
	e_1	e_2	e_3	e_4	
$\mathfrak{s}_1$	W	W	W	W	$I_1 + I_2 + I_3$
$\mathfrak{s}_2$	R	W	W	W	$I_2 + I_3$
$\mathfrak{s}_3$	W	R	W	W	$I_1 + I_3$
$\mathfrak{s}_4$	W	W	R	W	$I_1 + I_2$
$\mathfrak{s}_5$	S	S	S	R	
$\mathfrak{s}_6$	R	R	W	W	I_3
$\mathfrak{s}_7$	W	R	R	W	I_1
$\mathfrak{s}_8$	S	S	R	R	
$\mathfrak{s}_9$	R	W	R	W	I_2
$\mathfrak{s}_{10}$	R	S	S	R	
$\mathfrak{s}_{11}$	S	R	S	R	
$\mathfrak{s}_{12}$	R	R	R	S	
$\mathfrak{s}_{13}$	R	R	S	R	
$\mathfrak{s}_{14}$	R	S	R	R	
$\mathfrak{s}_{15}$	S	R	R	R	
$\mathfrak{s}_{16}$	S	S	S	R	

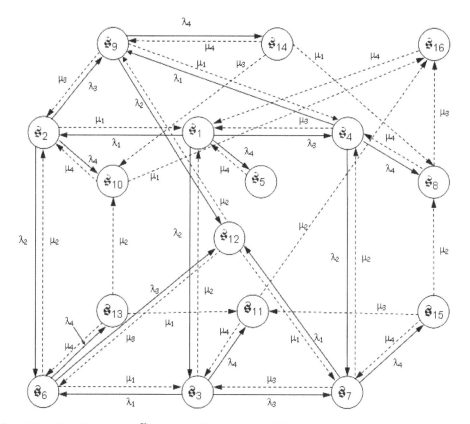

Figure 7.8.   Exploitation graph $\mathfrak{E}$ for an exploitation process of four-element chute system.

Having this matrix specified it is no problem to construct appropriate equations for determining limited probability states. The further part of the procedure is not interesting.   ◄

There are a number of interesting mine systems whose elements have exploitation processes that can be modelled by a Markov course of run. Let us now discuss an elementary system well-known in dressing plants.

■ **Example 7.3**
1. There is a system (Figure 7.9):

$$\mathfrak{S} : < e_i, i = 1, 2, 3; \lambda_i, \mu_i; \mathfrak{Q}_1 >$$

Element e_1 is a screen. The stream of mineral is divided into two different streams having specified diameters of particles.
  We wish to analyse this system.
2. Reduction of elements in series is not needed.
3. An exploitation repertoire $\mathfrak{E}$.
  The number of theoretically possible states is $L_t = 2^3 = 8$. In Table 7.3 all states are listed together with effects.
  By analysing the information contained in this table we can come to the conclusion that only four states are technically feasible and only one, $\mathfrak{s}_1$, is a work state of the system.
4. An exploitation graph for the system being considered is shown in Figure 7.10.
5. On the right-hand side of Figure 7.10 a matrix of the transitions between the states is shown.
  The exploitation graph as well as the matrix of the transitions between the states univocally indicates that this system is equivalent to a series system. This is true because the failure of any element means the failure of the system.

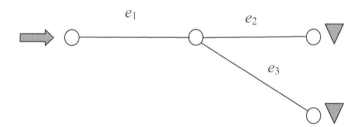

Figure 7.9.   Scheme of an elementary screening system.

Table 7.3.   Exploitation repertoire $\mathfrak{E}$ for three-element chute system.

| State of system | Element | | | Effect |
	e_1	e_2	e_3	
$\mathfrak{s}_1$	W	W	W	$I_1 + I_2$
$\mathfrak{s}_2$	R	S	S	0
$\mathfrak{s}_3$	S	R	S	0
$\mathfrak{s}_4$	S	S	R	0
$\mathfrak{s}_5$	R	R	W	(−)
$\mathfrak{s}_6$	R	W	R	(−)
$\mathfrak{s}_7$	W	R	R	(−)
$\mathfrak{s}_8$	R	R	R	(−)

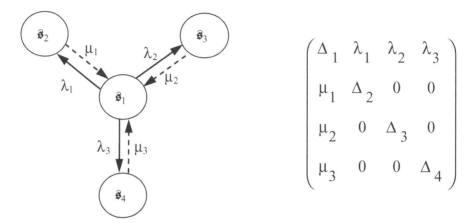

Figure 7.10.   Exploitation graph for an elementary screen system.

The calculation of this system is trivial.   ◄

Let us now analyse a different and interesting elementary system. In some publications it is termed as 'a pair of elements' and this system has been the point of interest several times (Gnyedenko 1964 and 1969, Gnyedenko et al. 1965, Kopociński 1973 for example). It can represent two parallel conveyor lines, one being a reserve for the second one. It can also stand for unit systems (hydraulic items) that are in use in some constructions of mine winders. One can find some other mine systems in which such an elementary system is a structural component.

- **Example 7.4**
1. There is a system

$$\mathfrak{S} : < e_i, i = 1, 2; \lambda_i, \mu_i; \mathfrak{Q}_1, \mathfrak{M}_u >$$

All notations are known except for the last component. Symbol $\mathfrak{M}_u$ here denotes the method of system utilisation being applied.

The scheme of the system is shown in Figure 7.11.

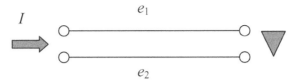

Figure 7.11.   Scheme of a pair of elements.

This system can be utilised in several different ways. Let us identify three of them.

a. A symmetric pair. One element executes its duties, the second one is held in reserve (a cold one). When a failure occurs in the working element, the second element commences its duties without delay. The first element is in a repair state. When the repair is finished—and a renewal occurs—the first element becomes the reserve. This situation exists until the moment when failure occurs in the second element. The situation is then reversed. A failure of the system occurs when a failure occurs in the working element during a repair of the second element.

b. A pair in order. One element executes its duties; a second one is in reserve (a cold one). When failure occurs in the working element, the second element commences its duties. The first element is in a repair state. When the repair is finished—and a renewal occurs—this repaired element restarts its duties again. The second element becomes the reserve once more—that is, it is a back-up element to be used when the first element is unavailable. Failure of the system is the same as in point (a). This system is a hierarchical one.

c. Elements half loaded. Let us presume that a stream of mineral is delivered to the system. Instead of fully loading one element, both elements carry half of the load. The idea of this solution is that a half-loaded element should have a higher reliability, perhaps a mean work time two times longer than that of one element working alone. Obviously, when one element is in failure the second one takes the full load. Failure of the system is the same as in (a) and (b).

An important question can be formulated here. Which solution is the best one? Or, some more in-depth questions might be posed. What kind of changes in the system parameters can be observed after the application of the reserve? What is the reliability of this system?

Let us discuss these modes of system operation taking into account the experience gained from mining practice.

The main idea of the last proposition (c) is that half-loaded elements will have a higher reliability. This higher reliability will pay for almost double element utilisation and, additionally, will earn a profit. Research has shown that this increase in reliability is usually small and the operational cost is almost doubled compared to the solution with a cold reserve. Therefore, this method is not recommended.

Utilisation of the system 'a pair in order' generates at least two problems. One element is in reserve and it does not work for the majority of the time. If a belt conveyor is in a standstill state for a long time some troublesome processes are observed. Re-starting generates problems. The intensity of failures during this operation is significantly higher than during regular transportation. This means that problems occur when they should not. A second troublesome property is connected with the fact that after a longer period of time one conveyor may become worn out, while the second—still almost new—becomes old but in a different sense. These two elements turn out to be different in the sense of their properties. They are not identical from a reliability point of view. Generally, this solution is impractical if the elements are mechanical ones. If the system consists of electronic items these annoying phenomena are not observed. But we are not analysing electronic systems here. For these reasons, this way of system utilisation is also not recommended.

The third solution—a symmetric pair—looks most practical at first glance. Elements wear out at the same intensity and over a long period of time the total work time will be approximately the same for both elements. However, for some pieces of equipment such as belt conveyors this method of utilisation is unsuitable because they are 'too reliable'. Failures occur rarely and the element being held in reserve is frequently in this state for a long time. If this happens, several annoying phenomena can be observed (greater belt sag between idlers, local belt deformations, etc.). Generally it is not good to keep a mechanical system in a standstill state for a prolonged period of time. For these reasons a fourth solution, a fourth method (d) of system utilisation is the best one to adopt—to switch an element from being in reserve to work without waiting for a failure to occur. If this action is repeated periodically with appropriate frequency, the reliability of both elements will be the same and failure problems connected with re-starting are eliminated to a great extent. It is worth noting that this method of system utilisation is equivalent to a symmetric pair in a reliability sense. Therefore, such a solution will be the focus into further considerations.

2. A reduction of elements in a series does not apply.

3. Determination of an exploitation repertoire $\mathfrak{E}$.
   The number of theoretically possible states is $L_t = 2^3 = 9$.
   Let us specify an exploitation repertoire of the system rejecting all technically unfeasible states, such as both elements are in reserve, one element is being repaired and the second one is in reserve. This repertoire is contained in Table 7.4.

4. All passages between states that are technically feasible are shown in Figure 7.12.

Table 7.4.   Exploitation repertoire $\mathfrak{E}$ for a symmetric pair of elements.

State of system	Element		Effect
	e_1	e_2	
$\mathfrak{s}_1$	W	S	I
$\mathfrak{s}_2$	W	R	I
$\mathfrak{s}_3$	S	W	I
$\mathfrak{s}_4$	R	W	I
$\mathfrak{s}_5$	R	R	0

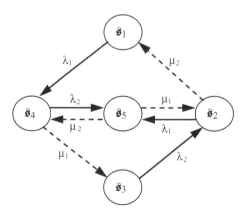

Figure 7.12.   Exploitation graph for a symmetric pair of elements.

5. A matrix of the transitions between states for the exploitation graph shown in Figure 7.12 is as follows:

$$\begin{bmatrix} -\lambda_1 & 0 & 0 & \lambda_1 & 0 \\ \mu_2 & -(\lambda_1+\mu_2) & 0 & 0 & \lambda_1 \\ 0 & \lambda_2 & (-\lambda_2) & 0 & 0 \\ 0 & 0 & \mu_1 & -(\lambda_2+\mu_1) & \lambda_2 \\ 0 & \mu_1 & 0 & \mu_2 & -(\mu_1+\mu_2) \end{bmatrix}$$

If we assume that both elements of the system are identical from a reliability standpoint—which usually holds—that is, $\lambda_1 = \lambda_2$ and $\mu_1 = \mu_2$ then the set of steady-state probabilities obtained from the matrix is:

$$P_1 = P_3 \quad P_2 = P_4 \quad P_3 = (\kappa^2 + 2\kappa + 2)^{-1}$$
$$P_2 = \kappa(\kappa^2 + 2\kappa + 2)^{-1} \quad P_5 = \kappa^2(\kappa^2 + 2\kappa + 2)^{-1}$$

(7.8)

where $\kappa = \lambda/\mu$ is the repair rate of the element.

Let us determine the steady-state availability of this system. Here we have:

$$A(\kappa) = 1 - P_5 = 2\frac{\kappa+1}{\kappa^2 + 2\kappa + 2}$$

(7.9)

Compare this availability measure with the steady-state availability of a single item which is determined by the formula:

$$A_1 = \frac{1}{1+\kappa}$$

An example plot of both functions is shown in Figure 7.13.

The advantage gained from the application of a spare unit is clearly visible. The problem here is whether the funds spent to buy, install and run this additional element will be at least covered

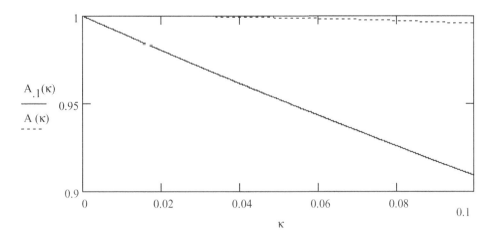

Figure 7.13.   The steady-state availability for a system without reserve $A_1(\kappa)$ and for a system with reserve $A(\kappa)$ as functions of repair rate κ.

by the profit gained due to its application. As a first attempt to answer this question this equation can be considered:

$$\Delta A \bar{I} \psi T = Z + kT \tag{7.10}$$

where:

ΔA is the steady-state availability increment
$\bar{I}$ is the mean stream of the mineral, t/d
ψ is the mean price of the mineral, €/t
Z is cost of reserve, €
k is the unit cost of operation of the reserve, €/d
T is search value, time, d.

Looking at both formulas used to determine steady-state availability, the following formula can be derived:

$$\Delta A = \frac{\kappa(\kappa+2)}{(\kappa+1)(\kappa^2 + 2\kappa + 2)} \tag{7.11}$$

At first glance formula 7.10 looks obvious. However, a more attentive analysis reveals that the magnitudes contained in this formula should be treated carefully. Let us pay attention to some of the parameters. Usually, time T is long and therefore the estimation of values k, ψ and $\bar{I}$ must be done precisely. Over a long period of time the price of the mineral being transported can change. Similarly, the unit cost of operation k can change. The last component $\bar{I}$ is more robust on possible changes due to obvious reasons.

Let us go a little further in our considerations. The method of system utilisation has an influence on the system's performance. It is better when the elements work and do not spend a long time in a standstill state. We can predict that with good management of the system, the elements will work with proper frequency. What does this mean for the system? There are two reliability indices in hand. The mean time of repair does not change and obviously the intensity of repair also stays unchanged. Thus, we may forecast that there will be a certain increase in the mean time of work. This means that the intensity of failures λ will be slightly lower.

Let us express the system availability as the function of both intensities. We have:

$$A(\lambda;\mu) = 2\mu \frac{\lambda + \mu}{\lambda^2 + 2\lambda\mu + 2\mu^2} \tag{7.12}$$

where μ is a parameter. Figure 7.14 shows example plots of this function for three different levels of intensity of repair $\mu_1 = 1$, $\mu_2 = 0.8$ and $\mu_3 = 0.6$.

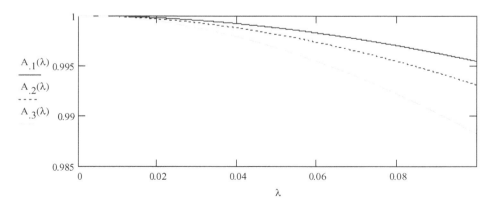

Figure 7.14.   The steady-state availability $A(\lambda)$ of a system of a symmetric pair as a function of the intensity of failures for different levels of intensity of repair μ; $\mu_1 = 1$, $\mu_2 = 0.8$ and $\mu_3 = 0.6$.

We can expect that with proper system utilisation, the intensity of failures λ will be slightly reduced and the corresponding value of A will be a bit higher resulting in a higher system output.

◄

7.4 SYSTEM CALCULATIONS

The considerations presented in the previous sections of this chapter mainly concerned comprehensive reliability analysis. The results of this analysis are used as the basis for system calculations, that is for determination of the basic exploitation parameters of these systems.

Usually, these are points of interest:

- calculation of the effective total work time of a system, subsystem and/or single element of the system
- calculation of the effective production of excavating machines and their systems, the hauling capacity of transporting means, and the dumping capability achieved by particular machines and their systems
- calculation of the number of failures and total repair times for the system and subsystem.

Moreover, in some cases considerations also comprise:

- analysis of the system under different methods of utilisation
- analysis of the system by applying elements with different parameters
- analysis of the system by applying different system configurations—that is, different reliability structures.

Calculation of a total work time $T_{j\Sigma}^{(w)}$ in a given time interval $(0, t)$ of element j is based on a simple equation:

$$T_{j\Sigma}^{(w)}(t) = \sum T_{jd} \sum_{j \in \mathbf{b}^{(w)}} P_j \qquad (7.13)$$

where:

ΣT_{jd} is total disposal time in the time interval $(0, t)$

$\Sigma_{j \in S^{(w)}} P_j$ is the sum of all probabilities connected with the set $\mathbf{b}^{(w)}$
$\mathbf{b}^{(w)}$ is the set of work states of the j-th element.

Notice, that this estimation will make sense if this interval is long. Also, the estimation of the probability contained in this formula is better for a longer period of investigation according to the well-known principle of mathematical statistics.

Similarly one can calculate the total time of a given state at the $(0, t)$ interval.

■ Example 7.1 (cont.)

Calculate an effective total work time $T_{j\Sigma}^{(w)}$ for elements e_1 and e_3 in the three-element chute system for a given total disposal time ΣT_{jd}

First, we have to identify a set of work states for these elements.

For element e_1 we have: $\mathbf{b}_1^{(w)}$: $\{\mathbf{s}_1, \mathbf{s}_3\}$ and for element e_3 we have: $\mathbf{b}_3^{(w)}$: $\{\mathbf{s}_1, \mathbf{s}_2, \mathbf{s}_3\}$.

Thus:

$$T_{1\Sigma}^{(w)}(t) = \sum T_{1d}(P_1 + P_3)$$

and

$$T_{3\Sigma}^{(w)}(t) = \sum T_{3d}(P_1 + P_2 + P_3)$$

◄

This line of investigation can be repeated in order to calculate the system's output. It is easy to construct a formula that takes into account the nominal output and the reliability of a given system branch or the whole system.

■ **Example 7.4 (cont.)**
Calculate the system output presuming that the method of system utilisation $\mathfrak{M}_u$ is equivalent to a symmetric pair. Presume $\kappa = 10^{-2}$ and the nominal output of the element is 3000 t/h.
We identify a set of work states: $\mathfrak{b}_s^{(w)}$: $\{\mathfrak{s}_1, \mathfrak{s}_1, \mathfrak{s}_3, \mathfrak{s}_4,\}$
Then compute the probabilities:

$$P_1 = P_3 = (\kappa^2 + 2\kappa + 2)^{-1} = 0.495 \quad P_2 = P_4 = \kappa \, (\kappa^2 + 2\kappa + 2)^{-1} = 4.95 \times 10^{-3}$$

Thus:

$$P_1 + P_2 + P_3 + P_4 = 0.9999$$

The system output: $3000 \times 0.9999 = 2999.7$ t/h. This means that only 0.1% is lost. ◄

Generally, a calculation of the number of failures of a given element, subsystem or the whole system is based on the limited probability distribution, while the normal one is determined by formula 5.26 after the appropriate modification of the value of time t (because the system is being investigated here).

Similarly, calculation of the total work time of a given element, subsystem or the whole system is based on the limited probability distribution, and the normal one is determined by formula 5.27 after the appropriate modification of the value of time t.

7.5   QUICK APPROXIMATION OF SYSTEM OUTPUT

As has been stated, an analysis of a system with many branches is tedious and complicated. If the point of interest is only system output, it can be estimated very easily. The basis for such an evaluation is the following statement:

An effective system output is not greater than the sum of the quotients of nominal output multiplied by the probability of the transportation of a mass from the point where this mass comes into the system to the point of its disposal, presuming that all streams of masses flow independently from each other.

This principle when converted into a formula is:

$$W_{ef}^{(S)} \leq \sum_j W_{ef}^{(j)} = \sum_j W_n^{(j)} \frac{1}{1 + \sum_{i \in j} \kappa_i} \qquad (7.14)$$

where:

$W_{ef}^{(S)}$ – the effective system output

$j - j$-th stream of mass

$\Sigma_i \kappa_i$ – this sum comprises all elements carrying the j-th stream of mass

$W_n^{(j)}$ – nominal output of first element receiving the j-th stream of mass.

This means that it is assumed that all streams of masses flow through systems of elements connected in a series and the steady-state availability of these basic systems should be calculated by applying the second principle of reduction.

Let us illustrate by an example.

■ **Example 7.1 (cont.)**

It was presumed that for the three-element chute system the first element was of 1200 t/h nominal output and the second element was of 1600 t/h productivity. Let us estimate the system output by applying the method of estimation outlined above.

We have two streams of masses. The estimation of output from the first working face is:

$$\frac{1200}{1+149.38\times10^{-3}+17.39\times10^{-3}}=1028 \text{ t/h}$$

and for the second place of generation:

$$\frac{1600}{1+142.73\times10^{-3}+17.39\times10^{-3}}=1379 \text{ t/h}$$

For these reasons the system output should be no greater than:

$$1028 + 1379 = 2407 \text{ t/h}$$

A precise calculation gives an output assessment of 2405 t/h. The error made by the estimation method is very small in this case. ◀

The statement allowing for a quick estimation of system output is extremely practical. However, there is no way to assess the accuracy of this evaluation. The reason for this vagueness is the fact that this error depends on the properties of the system concerned: on the number and reliability of elements, system configuration, method of utilisation and the method of system maintenance applied.

7.6 SEMI-MARKOV SYSTEMS[7]

It worth realising that there are other devices and some other mine systems that differ from those described up to this point in this chapter, and which can be encountered in continuously or almost continuously operating equipment. Here two cases can be enumerated, at least:

a. powered supports used in underground coal mining
b. systems comprising loading machine and crusher and continuous haulage of the belt conveyor type, which is used mainly in surface mining.

Perhaps, system (a) does not fit at first glance and it needs more attention to recognise its exploitation. Let us look at the method of operation of a single unit of a powered support. From the moment of its placement at a working face and from the moment when suitable pressure is placed on the hydraulic system, a powered support works at the face continuously. Sometimes some failures occur, sometimes repair is compulsory, but this does not nullify the fact that the support—in a stochastic sense—almost constantly executes its duties. And for this reason it can be counted in the category of devices being analysed. Unfortunately, the times of the states of a powered support are not exponential in many cases and due to this fact Markov processes cannot be applied to describe its operation in general.

Some additional explanation should also be given about system (b). There are some machinery systems of this type in operation, where material is loaded by a power shovel directly into a crusher on crawlers that follows the displacement of the loading machine. The situation here is a bit similar to a system of a winning machine—a continuous miner working underground plus a subsystem of armoured flight conveyors creeping laboriously on crawlers following the winning machine. System (b) is used in surface mining and loading has a cyclic character. Moreover, the

---

[7] The theoretical considerations in this chapter arose after consultations with Prof. Limnios and Prof. Bratiichuk.

subsystem of belt conveyors is fixed, continuously receiving a stream of bulk from the sizer. In some systems of this kind a beltwagon is applied as a link between the crusher and the belt conveyor subsystem. Though exploitation of belt conveyors in this system can usually be described by Markov processes, in some cases the operation of the crushers fits the Markov system, in some cases it does not. Similarly, the process of changes of the states of a power shovel sometimes is of a Markov type, but every so often the times of repair cannot be described satisfactorily by exponential distribution (Czaplicki's monograph 2009). For this reason Markov processes can frequently be useless. Therefore, the problem arises as to what kinds of mathematical tools should applied to analyse such systems. In these cases the Markov processes are too simple, and a more advanced method must be employed.

If the states in a given process are mutually independent in terms of times, the theory of semi-Markov processes can be applied.

Semi-Markov systems have the great advantage that they are much more general than the Erlangian systems that will be considered in Chapter 8.6; however, some connection with Markov processes remains. Before we begin to analyse these systems, we have to first define a supplementary term that will be required—a Markov chain. Its meaning can be explained in the following way.

There is given a set of states $\hat{s} = \{\hat{s}_1, \hat{s}_2, ..., \hat{s}_n\}$. The process starts in one of these states and moves successively from one state to another. Each move is called a *step* or *jump*. If the process is currently in state $\hat{s}_i$, then it moves to state $\hat{s}_j$ at the next step with a probability denoted by p_{ij}, and this probability does not depend on the states that the process was in before the current state (memoryless property). The probabilities p_{ij} are called transition probabilities. The process can remain in state i with probability p_{ii}. An initial probability distribution, defined on $\hat{s}$, specifies the starting state. In a continuous space the probability distribution that has a memoryless property is an exponential distribution, whereas in a discrete space it has a geometric distribution. The process is called a 'chain' because not only are the states discrete, but the moments when jumps occur are also discrete.

Having an idea of a Markov chain, a definition of a semi-Markov process can be given.

A semi-Markov process is a random process $X(t)$ with a finite or countable set of states s having stepwise trajectories with jumps at times $0 < \tau_1 < \tau_2 < ...$ such that:

- The values $X(\tau)$ at its jump times form a Markov chain with transition probabilities:

$$p_{ij} = P\{X(\tau_n) = j \mid X(\tau_{n-1}) = i\}$$

- The distributions of the jumps τ_n are described in terms of the distribution functions $F_{ij}(t)$ as follows:

$$P\{\tau_n - \tau_{n-1} \leq t, X(\tau_n) = j \mid X(\tau_{n-1}) = i\} = p_{ij} F_{ij}(t)$$

This process is called a semi-Markov one (see Korolyuk and Turbin 1976, Limnios and Oprişan 2001, Harlamov 2008).[8]

---

[8] Semi-Markov processes were introduced by Levy (1954) and Smith (1955). Takács (1954, 1955) investigated similar processes. The foundations of the theory of semi-Markov processes were mainly laid by Pyke (1961a, b), Pyke and Schaufele (1964), Çinclar (1969) and Korolyuk and Turbin (1976).

It is of practical importance to emphasise that states in this type of process are mutually independent.

If $F_{ij}(t)$ are exponential, then the semi-Markov process is a continuous-time Markov chain or in short a Markov process. If all the distributions degenerate to a certain point, the result is a discrete-time Markov chain. In analytic terms, the investigation of semi-Markov processes can be reduced to a system of integral equations (Korolyuk and Turbin 1976).

By analysing the definition it is easy to conclude that the class of semi-Markov processes has great value—the degree of generalisation is really quite high. However, the distance from theory to practice—as in many cases—is not short and is sometimes difficult. If there are many states and they fulfil all of the conditions indicating that their process of changes can be classified as a semi-Markov system, the analysis of such a process is quite tedious and time-consuming. In addition, such an analysis is mainly interesting for researchers who are involved in the specific investigations being considered; there is a little sense in publishing the analysis. If a certain problem having a connection with practice has been solved, it is short and relatively communicative; this has been pointed out in many publications (see, for example, Bousfiha et al. 1996, Bousfiha and Limnios 1997, Limnios and Oprişan 2001 or Brodi and Pogosjan 1973, Barlow and Proschan 1975, Grabski and Jaźwiński 2001).

Now, we can return to the modelling and analysis of one of the systems listed at the beginning of this chapter—a single unit of a powered support. The problem presented here is greatly simplified compared with the dilemma as originally formulated. The reasons for such an approach will be given later.

An exploitation investigation carried out on data taken from an underground operation of powered supports showed (Uzar 2001) that all failures which occurred in this piece of equipment could be divided into two separate states: full failure, which means repair, and partial failure where the support still operated. Therefore, an exploitation repertoire $\mathfrak{E}$ can be defined as:

$$\mathfrak{E} : < \mathfrak{s}_2, \mathfrak{s}_1, \mathfrak{s}_3 >$$

where:

- $\mathfrak{s}_1$ repair
- $\mathfrak{s}_2$ work
- $\mathfrak{s}_3$ partial failure/work.

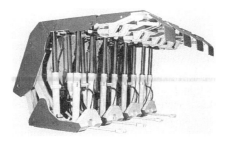

As empirical investigations have exposed, an object with a partial failure is not repaired as a rule. It operates until the moment when full failure occurs. Such an event is the signal to begin repairs: the support does not fulfil its duties, it does not work.

An exploitation graph illustrating the possible transitions between these states is shown in Figure 7.15. The number 1 near a given arrow means that only one possibility exists to pass from one particular state to a different one.

Because a support in a work state can either fail partially or fail entirely, corresponding probabilities were ascribed to these two events denoted by p_{21} if the passage was from state $\mathfrak{s}_2$ to state $\mathfrak{s}_1$ and by p_{23} if the passage was from state $\mathfrak{s}_2$ to state $\mathfrak{s}_3$. The probabilities of these two states are obviously closed to unity. It is also worth remembering that both states are work states.

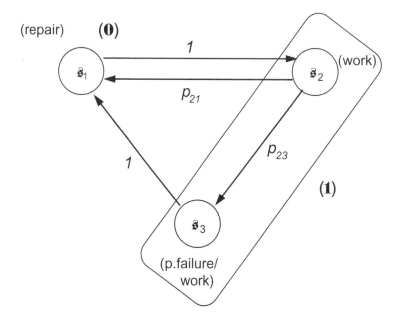

Figure 7.15.   An exploitation graph of a powered support—possible transitions between states.

Exploitation investigations of powered support failures (Uzar 2001, Markowicz 2003, Jaszczuk et al. 2004, for instance) allowed it to be proven that the times of these states can be satisfactorily described by the Weibull family of functions.

Recall, a random variable has a Weibull distribution if its probability density function can be described by this formula:

$$q(t) = \frac{\delta}{\eta^{\delta}} x^{\delta-1} e^{-\left(\frac{t}{\eta}\right)^{\delta}} \quad \text{if } t \geq 0 \tag{7.15}$$

Parameter δ is the shape parameter and η is the scale parameter. The Weibull distribution is denoted by $W(\delta, \eta)$. Plots of the Weibull distribution for different values of parameters are shown in Figure 7.16.

The final piece of information that is needed to identify the exploitation process of changes of states for a powered support as a semi-Markov process is the mutual independence of the component states.

Data gathered from several mines allowed a hypothesis proclaiming this regularity to be checked and there were no grounds to reject this supposition. Therefore, we can presume that the point of interest from now on is a semi-Markov process.

It is known from theory (for example, Gnyedenko and Kovalenko 1966, Korolyuk and Turbin 1976, Limnios and Oprişan 2001) that the semi-Markov process is well defined if its two characteristics are given:

- an initial probability distribution
- a semi-Markov kernel.

A semi-Markov kernel is defined as follows.

Let $\mathfrak{N} = \{1, 2, \ldots\}$, $\mathfrak{N}_0 = \{0, 1, 2, \ldots\}$, $\tilde{\mathfrak{N}}_+ = [0, \infty)$ and let $\mathfrak{s}$ denotes a finite or countable set of states.

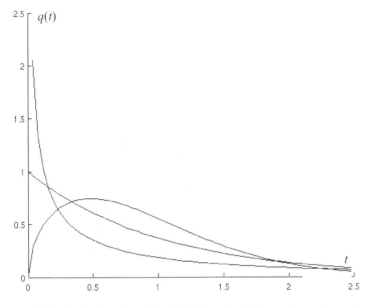

Figure 7.16. Probability density functions of the Weibull type for different values of parameters.

A semi-Markov kernel is a matrix:

$$\mathbf{Q}(t) = [\hat{Q}_{ij}(t) : i, j \in \mathbf{S} \times \mathbf{S}] \tag{7.16}$$

that has as its elements functions $\hat{Q}_{ij}(t)$, $t \in \tilde{\mathfrak{N}}_+$ which fulfil the following conditions:

i.   for all pairs $(i, j) \in \mathbf{S} \times \mathbf{S}$, $\hat{Q}_{ij}(t)$ are non-decreasing, right continuous real functions
ii. for all pairs $(i, j) \in \mathbf{S} \times \mathbf{S}$, $\hat{Q}_{ij}(0) = 0$ as well as for each $t \in \tilde{\mathfrak{N}}_+$ $\hat{Q}_{ij}(t) \leq 1$
iii. for each $i \in \mathbf{S}$ $\lim_{t \to \infty} \Sigma_{j \in S}$ $\hat{Q}_{ij}(t) = 1$.

If so, we are almost ready to have our semi-Markov kernel specified. The only remaining item is the determination of probabilities p_{21} and p_{23}.

Based on the total probability principle the following equation holds:

$$p_{21} = 1 - p_{23} = \int_0^\infty [1 - Q_{23}(t)] q_{21}(t) dt \tag{7.17}$$

Thus the matrix of transitions between states for a semi-Markov process for a powered support is:

$$\mathbf{Q}(t) = \begin{pmatrix} 0 & \hat{Q}_{12}(t) & 0 \\ \hat{Q}_{21}(t) & 0 & \hat{Q}_{23}(t) \\ \hat{Q}_{31}(t) & 0 & 0 \end{pmatrix} \tag{7.18}$$

where:

$$\hat{Q}_{12}(t) = Q_{12}(t) = 1 - \exp\left[-\left(\frac{t}{\eta_1}\right)^{\delta_1}\right] \tag{7.19a}$$

$$\hat{Q}_{21}(t) = p_{21}Q(t) \quad Q_{21}(t) = 1 - \exp\left[-\left(\frac{t}{\eta_{21}}\right)^{\delta_{21}}\right] \tag{7.19b}$$

$$\hat{Q}_{23}(t) = p_{23}Q(t) \quad Q_{23}(t) = 1 - \exp\left[-\left(\frac{t}{\eta_{23}}\right)^{\delta_{23}}\right] \tag{7.19c}$$

$$\hat{Q}_{31}(t) = Q_{31}(t) = 1 - \exp\left[-\left(\frac{t}{\eta_3}\right)^{\delta_3}\right] \tag{7.19d}$$

where δ_{12}, δ_{21}, δ_{23}, δ_{31}, η_{12}, η_{21}, η_{23}, η_{31} are structural functions parameters.

The probability distributions $Q_{ij}(t)$ concern the following random variables:

$Q_{12}(t)$—the powered support repair
$Q_{21}(t)$—the powered support work time up to the occurrence of full failure
$Q_{23}(t)$—the powered support work time up to the occurrence of partial failure
$Q_{31}(t)$—the powered support work time with partial failure up to the occurrence of full failure.

To obtain the fully identified semi-Markov process for a powered support, the initial probability distribution must be given. It can be presumed that:

$$\alpha = (\alpha_1 \ \alpha_2 \ \alpha_3) = (0 \ 1 \ 0) \tag{7.20}$$

which means that we presume that at the beginning of an operation the support was good—that is, in a work state without any failure—and the following notation is presumed:

$$\alpha = (\alpha_0 \ \alpha_1) \quad \alpha_0 = \alpha_1 \quad \alpha_1 = (\alpha_2 \ \alpha_3) \tag{7.21}$$

To illustrate the principle of transitions between states, we can have a look at Figure 7.17.

From a mathematical standpoint the determination of the process solves the problem but in the engineering field we are interested in getting an estimation of some basic parameters of the object, such as the mean time between failures, the mean time of repair, etc.

In order to construct the appropriate formulas allowing interesting estimations of these parameters to be achieved, two additional components taken from the theory of semi-Markov processes are needed, namely:

– the embedded Markov chain for the semi-Markov process being considered
– the ergodic[9] probability distribution of this chain.

Let $x(t)$, $t \geq 0$, be a time-homogeneous jump Markov process. Let τ_n, $n \geq 0$ be the jump times for which we have $0 = \tau_0 \leq \tau_1 \leq \cdots \leq \tau_n \leq \ldots$.

The stochastic process x_n, $n \geq 0$ defined by:

$$x_n = x(\tau_n), \quad n \geq 0$$

is called the embedded Markov chain of the Markov process $x(t)$, $t \geq 0$.

Now, we come to the construction of the embedded Markov chain for the semi-Markov process that is the point of our interest.

---

[9]In probability theory, a stationary ergodic process is a stochastic process that exhibits both stationarity and ergodicity. In essence this implies that the random process will not change its statistical properties over time (or other process parameter) and that its statistical properties (such as the theoretical mean and variance of the process) can be deduced from a single, sufficiently long sample (realisation) of the process.

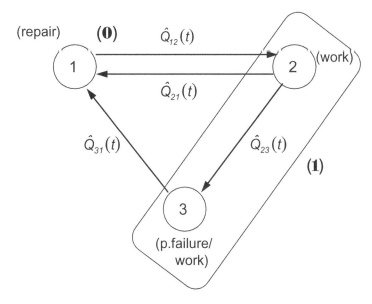

Figure 7.17. An exploitation graph of a powered support—principle of transitions between states.

Look at Figure 7.15. We have two states: **0** which means repair and **1** which means work and this state consists of two sub-states. The principle of the construction of this chain relies on the production of the following matrix of probabilities:

$$\mathfrak{P} = \begin{pmatrix} P_{00} & P_{01} \\ P_{10} & P_{11} \end{pmatrix} = \begin{pmatrix} \boxed{\text{From } \mathbf{0} \text{ to } \mathbf{0}} & \boxed{\text{From } \mathbf{0} \text{ to } \mathbf{1}} \\ \boxed{\text{From } \mathbf{1} \text{ to } \mathbf{0}} & \boxed{\text{From } \mathbf{1} \text{ to } \mathbf{1}} \end{pmatrix}$$

that yields to[10]:

$$= \begin{pmatrix} \boxed{\begin{array}{ccc} 0 & \boxed{1} & 0 \end{array}} \\ \boxed{P_{21}} \; \boxed{\begin{array}{cc} 0 & P_{23} \\ 1 \end{array}} \; \boxed{\begin{array}{cc} 0 & 0 \end{array}} \end{pmatrix} \tag{7.22}$$

where the fact that state **1** consists of two sub-states is taken into account.

Now, we are going to find the ergodic probability distribution—usually denoted by Π—for this Markov chain.

The following matrix equation holds:

$$\Pi\mathfrak{P} = \Pi \tag{7.23}$$

The probability distribution Π consists of three elements—probabilities—because we have three states of the process:

---

[10] This matrix is different to the matrixes considered in Chapter 7 where the main diagonal consisted of non-zero elements that represented the intensity of remaining at a given state.

$$\Pi = (\Pi_1\ \Pi_2\ \Pi_3)$$

Changing equation 7.23 into the coordinate form we have:

$$\begin{cases} -\Pi_1 + \Pi_2 p_{21} + \Pi_3 = 0 \\ \Pi_1 - \Pi_2 = 0 \\ \Pi_2\, p_{23} - \Pi_3 = 0 \end{cases} \tag{7.24}$$

Unfortunately, this set of equations is indeterminable. Therefore, it is necessary to reject one equation and replace it with the equation closing all probabilities to unity (see Chapter 7.1).

$$\Pi_1 + \Pi_2 + \Pi_3 = 1$$

Solving this set of equations, we have:

$$\Pi_2 = \Pi_1 = \frac{1}{2 + p_{23}} \qquad \Pi_3 = \frac{p_{23}}{2 + p_{23}} \tag{7.25}$$

The first set of interesting parameters that will be determined is a set of the average times of states. This set obviously consists of three equations:

$$\begin{cases} m_1 = m_{12} = \int_0^\infty x dQ_{12}(x)dx = \eta_{12}\Gamma\left(\dfrac{1}{\delta_{12}} + 1\right) \\[2mm] m_2 = p_{21}m_{21} + p_{23}m_{23} = p_{21}\int_0^\infty x dQ_{21}(x)dx + p_{23}\int_0^\infty x dQ_{23}(x)dx \\[2mm] \qquad = p_{21}\eta_{21}\Gamma\left(\dfrac{1}{\delta_{21}} + 1\right) + p_{23}\eta_{23}\Gamma\left(\dfrac{1}{\delta_{23}} + 1\right) \\[2mm] m_3 = m_{31} = \int_0^\infty x dQ_{31}(x)dx = \eta_{31}\Gamma\left(\dfrac{1}{\delta_{31}} + 1\right) \end{cases} \tag{7.26}$$

where m_1 is the expected value of the repair state, m_2 is the expected value of the work state and m_3 is the expected value of the work state but with partial failure.

We can construct a matrix of these mean values:

$$m = (m_1, m_2, m_3) = (m_0\ m_1) \quad m_0 = m_1 \quad m_1 = (m_2\ m_3) \tag{7.27}$$

that will be helpful in further reasoning.

Having values (Π_1, Π_2, Π_3) and (m_1, m_2, m_3) the ergodic probability distribution for the semi-Markov process can be determined. It consists of the following elements:

$$\rho_1 = \frac{\Pi_1 m_1}{M} \qquad \rho_2 = \frac{\Pi_2 m_2}{M} \qquad \rho_3 = \frac{\Pi_3 m_3}{M} \tag{7.28}$$

where:

$$M = \Pi_1 m_1 + \Pi_2 m_2 + \Pi_3 m_3 \tag{7.29}$$

Having these parameters of the semi-Markov process we can determine further parameters of interest for engineering practice.

The mean time to full failure (repair) is given by:

$$MTTFF = \alpha_1 (I - P_{11})^{-1} m_1 \tag{7.30}$$

where $I = \text{diag}(1, \ldots, 1)$. For this reason:

$$MTTFF = (\alpha_2, \alpha_3) \begin{pmatrix} 1 & p_{21} \\ 1 & 1 \end{pmatrix}^{-1} \begin{pmatrix} m_2 \\ m_3 \end{pmatrix} \tag{7.31}$$

The mean time to any failure occurrence is given by:

$$MTTF = \alpha_2 (I - P_{01})^{-1} m_2 = m_2 \tag{7.32}$$

In addition to the average times interesting reliability parameters are associated with the occurrence of failure. Here two measures are significant.

The intensity of the departure from state $\mathbf{\hat{s}}_2$:

$$\lambda_{2\to} = \frac{1}{m_{21} p_{21} + m_2 p_{23}} \tag{7.33}$$

and the intensity of failures of the process, which is determined by equation:

$$\lambda = \frac{1}{p_{21}(m_{21} + m_{12}) + p_{23}(m_{23} + m_{31} + m_{12})} \tag{7.34}$$

The denominator takes into account all possible passages from the work state $\mathbf{\hat{s}}_2$ to another state and a return to state $\mathbf{\hat{s}}_2$. In the case being considered it can be $\mathbf{\hat{s}}_2$–$\mathbf{\hat{s}}_1$–$\mathbf{\hat{s}}_2$ and it occurs with the probability p_{21}. The other possibility is the sequence $\mathbf{\hat{s}}_2$–$\mathbf{\hat{s}}_3$–$\mathbf{\hat{s}}_1$–$\mathbf{\hat{s}}_2$ and it occurs with the probability p_{23}. Thus, this intensity takes into account a full stochastic cycle of the semi-Markov process.

The steady-state availability, that is the probability that a powered support is in state $\mathbf{\hat{s}}_2$ or state $\mathbf{\hat{s}}_3$ is:

$$A = P\{\mathbf{\Psi(\mathfrak{S})} \subset \mathbf{\hat{s}}_2 \cup \mathbf{\hat{s}}_3 \} = P\{\mathbf{\Psi(\mathfrak{S})} \subset \mathbf{1}\} = \rho_2 + \rho_3 \tag{7.35}$$

Now, we will try to place this model into mining reality, which unfortunately—as usual—is much more complicated, especially where powered supports are concerned.

Let us commence our considerations from a statement that each working face has its own exploitation time. After a certain period, the life of a working face is terminated. At this moment the mine management must decide what to do with all the equipment that operated at this face. In the majority of cases the equipment is moved to a new working point using auxiliary means of transport, such as those that were described in Chapter 4.1. Some machines are transported intact, but some must be dismantled into pieces that are more easy to move. When these elements of dismantled machines reach their destination, they are reassembled. All pieces of equipment, before final installation, are surveyed, repaired if needed and adjusted. Some prophylactic actions are usually done.

From a theoretical standpoint, at this moment a regeneration of the exploitation processes of these machines takes place.[11] The simplest case is when the regeneration is total and in mining engineering—if regeneration is taken into account—this type of regeneration is considered almost exclusively. If this is the case, it can be presumed that the exploitation process starts from the beginning. If a given exploitation process has many points of regeneration, it can usually be presumed that we observe many stochastic copies of the same process.

Let us look more attentively at the wearing process of a powered support. This piece of equipment is constructed of many elements. Some of them can be replaced after failure, or can be treated as new after repair. But some have wearing processes of a different nature. An accumulation of the effects of wearing is observed. After repair, an item is not the same as a new one, but has slightly worse properties—that is, after repair the item is not as good as a new one. As a result, from a reliability point of view, the intensity of failures is not constant but is a function of time and it increases with time. However, the intensity of failures does not change directly proportionally with time, going continuously down, but it drops gradually but only to a certain level. This situation is a bit similar to that observed in the exploitation process of trucks (see Crawford 1979). Usually, mine management is not interested in using equipment with a very low reliability. However, in some poor countries some pieces of mine equipment of really low reliability are still in operation generating a lot of problems. In many countries the principle holds that a powered support remains in use until the end of operations at a given face, even when this piece of equipment is of poor quality. The rationale for this attitude is connected with the fact that general repair of a support requires a lot of time (sometimes even few days) and halts all operations at the face. Great losses are inevitable.

Let us continue with consideration of the operation of a powered support. Observation of mining practice indicates that a regeneration of an exploitation process is now noticed more frequently. In many mining countries a standard holds that after several days of operation one shift is devoted to wide-ranging maintenance of all sections of the powered support operating at a given working face. If this piece of equipment is new or almost new, this maintenance is done after longer intervals; for supports that have been in use for long time, this maintenance is done much more frequently. And here again, it can be presumed that regeneration takes place. For these reasons we can suspect that the model presented above holds for two neighbouring regeneration points only. Therefore, we can presume that here we have an intermittent exploitation process similar to that found in the operation of hoists. However, in the case of hoists, breaks are taken regularly and they are so frequent that it is presumed that their exploitation process runs continuously removing planned standstills and that their operational parameters, such as the intensity of failures, are constant. We can pose the question here whether the same approach can be adopted in the case of the operation of a powered support. Because the frequency of general maintenance does not occur as often as with hoists, we hesitate from answering this question. But, what happens when we presume that we cannot repeat the approach taken with hoists? This means that each copy of an exploitation process is a unique one and is never repeated. And it is so short in duration that an estimation of its basic parameters will be very poor due to the small size of the samples. Thus, the only solution is to presume that the intensity of failures of a powered support is constant in time, while being fully aware of fact that this is an approximation only. An analysis of the exploitation process of a support can be done in such a way.[12]

These frequent regeneration points have an influence on the manner of the evaluation of some parameters. The intensity of failures for instance, determined by equation 7.34, is valid for a stabilised period of the process (ergodic characteristic), that is for a prolonged period of

---

[11] Regeneration in the exploitation process of shovel-truck systems was investigated in Czaplicki's monograph (2009, pp. 92–93).

[12] To realise how important it is to keep sections of a powered support in a proper state it is enough to point out that in Poland we have about 130 longwalls composed of approximately 18,000 sections.

exploitation of the support. However, keeping in mind regeneration, in some cases the intensity, depending on time that is determined by the equation, can be more interesting:

$$\lambda(t) = \frac{-R'(t)}{R(t)} \tag{7.36}$$

where $R(t)$ is the reliability function at time $t \geq 0$.

Here some comments are necessary in connection with these two failure intensities. The function determined by equation 7.36—usually called the hazard function—concerns objects working until the first occurrence of failure. It is a conditional probability of failure occurrence in a unit of time provided that up to this moment there was no failure. Other intensities of failures, determined by equations 5.7 or 7.33 for instance, are unconditional probabilities of the occurrence of failure in a unit of time if this unit is appropriately small (Gnyedenko et al. 1965). For these reasons function 7.36 has a local character only because it concerns an object that is repairable.

We omit here the method of deriving a formula for $R(t)$ given by Limnios and Oprişan (2001 p. 5.4). For the case being considered, this is:

$$R(t) = \begin{pmatrix} 1 & 0 \end{pmatrix} \begin{pmatrix} 1 & -p_{23}Q(t) \\ 0 & 1 \end{pmatrix}^{-1} \begin{pmatrix} 1 - p_{21}Q_{21}(t) - p_{23}Q_{23}(t) & 0 \\ 0 & 1 - Q_{31}(t) \end{pmatrix} \begin{pmatrix} 1 \\ 1 \end{pmatrix}$$

$$= 1 - p_{21}Q_{21}(t) - p_{23}Q_{23}(t)Q_{31}(t) \tag{7.37}$$

The exploitation process of a powered support can even be much more complicated. If the operation at a given face progresses and the ore seam meets a geological fault, the exploitation conditions of the support change drastically. The acting forces can be radically different—and often much, much greater. Similarly, we can observe considerably huge changes in the acting forces on a powered support if the working face crosses an area under the influence generated by a stope being excavated above the face. Some diverse zones with different physical properties may exist in the surrounding rock masses, and these can generate significant changes in the forces acting on the support. All these events occurring during a support's operation change its working conditions dramatically and the intensity of failures transforms to a great extent; no doubt—it is not constant. And here again, the physical progress of the face is a better parameter than a time process.

If we additionally take into account that a given section is almost always not operating alone but in a system that involves a combination of other pieces of equipment working at a given face, we can easily come to the conclusion that the modelling of the exploitation process of a powered support is very difficult and a great challenge for reliability researchers.

Therefore, for the time being, let us do an analysis of the exploitation process of a powered support by making use of the model presented in this chapter.

▪ Example 7.5

Four identical powered supports were observed during six months of exploitation. All failures and repairs were noted. The times of all states were recorded. The data gathered created four sequences of realisations of random variables.

First, a statistical hypothesis was formulated stating that the data collected was not associated with the time of recording, that is the sequences are stationary ones.

A test using Spearman's rank correlation coefficient allowed the assumption that the sequences are not correlated with time.

The second phase of preliminary analysis was focused on the homogeneity of the data for each state separately. The Kruskall-Wallis test gave no grounds to reject the formulated hypothesis.

Therefore, the data was treated as homogeneous and the appropriate sequences were gathered together.

A synthesis of the data was started from an estimation of two basic statistical parameters: the mean and the standard deviation. Outcomes of the evaluation gave these results:

– average time of repair: $\bar{m}_{12} = 115$ min
– average work time to repair: $\bar{m}_{21} = 129910$ min
– average work time to failure: $\bar{m}_{23} = 40270$ min
– average work time with failure to repair: $\bar{m}_{31} = 70660$ min

The next step comprised searching for the theoretical probability distributions that described the empirical distribution of the observed random variables satisfactorily. The Weibull family of functions was selected and structural function parameters were estimated based on the method of moments—as the first step—and later the method of most likelihood was applied to find the final estimates of these parameters. The results of the investigations are shown in Figures 7.18 to 7.21.

These estimates of unknown structural parameters of the probability density functions are the basic information necessary to construct the matrix of the transition between states for the semi-Markov process. Two additional parameters are required—probabilities p_{21} and p_{23}.

Calculating we have:

$$p_{21} = 1 - p_{23} = \int_0^\infty [1 - Q_{23}(t)]q_{21}(t)dt = 0.1$$

The initial probability distribution is determined by equation 7.20 and all elements of the matrix $\mathfrak{P}$ are known.

Now calculate the ergodic probability distribution Π for the Markov chain. Making use of equations 7.25 we have:

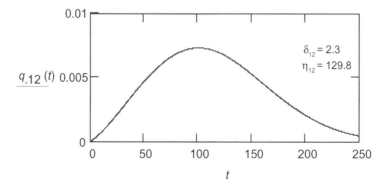

Figure 7.18.   The probability density function of repair times.

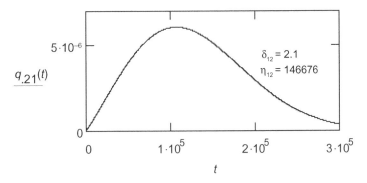

Figure 7.19. The probability density function of work times to repair.

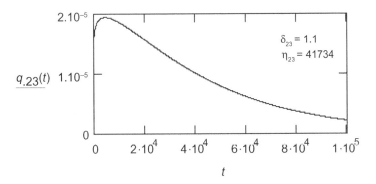

Figure 7.20. The probability density function of work times to failure.

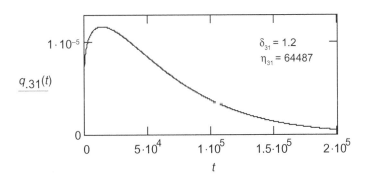

Figure 7.21. The probability density function of work in failure times to failure.

$$\Pi = (0.345 \quad 0.345 \quad 0.310)$$

In order to determine the set of average times of states, one mean needs an evaluation that is m_2. Taking into account the set of equations 7.26 we finally have:

$$m = (115 \quad 49234 \quad 70660)$$

The ergodic probability distribution for the semi-Markov process, applying formulas 7.28 and 7.29, is as follows:

$$\rho_1 = 0.001 \quad \rho_2 = 0.474 \quad \rho_3 = 0.525$$

The mean time to full failure (repair) determined by equation 7.31 gives:

$$MTTFF = 103\,802 \text{ min}$$

The steady-state availability for the support is:

$$A = \rho_2 + \rho_3 = 0.999$$

This is quite high, however partial failures were neglected.

The reliability function $R(t)$ determined by equation 7.37 looks similar to the one for exponential distribution and is shown in Figure 7.22 nonetheless the intensity of failures function 7.36 looks interesting, illustrated in Figure 7.23.

The intensity of failures for the ergodic case determined by equation 7.33 is:

$$\lambda_{2\rightarrow} = 1.744 \times 10^{-5} \qquad \blacktriangleleft$$

To get more experience in the modelling of mine systems by means of semi-Markov processes let us briefly consider the problem of the operation of a system comprising a power shovel working as a loading machine, a crusher[13] and a certain number of belt conveyors.

As the first step, a reduction of the elements connected in a series, including all belt conveyors up to the crusher, can be done provided that their processes of changes of states are of a Markov type. Presuming such a case, the original system is transformed into three-element one: loader—crusher—conventional unit (conventional 'conveyor').

An exploitation graph of this system is shown in Figure 7.24. Fig. 7.24a illustrates the possible passages between states and 7.24b illustrates the principle of transitions between the states of the system process.

Not resigning from general approach, it can be assumed that:

$$Q_{12}(t) = W(1, \eta_{12}) \quad Q_{13}(t) = W(\delta_{13}, \eta_{13}) \quad Q_{14}(t) = W(1, \eta_{14})$$
$$Q_{21}(t) = W(\delta_{21}, \eta_{21}) \quad Q_{31}(t) = W(\delta_{31}, \eta_{31}) \quad Q_{41}(t) = W(1, \eta_{41})$$

It is easy to see that three times exponential distributions were presumed (shape parameters equal 1) following information obtained from field investigations.

Obviously:

$$P_{12} + P_{13} + P_{14} = 1$$

The initial probability distribution can be specified as:

$$\alpha = (\alpha_1\,\alpha_2\,\alpha_3\,\alpha_4) = (1\,0\,0\,0)$$
$$\alpha_0 = (\alpha_2\,\alpha_3\,\alpha_4) \quad \alpha_1 = (\alpha_1)$$

We can calculate the average times of states using formula 7.26; however, the mean time for the $\hat{s}_1$ state must be weighted by all probabilities p.

---

[13]You can find more on the reliability of crushers for coal processing plants in Mishra and Pathak (2002); however, the reliability measures applied are characteristic for items working to the first failure occurrence. Additionally, only exponential functions are used. There is no information on repair times. This greatly limits the value of the information presented.

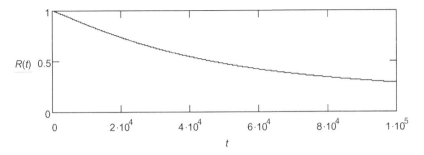

Figure 7.22.   The reliability function $R(t)$.

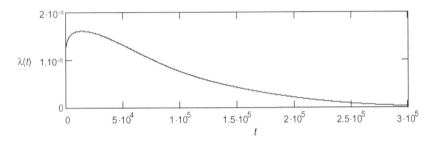

Figure 7.23.   The hazard function $\lambda(t)$.

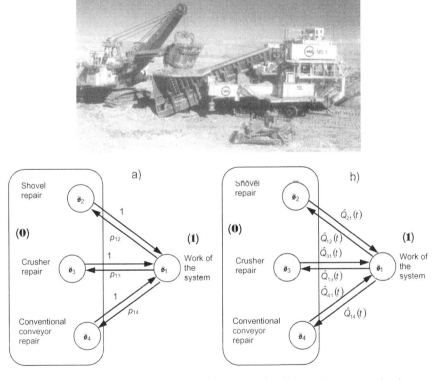

Figure 7.24.   An exploitation graph of a system: loader (power shovel)—crusher—conventional conveyor: a) possible passages between states, b) a principle of transitions between states.

Looking at Figure 7.23 it is easy to see that the system being considered has a series structure—it does not work if any of its elements are in a repair state.

If so, the embedded Markov chain for the semi-Markov process is:

$$
\mathfrak{P} = \begin{bmatrix} P_{00} & P_{01} \\ P_{10} & P_{11} \end{bmatrix} = \begin{pmatrix} \boxed{0} & \boxed{\begin{matrix} P_{12} & P_{13} & P_{14} \end{matrix}} \\ \begin{matrix} 1 \\ 1 \\ 1 \end{matrix} & \begin{matrix} 0 & 0 & 0 \\ 0 & 0 & 0 \\ 0 & 0 & 0 \end{matrix} \end{pmatrix}
$$

And the semi-Markov kernel is as follows:

$$
\mathbf{Q}(t) = \begin{pmatrix} 0 & \hat{Q}_{12}(t) & \hat{Q}_{13}(t) & \hat{Q}_{14}(t) \\ \hat{Q}_{21}(t) & 0 & 0 & 0 \\ \hat{Q}_{31}(t) & 0 & 0 & 0 \\ \hat{Q}_{41}(t) & 0 & 0 & 0 \end{pmatrix}
$$

Its construction indicates univocally that the system of interest is of a series type because only the first column and the first row are non-zero.

Based on the total probability principle the following equations hold:

$$
p_{12} = \int_0^\infty (1 - Q_{31}(t))(1 - Q_{14}(t)) d\hat{Q}_{12}(t)
$$

$$
p_{13} = \int_0^\infty (1 - Q_{14}(t))(1 - Q_{12}(t)) d\hat{Q}_{13}(t)
$$

The last probability to be determined is p_{14} and can be settled by constructing a similar equation to the one above or by using the principle to close all these probabilities to unity.

The ergodic probability distribution Π for the Markov chain of interest can be obtained by solving equation 7.23 which yields:

$$
\Pi_1 = \tfrac{1}{2} \quad \Pi_2 = p_{12} \Pi_1 \quad \Pi_3 = p_{13} \Pi_1 \quad \Pi_4 = p_{14} \Pi_1
$$

The ergodic probability distribution for the semi-Markov process consists of the following elements:

$$
\rho_1 = \frac{\Pi_1 m_1}{M} \quad \rho_2 = \frac{\Pi_2 m_2}{M} \quad \rho_3 = \frac{\Pi_3 m_3}{M} \quad \rho_4 = \frac{\Pi_4 m_4}{M}
$$

where:

$$
M = \Pi_1 m_1 + \Pi_2 m_2 + \Pi_3 m_3 + \Pi_4 m_4
$$

The steady-state availability of the system is:

$$
A = \rho_1
$$

The intensity of failures of the process is:

$$\lambda = \frac{1}{p_{12}(m_{12} + m_{21}) + p_{13}(m_{13} + m_{31}) + p_{14}(m_{14} + m_{41})}$$

A further method of modelling and analysis that could be adopted would be to use a similar approach to that shown for analysing a powered support. Readers may undertake this analysis according to their requirements. However, it is necessary to remember to include in the analysis of this system the accessibility of the power shovel and the crusher, both reducing the available time to execute the system's duties.

At the end of the considerations connected with semi-Markov models, two additional remarks must be made concerning the scope of their application and the possibility of analysing these systems in a different manner.

In the next chapter cyclic systems will be analysed, starting from the simplest case—Palm's model—up to the most advanced—the *G/G/k/r* model. After becoming familiar with these models, it will be easy to conclude that semi-Markov models can also be applied for modelling and analysing cyclic systems.

The second remark is that there are also different possibilities for analysing semi-Markov processes.

Limnios (Bousfiha and Limnios 1997, Limnios and Oprişan 2001) proposed applying ph-distributions for a reliability evaluation of semi-Markov systems. The idea of this method goes back to Cox (1955). It is possible to approximate a semi-Markov kernel by a phase-type semi-Markov kernel and then transform this into a Markov kernel by introducing additional states.

The initial stage consists of approximating the transition functions $Q_{ij}(t)$ of the original semi-Markov kernel by phase distributions, say, $B_{ij}(t)$. In order to do it in a proper way, an optimisation problem is formulated stating, for example, that:

$$\text{minimise} \int_{\Re_+} (Q_{ij}(t) - B_{ij}(t))^2 \, dt$$

under the constraints:

$$\int_{\Re_+} t^k dQ_{ij}(t) = \int_{\Re_+} t^k dB_{ij}(t) \quad \text{for } k = 1, 2, \ldots, r$$

Notice, these functions are statistical moments generating functions with regard to the supplied function. The idea here is identical to that used in a transition from a discrete type model to a continuous space and—after a solution is found in this space—in the return to the discrete space in the Sivazlian and Wang model. The main point is not to lose the statistical properties of the functions when a transformation or approximation is being applied.

For the Cox distribution family, whose Laplace-Stieltjes transform is of the form:

$$\sum_{s=1}^{n} a_1 \ldots a_k \prod_{i-1}^{s} \frac{\mu_i}{s + \mu_i}$$

we have to calculate the set of parameters $a_1, \ldots, a_n$; $\mu_1, \ldots, \mu_n$ and n.

As a rule, this approximation leads to a large number of parameters. Calculation is time consuming and tedious. Fortunately, Aldous and Shepp (1987) gave a lower limit for the number of phases required to represent a probability distribution Q with the mean μ and variance σ^2 on $\Re_+$. This boundary is the inverse of the square of the coefficient of the variation of Q, that is $(\mu/\sigma)^2$.

There is also one further approach to the analysis of semi-Markov processes proposed by Sing (1980) and called the equivalent rates method. Unfortunately, it is not an exact method. It consists in transforming a semi-Markov process into a Markov process by taking as the sojourn time in state i an exponential distributed time with parameter $(m_i)^{-1}$, where m_i is the mean sojourn time of the semi-Markov process in state i. The advantage of this method is that it gives the same limit distribution for the two processes, but the error for the transition probabilities in finite time is out of control. Nonetheless, we can apply the general method given in Chapter 4.6 of Limnios and Oprişan (2001) to bound the error of this method.

It seems that the method of analysis of semi-Markov processes presented here is quite practical and free of the artificial construction of additional states.

CHAPTER 8

Cyclic systems—selected problems

As was stated in Chapter 2.2, cyclic systems applied in mining can be divided into four types. The criterion for this division is the manner of operation of these systems, which determines to some extent which mathematical methods can be used to describe their exploitation processes.

In this chapter some problems of mine cyclic systems that can be described by queue theory will be considered. The only exception to this rule is a shovel-truck system. Its analysis, modelling and method of analytical calculation together with a vast literature review were done in Czaplicki's monograph of 2009. However, two combinations of a shovel-truck system with other pieces of equipment will be discussed in Chapter 9.

We commence our considerations by analysing a general model discussed in queue theory or—applying nomenclature used in Central and Eastern Europe—in the mass servicing theory.

8.1 A GENERAL MODEL OF QUEUE THEORY

This model is presented in Figure 8.1. Let us analyse its components.

A stream of arrivals (clients, calls) is directed to a service system. A call 'carries' a need for service. In a general case, the stream (A) can be treated as a stochastic process with the following properties:

- arrivals come to the system singly or in groups
- the group size is either deterministic or it is a random variable
- the moment of the occurrence of a call is deterministic or random
- clients come into the waiting room or they depart (because of a lack of free space in the waiting room or because there is a long queue in the waiting room for instance)
- the population from which arrivals come is either finite or not finite.

Usually, it is presumed that for a finite population, arrivals after service come back to the population and become potential calls for a new service. Systems with a finite source (population) are called *closed systems*. If, for example, a machinery system of a mine has a constant number of operating units, this system is a closed one until the moment when new pieces of equipment are purchased for the system or when a machine finishes its *life*.

Remember:
If one or more machines are introduced additionally to a given system, or one or more machines are withdrawn from the system or one or more pieces of equipment are replaced—the parameters and characteristics of this system will change.

Generally (A), the process of arrivals is a stochastic one; in a particular case, it can be characterised by a random variable or can be treated as deterministic one.

Waiting room (B) is in the system. These are its properties.

- A client in the waiting room waits for service when the service stand is occupied or in repair.
- Otherwise the client immediately goes to the stand and is serviced.
- The number of places in the waiting room is finite or not.

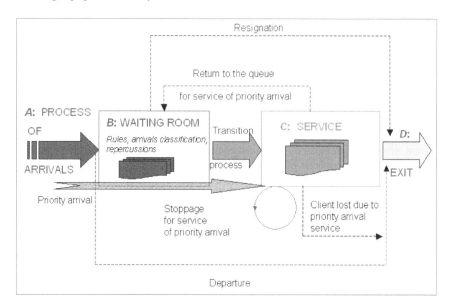

Figure 8.1.   Idea of a mass servicing system.

- If the call (or client) wants to wait for service (but it can usually resign at any moment in time), then it becomes the subject of the rules governing a queue (regulations determining how calls will be serviced). The most popular rules in this regard are:
    - *FIFO*—first-in first-out, arrivals are serviced in the sequence in which they occur in the system
    - *SIRO*—selection in random order, arrivals are serviced in a random sequence
    - *LIFO*—last-in first-out, the arrival that occurred last in the system is serviced first.

The first scheme is best understood intuitively. The second scheme, for example, holds when reserve parts are available. The collection of reserve parts is usually random. The third scheme holds in a situation involving the storage of reserve items. When there is limited access to the store, the item that was stored last is usually the one that is removed first.

In some queue systems, clients are classified—some clients are regarded as being more important than others. This means that such a system possesses a hierarchical organisation of servicing. Thus, some clients have priority. Usually, a client waiting the longest time from the class with the highest priority is serviced first. The classification of arrivals has an influence on the organisation of the service (*C*). This categorisation has consequences and therefore priority schemes are introduced.

There are two basic priority schemes.

- An absolute (ousting) priority, when a service in a process is halted to service the priority arrival; here are three cases to consider.
    - An absolute priority with losses; the service for the displaced client (or call) is lost as a result of the stoppage for the priority call.
    - An absolute priority with later continuation; following the priority service, after the stoppage the displaced client (or call) returns to its service stand and the halted service is continued.
    - An absolute priority with starting from beginning; after the stoppage of its service, the displaced client (or call) waits for the service of the priority call to be finished and then it returns to the stand, but service must be done from the beginning.
- A relative priority, when a running service is not stopped but the priority arrival is first to be serviced, thus jumping the queue in the waiting room.

There are also some other types of priority schemes. For example, if a repair is advanced and will soon be finished, then the service is frequently not stopped for a priority. The priority arrival waits.

In some mine systems, there can be a reverse situation. Some *service* points have priority, and clients are directed by a dispatching system to these points. For example, during the exploitation process of a shovel-truck system, shovels loading priority material should operate round-the-clock and therefore have priority.

A unique queue system is one that has no waiting room. This system is called *without waiting* or *with losses*. An arrival at the system when all service stands are occupied is lost. Otherwise, the system is *waiting*. Another particular case here is a system *without losses*, where there is no way to leave the system without receiving service.

Now consider service (**C**). Its properties can be characterised as follows.

- The service subsystem can be equipped with one stand or can have many stands.
- The organisation of stands can be:

  o parallel, that is all stands perform the same kind of service
  o non-parallel, that is some stands can perform a certain type of service.

- If the service subsystem has a number of parallel stands then:

  o a queue can be formed in front of each stand, and clients can select a queue or they are serviced according to the priority scheme
  o one general queue can be formed and stands call clients (in order) when they are free.

- If the service has a number of non-parallel stands, then some of these stands can create an organised subsystem where sequent phases of service can be performed.
- Clients can be serviced non-singly but in groups, and the size of the groups can be:

  o deterministic
  o limited
  o random.

- The service subsystem and, therefore, the whole system can be available:

  o continuously
  o periodically
  o with unpredicted breaks (because of breakdowns).

- In some systems, a client being serviced blocks the whole stand; in some effective systems, a queue can be formed in front of each stand.

Generally, servicing (**C**) is a process that is partly stochastic, the time of service is a random variable, but sometimes it can be treated as a deterministic one.

In this general model there are three incoming random variables:

- the time between two neighbouring arrivals (either single calls or groups ones)
- the number of arrivals at a given moment $\equiv$ group size
- the time of servicing a single client.

In this model there are also three parameters:

- the number of service stands, s_1
- the number of spaces in the waiting room $\equiv$ maximum queue length (s_2)
- the population size.

Assumptions of the model should denote:

- specification of random variable distributions
- mutual dependence of these random variables
- the queue rule, the gradation of arrivals and the consequences resulting from this categorisation.

8.2 MATHEMATICAL CLASSIFICATION OF CYCLIC SYSTEMS

In queue theory in order to identify what kind of a system one is dealing with the following notation is often used[1]:

$$A/B/s_1/s_2$$

where:

A is the probability distribution of the times between neighbouring arrivals to a mass servicing system
B is the probability distribution of times of service.

However, in the place of the letters *A* and *B* the following notation is used in this book:

D – deterministic distribution (identical time intervals)
M – exponential distribution (*M* after Markov)
G – general distribution
GI – general distribution of times between arrivals and general distribution of times of repair, independent of each other
E_k – Erlang distribution of the order *k*
K_n – χ^2 distribution with *n* degrees of freedom.

There is also some other notation that can be applied, but this is seldom used and it is not necessary to cite details here.

The general purpose of queue theory is to construct methods for the analysis of mass servicing systems in order to run them in rational way.

One can look at a queue system:

– from a client point of view
– from a management point of view.

A client is interested in following issues:

– selection of the moment of arrival to the system (to avoid rush hour, to select a convenient time for prophylactic action, etc.)
– selection of a system (perhaps, for example, a more expensive service but one offering better quality)
– selection of preferences (preference means privilege; priority in service).

The system management is interested in:

– attaining a more effective service, which means developing a more profitable system
– attaining a more attractive service,[2] perhaps through eye-catching deals, or making the service a bit cheaper.

Usually, the most important system characteristic is the limited distribution of system states that characterises the system after a long period of operation. The system is called *stable* if this distribution does not depend on the initial state of the system. A familiarity with this distribution allows many vital problems from a practical point of view to be solved.

---

[1] This classification is wider than the classical one presented by Kendall because of the element s_2. However, it neglects the expectation rule which is considered in the Lee notation. Mass servicing systems being considered here will be of *FIFO* type.
[2] Beautiful long-legged girls working in a service system could be '*an element* that makes the system more attractive'.

8.3   THE REPAIRMAN PROBLEM[3]

Consider a system which consists, from a functional point of view, of $m + r$ homogeneous machines, where m machines are directed to work and r machines are held in reserve. Spare units do not fail, which means they create a cold reserve. Machines operate in parallel (that is, independently of each other) and each unit can fail. A failed machine is directed to repair and a new unit is taken from the reserve provided that it is not empty. The number of repair stands (facilities) is k. After renewal a machine is directed to the reserve or set directly to work when the reserve is empty and less than m machines are in work. Repair completely restores the capability of the machine to work. Repair and work times are independent of each other and they have their own probability distributions (G, G). The operating scheme of the repairman system is shown in Figure 8.2. A further specification of the system's operation should detail the organisation of the operation of repairmen in relation to any failures that occur.

In mining practice many reliability investigations have proved that the work and repair times for a variety of pieces of equipment are independent of each other. If this independence assumption was not met—and that happened rarely—these were special cases and, for this reason, they will not be discussed here.

It will also be assumed that the waiting-room rule is a first in first out (*FIFO*) one. Again, mine investigations have shown that this holds in a majority of cases. Sometimes, situations have been noted where repair work was allocated on the principle that a short repair is cleared out first. But such cases were rare.

In the classical repairman problem the main issue is the determination of a mathematical expression that describes system output, understood here as the number of machines in a work state. Further system characteristics such as the mean system productivity, indices of the work times of machines or idleness of repairmen can be found after having specified the limited probability distribution of the number of machines in a work state. Moreover,

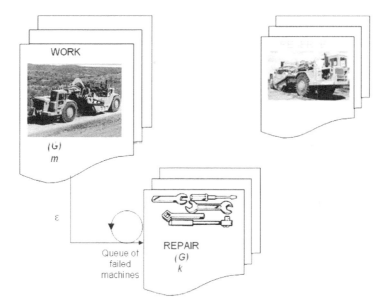

Figure 8.2.   Operating scheme of the repairman system.

---

[3] The repairman problem has been known for many years, probably starting with Palm (1947) and later in Feller (1957), Cox and Smith (1961) and Takács (1962).

the economic aspects of the operation of this system can also be considered by knowing this distribution.

Because there are a number of pieces of mine equipment for which both work and repair times can be satisfactorily described by exponential distributions, we will now discuss the Palm model.

8.4   THE PALM MODEL

Presume now that the probability distribution of repair times is exponential and the probability distribution of work times is also exponential. Denote as previously:

$$E(T_w) = \lambda^{-1} \quad E(T_r) = \mu^{-1} \quad \text{and} \quad \kappa = \lambda/\mu$$

The system we are interested in is: *M/M/k/r*

If so, the process $n(t)$ defined as the number of machines in repair at a given moment in time t is the birth-death process[4] of intensities of transition:

$$\mu_i = \begin{cases} i_\mu & \text{for } i = 0, 1, \dots, k \\ k_\mu & \text{for } i = k, k+1, \dots, m+r \end{cases} \tag{8.1a}$$

$$\lambda_i = \begin{cases} m\lambda & \text{for } i = 0, 1, \dots, r \\ (m+r-i)\lambda & \text{for } i = r, r+1, \dots, m+r \end{cases} \tag{8.1b}$$

The limited probability distribution of the number of machines in repair as a function of the number of machines in the reserve is:

- for $r = 0$

$$P_i^{(r)} = \begin{cases} \dfrac{m!}{i!(m-i)!} \kappa^i P_0^{(r)} & \text{for } i = 1, 2, \dots, k \\[4mm] \dfrac{m!}{k!k^{i-k}(m-i)!} \kappa^i P_0^{(r)} & \text{for } i = k, k+1, \dots, m \end{cases} \tag{8.2}$$

- for $r = 1, 2, \dots, k$

$$P_i^{(r)} = \begin{cases} \dfrac{m^i}{i!} \kappa^i P_0^{(r)} & \text{for } i = 1, 2, \dots, r \\[4mm] \dfrac{m^r m!}{i!(m+r-i)!} \kappa^i P_0^{(r)} & \text{for } i = r+1, r+2, \dots, k \\[4mm] \dfrac{m^r m!}{k!k^{i-k}(m+r-i)!} \kappa^i P_0^{(r)} & \text{for } i = k+1, k+2, \dots, m+r \end{cases} \tag{8.3}$$

---

[4] The Markov process determined on a finite or countable set of states characterised by the intensities of transitions is called a birth-death process if the states of this process can be ordered in such a way that only the intensities of transition between the neighbouring states can be non-zero.

- for $r = k + 1,\ k + 2, \ldots$

$$
P_i^{(r)} = \begin{cases}
\dfrac{m^i}{i!}\kappa^i P_0^{(r)} & \text{for } i = 1, 2, \ldots, k \\[2ex]
\dfrac{m^i}{k!\,k^{i-k}}\kappa^i P_0^{(r)} & \text{for } i = k+1, k+2, \ldots, r \\[2ex]
\dfrac{m^r m!}{k!\,k^{i-k}\,(m+r-i)!}\kappa^i P_0^{(r)} & \text{for } i = k+1, k+2, \ldots, m+r
\end{cases} \tag{8.4}
$$

where for each r the sum of all probabilities is closed to unity.

One can easily recognise that if the reserve size is zero and the number of machines in the system is identical to the number of repair stands, then the above probability distribution is confined to a binomial distribution:

$$
P_i^{(r)} = \binom{m}{i}(1-A)^i A^{m-i} \quad \text{for } i = 1, 2, \ldots, m
$$

And:

$$
A = \frac{\mu}{\lambda + \mu}
$$

which is the steady-state availability of machines.

For practical purposes the probability distribution of the number of machines in a work state is more interesting. Usually we are interested in its limited shape because only in readiness systems is the point of interest this probability distribution dependent on a given moment in time t.

Therefore defining a new random variable:

$$
m(t) = \min\left(m,\ m + r - n(t)\right)
$$

we can obtain the limited probability distribution of the number of machines in a work state:

$$
P_i^{(w)} = \begin{cases}
\displaystyle\sum_{j=0}^{r} P_j^{(r)} & \text{for } i = m \\[2ex]
P_{m+r-i}^{(r)} & \text{for } i = 0, 1, \ldots, m-1.
\end{cases} \tag{8.5}
$$

Based on the above, two important parameters for engineering practice can be defined, namely:

- the average number of machines in failure:

$$
L^{(r)} = \sum_{i=1}^{m+r} i P_i^{(r)} \tag{8.6}
$$

- the average number of machines in a work state:

$$
L^{(w)} = m \sum_{j=0}^{r} P_j^{(r)} + \sum_{i=1}^{m-1} i P_{m+r-i}^{(r)} \tag{8.7}
$$

Keeping in mind the system of machines, we can state that on average there will be a certain mean number of machines in a work state, a certain mean number of machines in failure and a certain mean number of units in reserve. In mining reality, at any given moment in time these three numbers are stochastic because they are random variables.

Consider a tutorial case.

■ Example 8.1

There are given two systems:

$$\mathfrak{S}_1 : < m = 24, k = 4, r = 4; A = 0.82; \mathfrak{M}_{gr} >$$

$$\mathfrak{S}_2 : < m = 24, k = 4, r = 4; A = 0.82; \mathfrak{M}_{ind} >$$

The first system consists of 24 machines directed to work and 4 machines in reserve. The steady-state availability of the machines is 0.82. There are 4 repairmen and the method $\mathfrak{M}_{gr}$ of their organisation is a group one.

The second system is identical to the first except for the method of the organisation of work of the repairmen; they operate independently of each other. Every repairman takes care of six pre-determined machines. This means that the second system can be divided into four identical subsystems:

$$\mathfrak{S}_2 = 4 \times < m = 6, k = 1, r = 1; A = 0.82 >$$

Compare the characteristics of both systems.

Let us begin by looking at both limited probability distributions. They are presented in Table 8.1. Consider now essential system parameters.

The average number of failed machines:

$$\mathfrak{S}_1 : L^{(r)} = 10.07 \quad \mathfrak{S}_2 : L^{(r)} = 4 \times 2.83 = 11.32$$

Table 8.1.   The limited probability distributions of the number of failed machines.

P_i	$\mathfrak{S}_1$	$\mathfrak{S}_2$
0	0.001	0.107
1	0.006	0.141
2	0.016	0.186
3	0.028	0.204
4	0.036	0.179
5	0.048	0.118
6	0.06	0.052
7	0.073	0.011
8	0.084	
9	0.092	
10	0.096	
11	0.095	
12	0.089	
13	0.078	
14	0.064	
15	0.049	
16	0.035	
17	0.023	
18	0.014	
19	0.008	
20	0.004	
21	0.002	

The average number of machines in a work state:

$$\text{\emph{S}}_1 : L^{(w)} = 17.85 \quad \text{\emph{S}}_2 : L^{(w)} = 4 \times 4.07 = 16.28$$

The average number of failed machines waiting for repair:

$$\text{\emph{S}}_1 : L^{(o)} = L^{(r)} - 4 = 6.07$$
$$\text{\emph{S}}_2 : L^{(o)} = L^{(r)} - 4 = 7.32$$

The average number of spare machines:

$$\text{\emph{S}}_1 : L^{(sp)} = m + r - L^{(r)} - L^{(w)} = 0.08$$
$$\text{\emph{S}}_2 : L^{(sp)} = m + r - L^{(r)} - L^{(w)} = 0.4$$

The results obtained indicate that group organisation is more advantageous than individual organisation, which is an obvious result. In the group organisation there situation cannot arise where a repairman is idle and a machine in failure is waiting for repair.

It is worth looking more closely at the interesting differences in some of the exploitation system parameters. Additionally, some relative exploitation parameters can be considered that characterise both system machinery and repairmen utilisation, that is:

$$\frac{\text{The mean number of machines in a work state}}{\text{The number of machines directed to work}}$$

$$\frac{\text{The mean number of busy repairmen}}{\text{The number of repairmen}}$$

Further interesting parameters are functions that determine how the basic system indices change for different sizes of the reserve. Using more in-depth reasoning, an interesting problem might be to consider for which reserve size does a further increase in the number of spare machines make no difference for the system exploitation parameters. If this problem is critical, it should also be considered on an economic basis.

Incidentally, if the steady-state availability of machines is 0.78 or lower, the average number of machines in reserve is zero. ◀

This problem was also considered by Palm (1947), Feller (1957), Gnyedenko and Kovalenko (1966).

8.5 A SYSTEM WITHOUT LOSSES

This model is a particular case of the Palm model. Let us discuss its properties by considering an example taken from mining practice.

■ Example 8.2

A certain type of failure in a mine was investigated. The mean number of this type of failure was found to be 1.04 failures per one working shift (8 hours). Further study showed that the times between the occurrences of two neighbouring failures can be satisfactorily described by the exponential distribution. Similarly, the times of repair can also be described by this type of distribution, with the mean equalling 2.7 hours. All failures are cleared out. The question is how many repairmen should be available during one production shift?

At first glance the problem looks like a trivial one—failure appears every $8/1.04 = 7.7$ hours on average and the mean repair time is 2.7 hours. One repairman should be enough.

However, look at this problem from the perspective of queue theory.

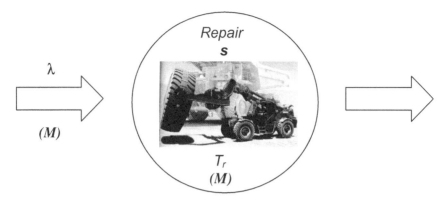

Figure 8.3.   Operating scheme of a system without losses.

Because there is no 'client' loss, the system can be noted as:

$$M/M/s/\infty$$

and Figure 8.3 illustrates the basic idea of the system.

For such a system the limited probability distribution of the number of failures is determined by the formula:

$$P_i = \begin{cases} (\kappa^i/i!)P_0 & \text{for } i = 1, 2,\ldots, s \\ (\kappa^i/s!s^{i-s})P_0 & \text{for } i = s, s+1,\ldots \end{cases} \tag{8.8}$$

where κ is obviously the repair rate. Because this formula has a recurrent character, an equation that the sum of all probabilities must be 1 has to be added.

For the system being considered $\kappa = 2.7/7.7 = 0.35$ and the limited probability distribution of the number of failures in relation to the number of repairmen is given in Table 8.2. The numbers in the last row are the expected values.

Looking at the outcomes of the calculations presented in this table, it is easy to notice that the distribution quickly tends to its limited form and there is no difference whether three of four repairmen are working. The difference between $s = 1$ and $s = 2$ is well recognised. We can hesitate about stating whether the difference between $s = 2$ and $s = 3$ is significant.

There should be no doubt that consideration ought to be transferred onto the economic aspects of the problem. Let us make a simple cost calculation.

Let us presume that the unit cost of a repairman is c_r €/h and the unit loss due to repair is c_l €/h.

Defining the average value of unit costs we have:

$$C_s = sc_r + c_l \sum_i iP_{i+s}$$

Therefore we search for such a value of s to get the minimum cost C_s.

Presume now, for instance, that $c_r = 40$ and $c_l = 500$. If so:

$$C_{s=1} = 133.6 \quad C_{s=2} = 85.35 \quad C_{s=3} = 120.4$$

For a further increase in the number of repairmen the total cost C_s increases permanently.

A simple conclusion can be formulated here that it is much better to have two repairmen rather than only one.

Table 8.2. The limited probability distributions of the number of failures P_i vs. the number of repairmen s.

P_i	$s = 1$	$s = 2$	$s = 3$	$s = 4$
0	0.65	0.702	0.705	0.705
1	0.223	0.246	0.247	0.247
2	0.080	0.043	0.043	0.043
3	0.028	0.008	0.005	0.005
4	0.010	0.001	0.001	
5	0.003			
6	0.001			
E	0.538	0.361	0.351	0.350

Remember that this conclusion holds for this set of values of the system parameters. For a different set the final conclusion may be different. ◄

Let us define some interesting system parameters.

- The average number of failures that have occurred in the system is:

$$L^{(r)} = \kappa \left[\frac{\kappa^s(\kappa s - \kappa^2 + s)}{s!(s - \kappa)^2} + \sum_{k=0}^{s} \frac{\kappa^k}{k!} \right] P_0 \tag{8.9}$$

- The average number of failures waiting for repair is:

$$L^{(o)} = \frac{\kappa^{s+1}}{(s - \kappa)^2(s - 1)!} P_0 \tag{8.10}$$

- The average number of free repairmen is:

$$L^{(f)} = s - \kappa \tag{8.11}$$

- The limited probability that all repairmen will be busy is:

$$P_B = \frac{\kappa^s}{(s - 1)!(s - \kappa)} P_0 \tag{8.12}$$

Some further system parameters can be constructed but they are of minor significance.

Generally, the literature from this field is actually quite rich and researchers have many positions to select from (see, for example, Cox and Smith 1961, Takács 1962, Cooper 1972, Lee 1966, Gross and Harris 1974, Tijms 2003, Wolf 2007).

8.6 ERLANGIAN SYSTEMS

We remain in the field of exponential probability distributions; however, now the level of generalisation is higher. The key to making our considerations more general is a property that has a random variable of Erlang distribution.

The Erlang distribution of density function:

$$f(t) = \begin{cases} \lambda(\lambda t)^{k-1} \exp \dfrac{-\lambda t}{(k-1)!} & \text{for } t \geq 0 \\ 0 & \text{for } t = 0 \end{cases} \qquad (8.13)$$

is the distribution of the sum of k independent random variables exponentially distributed with the same parameter λ.

If the probability distribution of the times between arrivals to a queue system can be described by an Erlang distribution and, similarly, the probability distribution of times of service can be satisfactorily described by an Erlang distribution, then—making an appropriate conventional decomposition—considerations can be shifted to the well-known and convenient Markov theory field.

If, for example, the probability distribution of the times between calls to the service system is of the order s, then we can decompose the time into s succeeding phases, each one of an exponential distribution. Analogically, a random variable—time of service—can be decomposed into, say, v phases, each one of an exponential distribution. Therefore, analysis of the system:

$$E_s/E_v/k/r$$

can be converted into an analysis of the Markov processes.

The analysis of an Erlangian system can be done using the following scheme:

1. Original system decomposition to identify the phases exponentially distributed
2. Identification of the intensities of the transitions between phases
3. Identification of the exploitation repertoire
4. Construction of a matrix of intensities of passages between phases
5. Construction of equations determining the limited probabilities of phases
6. Calculation of probabilities, lumping appropriate phases, and final calculation of parameters of states.

It is easy to recognise that some steps are similar or identical to those in the methodology for the analysis of continuous systems.

Let us discuss an example.

■ Example 8.3

There are two dumpers in a certain quarry delivering excavated rock to a crusher. Due to the high demand for the crushed material, and the not so high reliability of the trucks, one additional truck was purchased to operate as a reserve unit and the decision was made to work 18 h/d.

To maintain the trucks in a proper condition two repair stands were arranged. Figure 8.4 illustrates this system.

A reliability investigation showed that the work times of the trucks are of the Erlang distribution of the order 2 with a mean of 10 hours. Repair times were also described by the Erlang distribution of the order 2 but with an average value of 3.1 hours. Figure 8.5 shows both probability density functions.

Let us analyse this system and try to find out whether the decision to purchase one more dumper is a rational one.

Let us start from an identification of the system.

The system discussed can be determined as follows:

$$\mathfrak{E} : <m = 2; k = 2; r = 1; A = 0.763>$$

and its notation is:

$$E_2/E_2/2/1$$

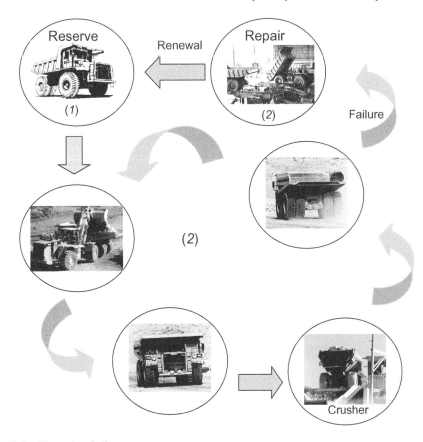

Figure 8.4.   The system in the quarry.

$$k := 2 \quad f(x) := \frac{\lambda^k \cdot x^{k-1}}{(k-1)!} \cdot e^{-\lambda \cdot x} \quad g(x) := \frac{\mu^k \cdot x^{k-1}}{(k-1)!} \cdot e^{-\mu \cdot x}$$

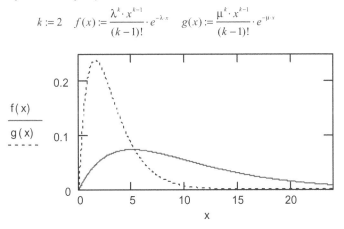

Figure 8.5.   The probability density functions for trucks in the analysed system.

In order to transfer our consideration to the Markov theory field we presume that the work time is a two-phase one. Let the first phase be called 'with great potential' and the second one 'with low potential'. The repair time will be treated similarly. The first phase may be termed as the 'beginning of repair' and the second one the 'finishing of repair'. All four random variables are obviously exponential.

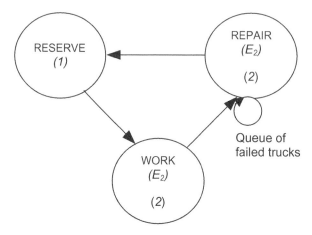

Figure 8.6. Operating scheme of the system.

We need to calculate the intensities of transition between phases. These intensities are parameters of their exponential distributions.

The Erlang distribution of the second order with an average 10 h means that the intensity
$\lambda = 2/10 = 0.2$. Similarly, the Erlang distribution of the second order with an average 3.1 h means
the intensity $\lambda = 2/3.1 \cong 0.65$.

Determination of an exploitation repertoire $\mathfrak{C}$.

The situation here is different to that in an analysis of continuous systems. The states here are
divided into phases and the process of changes of phases is identified. These phases are stipulated
and unrealistic ones. When this is recognised, the phases are lumped and in this way a return to
reality is achieved.

Let us introduce a notification of the states of the phases of system being considered as:

$$(a, b, c, d)^5$$

where each letter means the number of machines at a given phase. All these numbers are obviously random. In our case the letters can take the values: 0, 1, 2. Notification will be according to
the rule that is assumed in theory of queues. A given machine starts duty (work) in s-phase and
finishes in the first phase. Similarly, repair commences in v-phase and is terminated in the first
phase. Transitions are running from the right to the left. Saying this in different way—a phase
number equals the number of phases required to finish work/repair. For better understanding let us
consider—for example—the following states of the phases: (2000), (0201) and (0011).

The state of phases (2000) means that 2 machines are at the phase of work with low potential
(say, just before failure), there is no machine at the phase of work with high potential, there is no
machine at the second phase of repair and at the first phase of repair. An inference: one machine
is in the reserve.

The state of phases (0201) means that there is no machine at the phase of work with low potential, there are 2 machines at the phase of work with high potential, there is no machine which
repair is about to finish, and there is a machine at the second phase of repair (repair started). An
inference: there is no machine in the reserve.

The state of phases (0011) means that there is no machine at the phase of work with low
potential, there is no machine at the phase of work with high potential, there is a machine at the

---

[5] Usually, the number of phases are not divided by commas. This notification is an exception in order to
indicate that all numbers are separate.

first phase of repair and also there is a machine at the second phase of repair. An inference: all machines are down and one is waiting for repair.

The analysed system can be at 18 states of phases. They are enumerated in a matrix (shown below) of the intensities of the transitions between states. It is presumed that time is a continuous variable and for this reason only one transition is possible at a given moment in time. In the matrix only non-zero elements are inserted. Because this matrix concerns the Markov process then the sum of all intensities in a given row must be zero. In such a way the main diagonal elements are defined.

The last case to explain is how to find the passages between the phases of states. Let us analyse, for instance, the state (2000). We know what it means. Consider now, what kind of events might happen in the near future. One possibility is that one of working machines can fail and the state will be (1101). The number 1 has been removed from note 2 and this number occurred at the last place in the notation (repair started). The machine failed—intensity λ. But in the place of the failed machine a new unit appeared from the reserve. This fact is indicated in bold. Because this situation can concern both working machines, the number 2 occurred before the intensity λ in the matrix. Because there are no other possibilities, the rest of intensities are zero.

If the matrix is a specified, construction of equations determining the limited probabilities of states is according to the well-known principles (see Chapter 7.3). Rejecting one equation and replacing it with the equation that the sum of all probabilities must be closed to unity allows estimations of unknown probabilities to be obtained. They are shown in Table 8.3.

Matrix of transitions between the states for the system being considered

From state \ To state	(2000)	(1100)	(0200)	(2001)	(2010)	(1101)	(1110)	(0201)	(0210)	(1020)	(1011)	(1002)	(0120)	(0111)	(0102)	(0020)	(0011)	(0002)
(2000)	Δ_1					2λ												
(1100)	λ	Δ_2						λ										
(0200)		2λ	Δ_3															
(2001)				Δ_4	μ							2λ						
(2010)	μ				Δ_5						2λ							
(1101)				λ		Δ_6	μ								λ			
(1110)		μ			λ		Δ_7							λ				
(0201)						2λ		Δ_8	μ									
(0210)			μ				2λ		Δ_9									
(1020)							2μ			Δ_{10}						λ		
(1011)						μ				μ	Δ_{11}					λ		
(1002)											2μ	Δ_{12}						λ
(0120)									2μ	λ			Δ_{13}					
(0111)							μ				λ		μ	Δ_{14}				
(0102)												λ	2μ		Δ_{15}			
(0020)													2μ			Δ_{16}		
(0011)														μ		μ	Δ_{17}	
(0002)																	2μ	Δ_{18}

Now it is time to lump the determined states to get interesting limited probabilities. Making use of the information contained in Table 8.3 the probability distribution of number of machines in work state is:

$$P_0^{(w)} = 0.009$$
$$P_1^{(w)} = 0.094$$
$$P_2^{(w)} = 0.897$$

The mean value is: $L^{(w)} = 1.89$.

Following this method we can determine that the probability distribution of the number of machines in repair is:

$$P_0^{(r)} = 0.521$$
$$P_1^{(r)} = 0.376$$
$$P_2^{(r)} = 0.094$$
$$P_3^{(r)} = 0.009$$

The mean value is: $L^{(r)} = 0.591$.

A system without a spare unit has the probability distribution of the number of machines in a work state:

$$P_0^{(w)} = 0.055$$
$$P_1^{(w)} = 0.36$$
$$P_2^{(w)} = 0.585$$

with the mean:

$$L^{(w)} = 1.53.$$

Table 8.3.   Limited probability distribution of the states of the phases of the system.

State number	State of phases	Number of failed machines	Number of machines in work	Probability
1	(2000)	0	2	0.1764
2	(1100)	0	2	0.2522
3	(0200)	0	2	0.0926
4	(2001)	1	2	0.0198
5	(2010)	1	2	0.031
6	(1101)	1	2	0.104
7	(1110)	1	2	0.0982
8	(0201)	1	2	0.0664
9	(0210)	1	2	0.057
10	(1020)	2	1	0.01
11	(1011)	2	1	0.0186
12	(1002)	2	1	0.007
13	(0120)	2	1	0.0128
14	(0111)	2	1	0.0296
15	(0102)	2	1	0.0156
16	(0020)	3	0	0.004
17	(0011)	3	0	0.004
18	(0002)	3	0	0.001

Therefore the increase in the system output because of one additional truck as a spare unit is 1.89 − 1.53 = 0.36 truck output. Taking into account the effective truck output and the price of crushed rock we can approximately evaluate whether the decision to purchase this new dumper is a rational one. If we would like to consider the problem more carefully, the economic aspect of repairs plus probable fluctuations in the price of crushed rock should be taken into account. ◄

At the end of the consideration of Erlangian systems some remarks ought to be made. From a mathematical standpoint the methods of analysing these systems has been well-known for years (cf. Kopociński 1973) and they are easy to conduct. Probability distributions in these systems are not exponential ones and that is a fundamental advantage. However, problems occur when the number of machines in the system being considered is large, when the number of repair stands increases or the order of Erlang distribution increases. The matrix of intensities of transitions between states increases to a great extent and the set of equations used to determine the limited probabilities of states can comprise a few hundred equations or even more. Taking into consideration that each probability distribution is always only a certain approximation of reality, especially because one parameter is a natural number, and computers solving such a great number of equations will round outcomes, the results of computations can have a low likelihood. Besides, a careful analysis of a complicated system using the Erlangian scheme is quite a tedious and time-consuming job.

8.7   THE TAKÁCS MODEL

This model dating back to 1962 is interesting from a mining point of view because its application to mining engineering problems was considered in 1968 by Kopocińska. This model falls into the repairman problem field but it is more general than the Palm model. In the Palm system both probability distributions are exponential, in the Takács model one distribution is released from its exponential fetters. This model is without a reserve and without losses.

The description of the model is as follows. A system consists of *m* working machines and the probability distribution of times between services is an exponential one. A time of service can be determined by any type of distribution—which means it is a general case. The service subsystem consists of only one unit. All machines are serviced according to the *FIFO* principle and a queue is created in front of the service stand if it is busy. Thus the system notation is:

$$M/G/1/m$$

Figure 8.7 illustrates the operating scheme of the system.

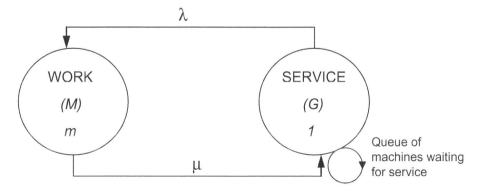

Figure 8.7.   Operating scheme of the Takács system.

Analysis of this system will be done by considering an example quite similar to one analysed by Kopocińska (1968). However, the approach to some engineering aspects of the problem will be a bit different.

▪ Example 8.4

The operation of a gravel mine is being analysed where one power shovel loads gravel onto dumpers. The gravel is delivered to a plant producing prefabricated concrete elements. The gravel is unloaded at the plant and the truck returns to the mine. The loading time of the truck is a random variable of a certain probability distribution $G(t)$ with T_z mean and a small variance. The total time of truck travel, that is for the cycle haul–dump–return is a random variable exponentially distributed with the parameter λ. A representation of the system is shown in Figure 8.8.

Based on a simplified economic analysis we will try to find the optimal number of trucks for the system.

Notice at the very beginning that instead of a classical repair facility we have here a different service point—a loading point. All appears to be correct but we have 'lost' the machine's reliability. We will return to this issue later on.

In this simplified economic analysis, we will neglect the constant cost. Let us consider an operation cost that will be the sum of two components:

1. the cost proportional to the number of trucks in the system—$k_1 m$
2. the cost proportional to the average number of trucks in operation—$k_2 \Sigma i q_i$

where:

k_1, k_2 – proportional coefficients; $k_2 \gg k_1$ due to obvious reasons
$\Sigma i q_i$ – the average number of travelling trucks
q_i – probability that i trucks are travelling.

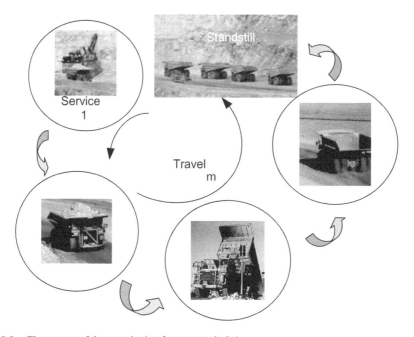

Figure 8.8.   The system of the gravel mine from example 8.4.

The mine income is proportional to a fraction of the work time of the shovel. Denote this proportional coefficient by k. Thus, the mine profit can be expressed by the formula:

$$Z = k(1 - q_m) - k_1 m - k_2 \sum_{i=1}^{m} i q_i \tag{8.14}$$

The mine profit depends on the number of machines being used, that is obvious, and depends heavily on the reliability of the equipment involved (as with the shovel-truck systems, which was clearly proved in Czaplicki's monograph 2009). Unfortunately, this latter issue is lacking in the model being considered.

Let us discuss the outcomes obtained by Takács.

We commence our consideration from the Laplace transform of the times of loading. Recall, the Laplace transform of a function $f(t)$ is an integral:

$$f^*(s) = \int_{0}^{\infty} e^{-st} f(t) dt$$

if the function is determined for $t \geq 0$, it is continuous and increases no greater than a certain exponential function $C\exp(at)$.

It is a well-known fact that times of loading in practice can be satisfactorily described by the normal distribution $N(T_z, \sigma_z)$ but this distribution is determined on the whole real numbers axis. This does not correspond with reality. Thus, an assumption is usually made that $\sigma_z \ll T_z$ (this often holds in practice), which means that the mass of probability over the negative part of axis can be neglected. The Laplace transform of the normal distribution is:

$$f^*(s) = \exp\left(-T_z s + \frac{\sigma_z^2 s^2}{2}\right)$$

Because the second component is of a lower order, it can be neglected and for this reason it can be assumed that:

$$f^*(s) \approx \exp(-T_z s) \tag{8.15}$$

Having defined this function, the following set of auxiliary functions and probability distributions should be determined:

$$C(r) = \prod_{s=1}^{r} \frac{f^*(s\lambda T_z)}{1 - f^*(s\lambda T_z)} \quad C(0) = 1 \quad r = 1, 2, \ldots, m - 1 \tag{8.16}$$

$$B(r) = C(r) \frac{\sum_{j=r}^{m-1} \binom{m-1}{j} C^{-1}(j)}{C(0) + \sum_{j=r}^{m-1} \binom{m-1}{j} C^{-1}(j)} \quad B(0) = 1 \tag{8.17}$$

$$P(j) = \sum_{r=j}^{m-1} (-1)^{r-j} \binom{r}{j} B(r) \quad P(0) = 1 + \sum_{r=1}^{m-1} (-1)^r B(r) \tag{8.18}$$

If so, the limited probability distribution of the number of travelling trucks $\{q_i\}$ is determined by formula:

$$q_i = \frac{mP(i-1)}{i(m\lambda T_z + P(m-1))} \quad i = 1, 2, \ldots, m \quad q_0 = 1 - \sum_{i=1}^{m} q_i \qquad (8.19)$$

Thus, the limited probability distribution of the number of trucks at the shovel is given by:

$$Q_i = q_{m-i} \quad i = 0, 1, \ldots, m \qquad (8.20)$$

Let us define two basic exploitation system parameters.

• The average number of trucks at the shovel is:

$$L^{(0)} = \sum_{i=1}^{m} iQ_i \qquad (8.21)$$

• The average number of travelling trucks is:

$$L^{(w)} = \sum_{i=1}^{m} iq_i = \lambda T_z(1 - q_m) \qquad (8.22)$$

Presume that the average time of truck loading is 3.5 minutes and the average time of truck travel is 38 minutes. Let the proportional coefficients be $k = 1000$, $k_1 = 10$ and $k_2 = 60$. Calculate the limited probability distributions of the number of trucks at the shovel for the different number of dumpers being used: $m = 3, 5, 7, 9, 11, 12, 13, 14$.

The results of the calculations are given in Table 8.4.

Let us make some comments in relation to the outcomes contained in this table. Some information has a general character.

Table 8.4. The limited probability distributions of the number of trucks at the shovel, and the average number of trucks at the shovel and an economic evaluation.

Q_i	Number of trucks in the system								
i	3	5	7	9	11	12	13	14	15
0	0.749	0.587	0.431	0.287	0.164	0.114	0.073	0.043	0.023
1	0.227	0.327	0.371	0.352	0.273	0.218	0.161	0.107	0.064
2	0.023	0.077	0.152	0.223	0.251	0.238	0.206	0.161	0.111
3	0.001	0.009	0.039	0.097	0.169	0.195	0.203	0.188	0.153
4			0.007	0.031	0.09	0.128	0.162	0.182	0.177
5				0.007	0.037	0.067	0.106	0.145	0.171
6				0.001	0.012	0.028	0.056	0.095	0.137
7					0.003	0.009	0.023	0.05	0.089
8						0.002	0.007	0.021	0.047
9							0.002	0.007	0.02
10								0.001	0.006
11									0.001
$\Sigma i q_i$	0.3	0.5	0.8	1.3	1.9	2.4	2.9	3.6	4.4
Z	57.5	94.1	128.4	158.5	181.4	188.9	193	193.5	190.6

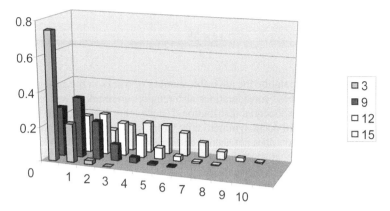

Figure 8.9. The probability distributions of the number of trucks at the shovel for a different number of trucks in the system.

Let us notice, at first, that when the number of haulers in the system increases, the probability distribution of the number of trucks at the shovel changes its character. When there is a distinct shortage of trucks, the distribution has an exponential character. With an increasing number of haulers in the system, the distribution loses its exponential nature, the mass allocation becomes flatter and this mass moves away from zero. The most probable situation is also shifted away from zero (see Figure 8.9). All these changes have a meaning in engineering. When the mass of probability is located near zero, it means that the power shovel operator has a lot of time free.

When the mass of probability moves away from the origin of coordinates, this operator becomes busier and busier. However, notice some subtleties. When there are 9 trucks in the system, the average number of haulers at the shovel is more than one (precisely, 1.3) but in almost 30% cases the shovel operator will be waiting for a truck to be loaded. When the number of dumpers reaches 14 he is 'deadly' busy—about 5% of time will be free and there will be 3.6 units at the shovel, on average.

All these remarks neglect the economic aspects. Now, in the last row of Table 8.4, the results of the financial calculations are given, showing that the most profitable solution is to have 14 trucks in the system. Remember that for a different set of financial parameters the optimum can be located at a different point. ◀

One item must be discussed at the end of this part of our considerations.

In many manuals and other publications, a recommendation is given that the optimal number of trucks in a system is determined by the formula[6]:

$$\frac{\text{Average time of loading} + \text{Average time of travel}}{\text{Average time of loading}}$$

In no event should this advice should be followed. At most, it may be treated as a suggestion (a hint) and the number that results may be the starting point to search for a proper value. Let us list the errors associated with this formula:

a. It completely neglects the economic side of the problem
b. The reliability of the equipment it is not taken into consideration
c. It is based on deterministic values (the average is a deterministic value), whereas the times of loading and travel are random variables.

---

[6] This formula is probably taken from analysing results of deterministic simulation. You can find examples of this method of analysis in Bise 2003, however it is not stated that this is a deterministic simulation.

Notice that applying this recommendation we have:

$$\frac{3.5 + 38}{3.5} \cong 12 \text{ haulers}$$

but according to our calculation the advisable number of trucks in the system is 14 units.

One further issue in these considerations also requires a remark. It was assumed that the times of truck travel have an exponential distribution. Usually, this does not hold in practice. Empirical distributions in this regard occur symmetrically with a shape similar to a normal distribution. For this reason, the assumption does not accord with practice.

Let us now look briefly at what should be taken into account when the reliability of the equipment in the system appears in the analysis.

Let us start from the reliability of a power shovel. If the reliability of this loading machine decreases, the 'stream' of gravel will become more and more narrow. Therefore, fewer hauling units will be needed for the system. Moreover, almost every shovel repair that takes longer than the average time of truck travel will create a regeneration of the exploitation process of the system and the process will 'start from the beginning'. All good trucks will come back from the unloading point. Notice, that if the process of loading commences again, the exploitation characteristics will be different from those that are found when the exploitation process is stabilised.

Let us now discuss the problem of truck reliability. If the reliability of hauling machines is to be taken into account, it means that randomisation has to be done. Previously, the number of trucks was a deterministic value, now it becomes a random variable. It is necessary to identify its probability distribution. Obviously, the lower the truck availability, the greater the number that is required in the system.

Additionally, we should take into account that almost always for larger systems of this kind a truck reserve is arranged. The occurrence of a reserve changes the method of truck system calculation. Going further along this line of reasoning, the existence of a truck maintenance/repair shop must be taken into account as well. If this shop has a low capability of making repairs, a queue of failed trucks will appear in front of the shop. For this reason more units will be needed. If this analysis is conducted in even more depth, the reliability of the repair stands should also be considered. The analysis becomes complicated but it can be done successfully. All these problems were discussed in Czaplicki's monograph of 2009.

8.8   THE MARYANOVITCH MODEL

A given system consists of m homogeneous machines directed to work and r machines in cold reserve (that is, the intensity of failures of machines in reserve is negligible). It is assumed that working machines can fail with a constant intensity ε. The workshop in the system contains an adequate number of repair stands so that all failed machines can be repaired simultaneously (that is, the number of repair stands is $m + r$). Therefore, there is no queue of machines waiting for repair. It is also assumed that repair times are independent random variables characterised by a general probability distribution $G(x)$. The expected value of this variable is known and equals T_n and for this reason the intensity of repair is $\eta = T_n^{-1}$. Additionally, three classical assumptions for queue systems are fulfilled, namely:

a. machines are served in a *FIFO* (first in, first out) regime
b. repair totally restores a machine's ability to work
c. random variables of the model—the work and repair times—are independent of each other.

A graphical representation of the Maryanovitch model is shown in Figure 8.10.
According to Kendall's notation, the Maryanovitch model can be given as:

$$M/G/m + r/r$$

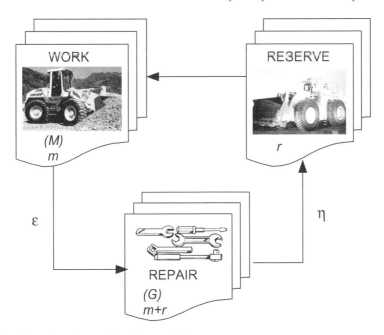

Figure 8.10.   Operating scheme of the Maryanovitch system.

It has been proved (see, for example, Maryanovitch 1961, Kopociński 1973) that the system described here has the probability distribution of a number of failed machines (that is, the machines in repair state) described by the formula:

$$P_k^{(n)} = \begin{cases} P_0^{(n)}(\kappa^k/k!)m^k & \text{for } k = 1, 2, \ldots, r \\ P_0^{(n)}(\kappa^k/k!)m^r m(m-1)\ldots(m-k+r+1) & \text{for } k = r+1, \ldots, m+r \end{cases} \qquad (8.23)$$

where $\kappa = \varepsilon T_n$ is the repair rate and obviously:

$$\sum_{k=0}^{m+r} P_k^{(n)} = 1$$

■ Example 8.5

The operation of a quarry is analysed and the demand is to produce X m³ of an aggregate stone. The mine operates T_d h/day and the utilisation of this time is ϕ, $0 \leq \phi \leq 1$. The crushed rock is loaded by the mine front-end loaders onto trucks delivered by a customer. The output of a single loader is W m³/h. The probability distribution of work times of loaders is exponential with parameter λ. The mean time of repairs of these machines is $T_r = \mu^{-1}$. Exploitation conditions are such that due to the confined space at the mine (the mine is located in the mountains) only m loaders can operate simultaneously, at most. Otherwise, they disturb one another. The problem is to find the number of loading machines to fulfil the customer's demand.

The object of interest is the system of front-end loaders. Keeping in mind information that only m of these machines can operate at the same time we can predict that a few machines should be in reserve because the demand is high. Knowing that the maintenance system of the mine is good, it can be presumed that each loader in failure will be repaired without delay. Therefore the notation of the system is:

$$M/G/m + r/r$$

and it is a Maryanovitch type system.

The limited probability distribution of the number of failed machines is determined by formula 8.23.

It is easy to see that the number n of loaders required to fulfil the customer's demand can be obtained from the equation:

$$n = \nabla\left(\frac{X}{T_d \phi W A}\right) \quad n \in \mathfrak{R} \tag{8.24}$$

where:

$\nabla(y)$ means that the real number y should be rounded up to the nearest natural number
A is the steady-state availability of loading machine.

The effective mine output per day is determined by the formula:

$$G_r = T_d \phi W\left[m\sum_{k=0}^{r}P_k^{(n)} + \sum_{k=1}^{m}(m-k)P_{r+k}^{(n)}\right]. \tag{8.25}$$

This formula is quite interesting. The product of the first two components is the effective number of hours of work of a single machine per day—that is, the time of work that we can expect in the long run, on average. This product multiplied by the nominal loader output indicates what should be the effective output of the loader in the long term. Components in the square brackets refer to three items: the steady-state availability of the machines, their number and the existence of a reserve. The first component in this bracket shows the steady-state availability of the system with a full output—all failed machines are replaced by units from the reserve. The second component determines the steady-state availability of the system when the system output decreases because the number of working machines decreases—that is, when there are no spare machines and those in operation are going into a down state. It can be stated that the steady-state availability of the system is determined in the square brackets.

If $G_r > X$ it means that the customer's demand is fulfilled.

Let us presume this data: $T_d = 16$ h $\phi = 0.82$ $X = 38{,}000$ m³/d
Some loaders with these parameters:

$$W = 410 \text{ m}^3/\text{h} \quad \lambda = 0.12 \text{ h}^{-1} \quad T_r = 2.8 \text{ h}$$

were at the mine's disposal immediately. The maximum number of working machines of this type is $m = 7$.

Calculate:

- the repair rate $\kappa = \lambda T_r = 0.336$
- the steady state availability of loaders $A = 0.749$

Thus:

$$n = \nabla\left(\frac{X}{T_d \phi W A}\right) = \nabla\left(\frac{38000}{16 \times 0.82 \times 410 \times 0.749}\right) = \nabla(9.4) \Rightarrow 10$$

It seems quite obvious that a fleet of 10 trucks, with 7 working units and 3 in cold reserve, will be not enough. However, calculate such a system to determine how far we are from the customer's demand.

We have:

$$G_{r=3} = 16 * 0.82 * 410\left[7\sum_{k=0}^{3}P_k^{(n)} + \sum_{k=1}^{7}(7-k)P_{k+3}^{(n)}\right] \cong 32500 \text{ m}^3/\text{day}$$

because the limited probability distribution of the number of trucks in repair is:

$$P_0^{(n)} = 0.180 \qquad P_1^{(n)} = 0.23 \qquad P_2^{(n)} = 0.27 \qquad P_3^{(n)} = 0.211$$

$$P_4^{(n)} = 0.124 \qquad P_5^{(n)} = 0.05 \qquad P_6^{(n)} = 0.014 \qquad P_7^{(n)} = 0.003$$

Further probabilities are below 0.000.

Let us discuss all the possibilities to increase the system output. They are:

a. an increase in the reliability of machines
b. an increase in the daily disposal time
c. usage of machines of a greater output.

The last possibility in this regard can be a careful analysis of whether, perhaps, $m + 1$ machines can load simultaneously. This idea was rejected given the type of loaders. It may be reconsidered if smaller machines are to be used at the mine.

Consider case (a). An increase in the reliability of the machines can be achieved in a two ways—by changing the policy of the maintenance of the machines or by buying new and better pieces of equipment. Changes in the method of maintenance could be directed towards an improvement of the repair actions and to reducing the mean time of repairs of the machines. It could comprise better diagnostic procedures, better equipment in the repair shop, a better quality of the repairs executed and fewer (or no) problems with spare parts. Sometimes some prophylactic actions can be implemented to decrease the intensity of failures of machines. Generally, if a maintenance system is of low quality, there are possibilities to improve the situation. If a maintenance system is good, possibilities to improve the situation are few. However, the mine assessed that this method would not give a significant increase in the mine's production.

There are two ideas that could increase the daily disposal time (b). The first idea is to reduce time losses connected with the accomplishment of the loading task. Looking at the figure $\phi = 0.82$, we are inclined to evaluate that this is rather a high number and no spectacular achievements can be expected in this area. A second idea is to seek an extension of the disposal time; a bit more than 16 h/day. This is extremely hard to achieve. There are two crucial problems. First, if an extension of work time is suggested, the trade union will be against it. This opposition may change if extra good pay is provided. This is usually very costly. Second, it requires the extension of the work time of the truck fleet (connected with the customer) and again an extra cost. For these reasons this method looks ineffective.

Let us now discuss the third possibility (c). This solution requires extra funds. New machines would need to be purchased; however, some recovery can be expected from selling the fleet of existing machines that are on hand at the mine. A dealer of loaders that usually delivers machines to the mine proposed the use of loaders with a greater output—590 m³/h. These machines had a larger loading bucket, greater power and slightly larger dimensions. Because the productivity of these new pieces of equipment is greater by 44% a system of 7 working machines with one in a cold reserve was considered. This would yield the following results.

- The limited probability distribution of the number of trucks in repair is:

$$P_0^{(n)} = 0.112 \qquad P_1^{(n)} = 0.262 \qquad P_2^{(n)} = 0.307 \qquad P_3^{(n)} = 0.206$$

$$P_4^{(n)} = 0.086 \qquad P_5^{(n)} = 0.023 \qquad P_6^{(n)} = 0.004$$

and further probabilities are below 0.000
- The effective mine output is:

$$G_{r=3} = 16 * 0.82 * 590 \left[7 \sum_{k=0}^{1} P_k^{(n)} + \sum_{k=1}^{7} (7 - k) P_{k+1}^{(n)} \right] \cong 39700 \ \text{m}^3/\text{day}$$

This outcome is 10% greater than is needed. It is good result. Even if some shortcomings are discovered with the new trucks, the demand should be fulfilled. If some more perturbations occur, an increase in the number of spare units should solve the problem.

It looks at first glance as though the problem is finally solved, but not entirely. Some additional issues should be carefully considered.

First, these new machines have larger dimensions than the ones they replace. It should be determined whether 7 units can still operate without disturbances. If, perhaps, only 6 machines can be utilised, the calculations must be repeated taking into account this new situation.

Second, these new machines have a larger bucket. It is recommended that the ratio of the bucket capacity to the truck box capacity be checked. If these new loading tools are of very high volume in relation to the truck box volume, some new problems might appear. A greater spillage of the crushed rock being loaded will be likely causing a longer average time of loading. Additionally, the heavier loads may mean that the increased impact during loading may reduce the durability of some assemblies of trucks (suspension, box lining, etc.). All these should be verified in practice.

Third, the economic aspect of purchasing new machines should be investigated. The main question is whether the demands of the purchaser of the crushed rock is for a long enough period of time that the money spent to buy new loaders will be recouped and some profit will be obtained.

One other important problem has been ignored till now—the trucks arrival to load the aggregate stone. It can be stated with a probability near one that some periods will occur where there will be a shortage of dumpers. Loaders will be waiting for hauling machines. A high number of these shortages can change the economic calculations. In a comprehensive system calculations of this situation have to be analysed.   ◀

8.9   THE RANDOMISED MARYANOVITCH MODEL

It looks as though the problem in Example 8.5 has been investigated thoroughly. However, one point, quite a sensitive one has not been analysed with the appropriate attention. The dealer offering the new loaders stated that the reliability of the new machines should be approximately identical to the ones they replaced. Will this be so?

For the mine, the proposed machines do not exist—in the sense of the machines having any history of exploitation. If these machines are purchased second-hand we can expect that some 'history of operation' will be available and some reliability indices can be roughly evaluated based on this history. But in the case being considered these machines will be delivered new and we can only predict the values of their reliability parameters. This means that our investigations should be shifted towards the theory of prediction. How to do this?

The main parameter of interest is the steady-state availability of the machines. It is well-known from practice that when buying a number of machines of a given type, each unit will have a

slightly different value of this availability parameter in practice. Therefore, it is quite rational that we can treat it as a random variable.

If so, two important questions arise:

- What is the type of the random variable?
- How is this information to be included into the procedure of estimation of the number of loaders required to fulfil the customer's demand?

Let us start to answer the first question. A random variable A is determined over a [0, 1] interval. The most popular theoretical random variable supported in such a space is the beta one. This probability distribution has been used for years in reliability theory and the theory of exploitation to determine steady-state availability. It concerns a set of several pieces of equipment of the same type with identical geometric and mechanical parameters.[7] It was also applied to describe the distribution of the utilisation of pieces of equipment (for example, Czaplicki 1977).

Recall from Chapter 5 that the beta probability distribution[8] is a family of continuous probability distributions defined over the interval [0, 1] parameterised by two positive shape parameters and in which the probability density function is determined by formula 5.31:

$$f(x) = \frac{\Gamma(p+q)}{\Gamma(p)\Gamma(q)} x^{p-1}(1-x)^{q-1} \quad 0 \le x \le 1, \, p > 0 \, q > 0$$

Frequently, instead of gamma functions the beta function is used here which is determined as:

$$B(p,q) = \int_0^1 x^{p-1}(1-x)^{q-1} dx \tag{8.26}$$

and:

$$B(p,q) = \frac{\Gamma(p)\Gamma(q)}{\Gamma(p+q)} \tag{8.27}$$

The expected value and the standard deviation are given by Equations 5.32.

For $p > 1$ and $q > 1$ the density function has its maximum at point $x = (p-1)/(p+q-2)$.

As is known from the theory, the graph of this function can create different shapes, as Figure 8.11 shows.

Let us now specify what kind of properties the probability density function of random variable A should have. Our expectations can very likely be meet by the function shown in Figure 8.12 because:

- this function has one maximum
- its mode and the mean value are near one (actually machines are of high reliability).

Transforming the above stipulations into a mathematical form articulated by function parameters, we can say that we are looking for the density function for which the following inequality holds:

$$\frac{\kappa\left[\kappa - S_a^2(1+\kappa)^2\right]}{S_a^2(1+\kappa)^3} > 1 \tag{8.28}$$

---

[7] The problem of the homogeneity of machines is analysed in Chapter 10.
[8] This is a special case of the Dirichlet distribution with only two parameters and it is also a special case of the Euler's distribution, again with only two parameters.

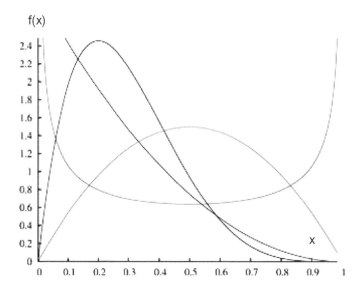

Figure 8.11.   Beta probability density functions for different values of parameters.

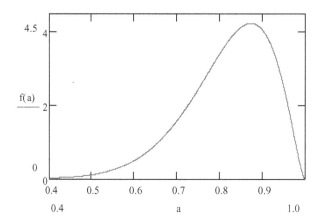

Figure 8.12.   An example of the probability density function of steady-state availability treated as a random variable.

where:

$\kappa = \dfrac{1-\bar{a}}{\bar{a}}$ and, obviously, it is the repair rate of machine

$\bar{a}$ is the mean of steady-state availability A

S_a^2 is the variance of the steady-state availability A.

As an alternative probability density function the following function can be applied:

$$z(a) = \frac{\gamma^c \delta^d \left[a^{d-1}(1-a)^{c-1} \right]}{B(c,d)\left[\delta a + \gamma(1-a) \right]^{c+d}} \quad 0 \le a \le 1 \;\; c > 0 \;\; d > 0 \;\; \gamma > 0 \;\; \delta > 0 \tag{8.29}$$

This is the Euler's probability density function. It is easy to notice that for $\gamma = \delta$ this function becomes the beta density function. Euler's density function was applied for the first time in reliability investigations in the 1960s in connection with investigations into availability measures of various types. This function should give a more adequate description of the empirical data because it has four parameters. However, such a function needs much more data to get an appropriate estimation of these parameters and for this reason it has low applicability.

Let us now make a list of the information which will be useful for our machinery investigation. We should have details on:

- the average value $\bar{a}$ of the steady-state availability of the machines
- the average repair time of the machines T_n
- the average work time of the machines T_p
- the standard deviation S_a of the steady-state availability of the machines.

Nevertheless, if we pay closer attention we may add one more piece of information. Presuming that the machine's manufacturer is well-known on the market—that is, its products are of a high quality—we can presume that there should be a certain reliability level of machines A_M (minimum one) and that the new loaders being offered should have at least this level. If so, the probability distribution of random variable A is a truncated one determined over $[A_M, 1]$. Thus, the probability density function is given by:

$$f_t(a) = \frac{f(a)}{\int_{A_M}^{1} f(x)dx} \quad a \in [A_M, 1] \tag{8.30}$$

where $f(a)$ is determined by pattern (8.26).

If this is so, the two most important parameters are determined by the equations:

$$\bar{a} = \frac{\int_{A_M}^{1} xf(x)dx}{\int_{A_M}^{1} f(x)dx} \quad S_a = \frac{\int_{A_M}^{1} (x - \bar{a})^2 f(x)dx}{\int_{A_M}^{1} f(x)dx} \tag{8.31}$$

If these parameters are known we can solve the set of equations to obtain estimations of parameters p and q for the truncated distribution 8.30.

Now, it is time to include this information in the calculation procedure. Let us recall formula 8.25 for the effective system output, indicating the random variable of interest. Here we have:

$$G_r(A) = T_d \phi W \left[m \sum_{k=0}^{r} P_k^{(n)}(A) + \sum_{k=1}^{m} (m-k)P_{r+k}^{(n)}(A) \right] \tag{8.32}$$

The probabilities in this formula are conditional ones. To release them from this property, it is necessary to take into account all possibilities determined by the probability distribution. Therefore the probability that 0 machines will be in repair is given by the formula:

$$P_0' = \left\{ \int_{A_M}^{1} \left[1 + \sum_{k=1}^{r} \left(\frac{1-a}{a} \right)^k \frac{m^k}{k!} + \sum_{k=r+1}^{m+r} \left(\frac{1-a}{a} \right)^k \frac{m!m^r}{k!(m-k+r)!} \right] f_t(a)da \right\}^{-1} \tag{8.33a}$$

Further probabilities are determined by:

$$P'_c = \int\limits_{A_M}^{1} P'_0 \left(\frac{1-a}{a}\right)^c \frac{m^c}{c!} f_i(a)da \quad \text{for } c = 1, 2, \ldots, r \tag{8.33b}$$

$$P'_c = \int\limits_{A_M}^{1} P'_0 \left(\frac{1-a}{a}\right)^c \frac{m^r m!}{c!(m-c+r)!} f_i(a)da \quad \text{for } c = r+1, \ldots, r+m \tag{8.33c}$$

■ **Example 8.6**

Let us presume that a quarry operates $T_d = 16$ h/d but that the utilisation of this time is $\phi = 0.76$. The output of a single loader is $W = 590$ m³/h and the customer's demand is for $X = 40,000$ m³/d. A machines dealer guaranties that the minimum reliability/availability level is $A_M = 0.5$. He claims that the average steady-state availability of the loader is 0.76 with a standard deviation of approximately 0.10. Let us repeat our calculations.

First, check condition 8.28. Here we have:

$$\frac{\kappa \left[\kappa - S_a^2(1+\kappa)^2 \right]}{S_a^2(1+\kappa)^3} = 4.14 > 1$$

Let us now estimate the unknown parameters of the distribution. Solving the set of equations 8.31 we obtain:

$$p = 11.3 \quad q = 3.68$$

These are the parameters of the truncated distribution supported over a [0.5, 1] interval. The probability distribution of the steady-state availability of loaders offered by a given producer is shown in Figure 8.13.

Let us note that the average value—if the information on the minimum level of availability is neglected—is 0.75 and the corresponding standard deviation is 0.24. Changes in the values of the basic statistical parameters are clearly visible.

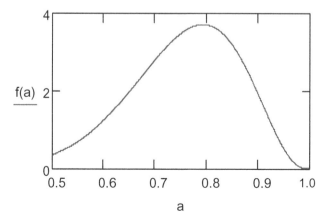

Figure 8.13.   The probability density function of the steady-state availability of loaders offered by the producer.

Now assess the mass of probability that is cut off because of truncation. Integrating the density function over a [0, 0.5] interval we get 0.019. This outcome indicates that we may neglect the truncation and the generated error will not be great. Notice that such attitude is more sceptical (more careful).

The probability density function of the beta type with these properties also has a property that:

$$E(A) < Me(A) < Mo(A). \tag{8.34}$$

In the case being considered:

$$E(A) = 0.754 < Me(A) = 0.766 < Mo(A) = 0.794$$

where *Me* means here the median and *Mo* means the mode.

This property is good for the customer. It means that if we make a selection of a machine of high reliability and we know nothing apart from the information on the mean reliability/availability level, we are in a good situation. We can expect that the selected machine will probably be of higher reliability than the average. Two issues should be taken into account. First, if the prediction is being made only once, a recommended principle is to select the mode of random variable because it is the most probable value and this mode is obviously greater than the mean. Second, it can be expected that in more than 50% of cases the selected machine will have a greater reliability than the mean because $E(A) < Me(A)$.

Now we can calculate the number of machines needed to fulfil the customer's demand. Considering formula 8.24 we have:

$$n = \frac{X}{WT_d\phi} = 5.6$$

This is *the mean number of loaders in a work state* that is needed at the mine. Taking into account the loader availability we have:

$$V = \frac{n}{A_M} = 7.3 \Rightarrow 8$$

This means that 8 loaders of the determined steady-state availability are required at the mine. But, if it is known that only *m* machines can operate simultaneously and $m < V$, we have to split the loader fleet into two sets of powers: *m* and *r*. Presume, as in the previous example, that $m = 7$. To compute the number of machines in a work state for such a system the Maryanovitch model can be applied giving the following results.

- The limited probability distribution of the number of trucks in repair is:

$$P_0^{(n)} = 0.125 \quad P_1^{(n)} = 0.277 \quad P_2^{(n)} = 0.306 \quad P_3^{(n)} = 0.193$$

$$P_4^{(n)} = 0.076 \quad P_5^{(n)} = 0.019 \quad P_6^{(n)} = 0.003$$

  and further probabilities are below 0.000.
- The average number of loaders in a work state is approximately 6, with 1.89 in failure and 0.11 in reserve.

If the steady-state availability of the loaders is presumed to be a random variable the results of calculations are as follows.

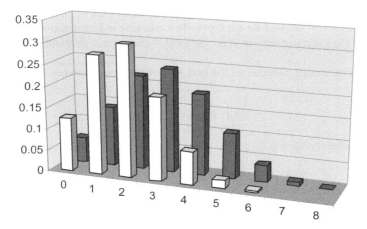

Figure 8.14. The probability distributions of the number of loaders in failure determined by the Maryanovitch model (light colour) and by the randomized Maryanovitch model (dark colour) for a given system.

- The limited probability distribution of the number of trucks in repair is:

$$P_0^{(n)} = 0.058 \quad P_1^{(n)} = 0.139 \quad P_2^{(n)} = 0.219 \quad P_3^{(n)} = 0.24$$

$$P_4^{(n)} = 0.189 \quad P_5^{(n)} = 0.105 \quad P_6^{(n)} = 0.039 \quad P_7^{(n)} = 0.009 \quad P_8^{(n)} = 0.001$$

- The average number of loaders in a work state is approximately 5.05, with 2.88 in failure and 0.7 in reserve.

Both probability distributions of the number of loaders in failure are shown in Figure 8.14. Recall, the assumption was made that there is no queue of failed machines waiting for repair.

Looking at Figure 8.14 it is easy to see that randomisation caused the probability distribution to become flatter and the mass of probability to be shifted a little away from zero.

Compare now the system output based on the above probability distributions:

- for the constant steady-state availability of loaders (no randomisation) the daily productivity is increased to 36,650 m³
- for the steady-state availability of loaders treated as a random variable the daily productivity is increased to 36,230 m³.   ◀

The results achieved here require some comments. By making some generalisations and analysing the outcomes obtained, we can make the following statements.

- Usually, if a given deterministic value is treated as a random variable, the uncertainty of the inference based on it increases.
- If it is assumed that the steady-state availability of loaders is a random variable, the productivity of the loader system is frequently estimated to be lower than that estimated under the assumption that the steady-state availability is a deterministic value, even when a stipulation is made that the availability of loaders is above a certain level (unless this level is really very high).
- If the stipulated minimum level of availability is estimated high, the mass of probability that is cut off grows and it cannot be neglected; however, if other system parameters remain unchanged, the estimated system productivity increases.
- The deterministic approach gives a higher value for the number of machines in a work state and a lower value for machines in failure, both compared to randomised approach, on average.

- The fewer the number of machines in the system, the smaller is the difference between the daily output estimated using both approaches.
- The more machines in reserve, the smaller is the difference between the daily output estimated using both approaches.
- If there is a greater uncertainty about the steady-state availability of a machine (that is, greater than its standard deviation), there is a greater difference between the daily output estimated using both approaches.

Generally, randomisation—such as in the case considered in Example 8.6—means that considerations become more cautious, perhaps underestimating the real system achievements and, perhaps, this is a not very bad approach.

In some cases when the production of machines is of very high quality, which means there is a small dispersion of availability, the probability distribution of the empirical steady-state availability can be approximated by the normal distribution, which has the whole mass located near the mean value.

8.10 THE *G/G/k/r* MODEL FOR HEAVY TRAFFIC SITUATION

This model is usually known as the Sivazlian and Wang model after the theory developed by these authors (1988, 1989). It is a more general model than the classical repairman problem (Chapter 8.3) because it assumes that the machines in reserve can also fail.

There is one limitation in this model. The exploitation situation in the system must fulfil the so-called *heavy traffic situation*. This allows a description of the system using the method of diffusion approximation that is based on the assumption that queues of failed machines in the repair shop are almost always non-empty. If this condition is fulfilled, one can replace the discrete type of queue process with those of a continuous type process. This change must be done in such a way that the characteristics of the original process will not be lost. In contrast to the discrete space, functions written continuously can be modified to give interesting characteristics. The result obtained is then transferred to the discrete space and the solution can be presented in an explicit form.

Many of the exact solutions to queuing problems with interarrival times or service times distribution of the general type have not been found. It is extremely difficult to obtain explicit equations, such as the steady-state probability mass function and the mean of the number of arrivals in the system, for a *G/G/k/r* queuing system.

The *G/G/k/r* system is characterised as follows.

A system of $m + r$ homogeneous machines is given. As many as m units may operate simultaneously in parallel (see Chapter 6). The size of the reserve is r machines, but these spares are warm-standby units.

A standby component is called a *warm standby* or a *lightly loaded* one if its intensity of failures α is non-zero but less than the intensity of failures of an operating unit δ, $0 < \alpha < \delta$. For $\alpha = 0$ the reserve becomes cold (unloaded) and for $\alpha = \delta$ it becomes hot (fully loaded). The repair is done with the intensity γ.

It is assumed that the work times probability distribution is a general one (*G*), the probability distribution of times to machine failure in reserve is general (*G*), and the repair times probability distribution is also general (*G*). The intensities of the transitions between states α, γ, and δ are known.

Furthermore, the three classical assumptions of queue systems listed previously (in Chapter 8.7) remain valid.

Figure 8.15 shows a graphical representation of the *G/G/k/r* model.

Using the application of the diffusion approximation to find the mass probability function of the number of failed machines, Sivazlian and Wang came to the conclusion that three pairs of

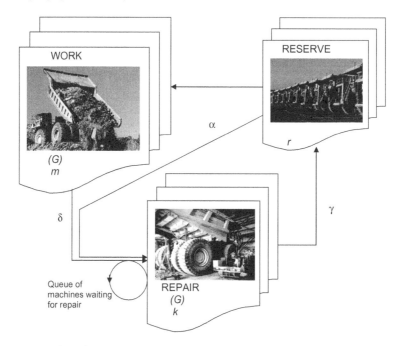

Figure 8.15.   Operating scheme of the *G/G/k/r* system.

statistical parameters are necessary. Apart from the named intensities α, γ, and δ, the following standard deviations should be known:

– σ_p the standard deviation of work times
– σ_s the standard deviation of times that the machine spent in reserve up to the moment when failure occurred
– σ_n the standard deviation of repair times.

After transformation from the discrete space to the continuous one, the function that we are looking for is the probability density function $h(x)$ of the number of failed machines.

Let us denote:

$$C_M = (\delta\sigma_p)^2 \quad C_S = (\alpha\sigma_s)^2 \quad C_R = (\gamma\sigma_n)^2 \tag{8.35}$$

and:

$$\xi = \delta/\gamma \quad \theta = \alpha/\gamma \tag{8.36}$$

Parameters C_M, C_S and C_R are the square coefficients of the variation of the succession of the uptimes of the working machines, the uptimes of the spare machines and the repair times, respectively. All these parameters are dimensionless.

Parameters ξ and θ are the repair rates for machines in a work state and in the reserve. They are also dimensionless.

The system being investigated can be described in following manner:

$$\mathfrak{S} : <\mathfrak{M}: m \geq 1, b = 1, 2, ..., m; \lambda, \sigma_M^2; r \geq 0, i = 0, 1, ..., r, \alpha, \sigma_S^2; \mathfrak{O}: k \geq 1, \mu, \sigma_R^2> \tag{8.37}$$

The probability density function of the number of failed machines x in the system expressed as the continuous function $h(x)$ is the sum of three components. They are defined as follows.

- For the reserve not lower than the number of repair stands, $k \leq r$

for $0 \leq x < k$

$$h_1(x) = \frac{K_1}{m\xi C_M + (r-x)\theta C_S + xC_R}\left(\frac{m\xi C_M + (r-x)\theta C_S + xC_R}{m\xi C_M + r\theta C_S}\right)^{\beta_1}$$

$$\times \exp\left(\frac{-2(\theta+1)x}{C_R - \theta C_S}\right) \quad \text{for } C_R - \theta C_S \neq 0,$$

$$h_1(x) = \frac{K_1}{m\xi C_M + r\theta C_S}\exp\left(\frac{2(m\xi + r\theta)x}{m\xi C_M + r\theta C_S} - \frac{(\theta+1)x^2}{m\xi C_R + r\theta C_S}\right) \quad \text{for } C_R - \theta C_S = 0 \quad (8.38a)$$

for $k \leq x \leq r$

$$h_2(x) = \frac{K_2}{m\xi C_M + (r-x)\theta C_S + kC_R}\left(\frac{m\xi C_M + (r-x)\theta C_S + kC_R}{m\xi C_M + (r-k)\theta C_S + kC_R}\right)^{\beta_2}\exp\left(\frac{2(x-k)}{C_S}\right) \quad \text{for } C_S > 0$$

$$h_2(x) = \frac{K_2}{m\xi C_M + kC_R}\exp\left(\frac{2(m\xi - k)(x-k)}{m\xi C_M + kC_R}\right) \quad \text{for } C_S = 0 \quad (8.38b)$$

for $r \leq x \leq m + r$

$$h_3(x) = \frac{K_3}{(m+r-x)\xi C_M + kC_R}\left(\frac{(m+r-x)\xi C_M + kC_R}{m\xi C_M + kC_R}\right)^{\beta_3}\exp\left(2(x-r)/C_M\right) \quad \text{for } C_M > 0$$

$$h_3(x) = \frac{K_3}{kC_R}\exp\left(\frac{2(m\xi + r\xi - k)(x-r)}{kC_R} - \frac{\xi(x^2 - r^2)}{kC_R}\right) \quad \text{for } C_M = 0 \quad (8.38c)$$

where:

$$\beta_1 = \frac{2m\xi[C_R - \theta C_S + (\theta+1)C_M] + 2r\theta[C_R - \theta C_S + (\theta+1)C_S]}{(C_R - \theta C_S)^2}$$

$$\beta_2 = \frac{2m\xi[1 - (C_M/C_S)] - 2k[1 + (C_R/C_S)]}{\theta C_S} \qquad \beta_3 = \frac{2k[1 + (C_R/C_M)]}{\xi C_M} \quad (8.38d)$$

- For a reserve lower than the number of repair stands, $r < k$

for $0 \leq x < r \quad h_4(x) = h_1(x)$ $\qquad\qquad\qquad\qquad\qquad\qquad\qquad\qquad (8.39a)$

for $r \leq x \leq k$

$$h_5(x) = \frac{K_5}{(m+r-x)\xi C_M + xC_R}\left(\frac{(m+r-x)\xi C_M + xC_R}{m\xi C_M + rC_R}\right)^{\beta_5}$$

$$\times \exp\left(\frac{2(\xi+1)(x-r)}{\xi C_M - C_R}\right) \quad \text{for } \kappa C_M - C_R \neq 0$$

$$h_5(x) = \frac{K_5}{(m+r)\xi C_M} \exp\left(\frac{2(x-r)}{C_M} - \frac{(\xi+1)(x^2-r^2)}{(m+r)\xi C_M}\right) \quad \text{for } \kappa C_M - C_R = 0 \qquad (8.39b)$$

for $k \le x \le m+r$

$$h_6(x) = \frac{K_6}{(m+r-x)\xi C_M + kC_R}\left(\frac{(m+r-x)\xi C_M + kC_R}{(m+r-k)\xi C_M + kC_R}\right)^{\beta_5} \exp\left(2(x-k)/C_M\right) \quad \text{for } C_M > 0$$

$$h_6(x) = \frac{K_6}{kC_R} \exp\left(\frac{2(m\xi + r\xi - k)(x-k)}{kC_R} - \frac{\xi(x^2-k^2)}{kC_R}\right) \quad \text{for } C_M = 0 \qquad (8.39c)$$

where:

$$\beta_5 = \frac{2(m+r)\xi(C_R + C_M)}{(C_R - \kappa C_M)^2} \qquad (8.39d)$$

The probability density function $h(x)$ of the number of failed machines is determined by the formula:

$$h(x) = \begin{cases} h_1(x) + h_2(x) + h_3(x) = K_1 g_1(x) + K_2 g_2(x) + K_3 g_3(x) & \text{for } k \le r \qquad (8.40a) \\ h_4(x) + h_5(x) + h_6(x) = K_4 g_4(x) + K_5 g_5(x) + K_6 g_6(x) & \text{for } k > r \qquad (8.40b) \end{cases}$$

The unknown constants K_i, $i = 1, 2, \ldots, 6$ are used in the construction of functions 8.38 and 8.39. Therefore, it is necessary to build six equations that allow these constants to be determined.

Function $h(x)$ must fulfil two conditions.

The first one is the normalisation of the probability mass to unity.

Thus:

$$\int_0^{m+r} h(x)dx = K_1\int_0^k g_1(x)dx + K_2\int_k^r g_2(x)dx + K_3\int_r^{m+r} g_3(x)dx = 1$$

and:

$$\int_0^{m+r} h(x)dx = K_4\int_0^r g_4(x)dx + K_5\int_r^k g_5(x)dx + K_6\int_k^{m+r} g_6(x)dx = 1 \qquad (8.41)$$

The second condition assumes the continuity of $h(x)$. The following equations then hold:

$$K_1 g_1(k) = K_2 g_2(k) \quad \text{and} \quad K_2 g_2(r) = K_3 g_3(r)$$

as well as:

$$K_4 g_4(r) = K_5 g_5(r) \quad \text{and} \quad K_5 g_5(k) = K_6 g_6(k) \qquad (8.42)$$

Before we convert these functions into the discrete space (the number of failed machines is part of the set of natural numbers plus zero), it is worth considering two cases.

The first case is when $C_M = 0$. This case was the result during the mathematical analysis. Let us translate it into engineering language: it means that the appropriate standard deviation of the work times of the machines is zero. This does not hold in practice. Therefore, this case will be excluded from further investigations.

The second case is the assumption that the machines in reserve can fail. Fortunately, this is a rare occurrence. Even if such events can sometimes occur, the value of the corresponding intensity of failures is very small. We can ignore it, and for this reason we further assume that $\alpha = 0$.

Now, we return to our procedure, and we will obtain the corresponding function in the discrete space. There are several different methods of discretisation. We repeat here the procedure suggested by Sivazlian and Wang following Halachmi and Franta's (1978) suggestion. According to this procedure, the steady-state probabilities P_j, $j = 0, 1, \ldots, m + r$ are given by following set of formulas:

$$P_j = \int_{j-0.5}^{j+0.5} h(x)dx, \quad \text{for } j = 1, 2, \ldots, m + r - 1$$

$$P_0 = \int_0^{0.5} h(x)dx \quad \text{and} \quad P_{m+r} = \int_{m+r-0.5}^{m+r} h(x)dx \tag{8.43}$$

where P_j is the probability that j machines failed.

However, using these formulas it is necessary to pay special attention to the limits of the determination of particular functions $h(x)$.

To close our discussion, a final problem needs our attention. This is the criterion that must be fulfilled in order to validate the system's operation under the heavy traffic condition.

The first idea is that the probability that the repair shop is empty should be small. This condition is impractical. Our service station possesses k stands and the heavy traffic situation requires occupancy of the majority of these units almost continuously. Let us express this condition by means of system parameters. According to Halachmi and Franta (1978) the decisive factor is the fulfilment of the following inequality:

$$\rho = \frac{\Lambda}{k\gamma} \geq 0.75 \tag{8.44}$$

where Λ is the intensity of arrivals to the service station and ρ is the coefficient of flow intensity in the station (see, for example, Gross and Harris 1974). It is worth noting here that this condition has no precise meaning. It has been stated that if the coefficient tends to decrease below 0.75 the assessment of probabilities given by equation 8.43 becomes poorer. Let us end by noting that the coefficient of flow intensity in the service station should also fulfil the inequality:

$$\rho < 1 \tag{8.45}$$

which means that the intensity of arrivals to the service system should be less than the intensity of service in the system. If ρ is appreciably above 1 an almost permanent queue will be observed, clients will be dissatisfied, and trouble will be inevitable.

Let us define some basic system parameters.

- The expected number of machines in failure is:

$$E(X) = K_1 \int_0^k xg_1(x)dx + K_2 \int_k^r xg_2(x)dx + K_3 \int_r^{m+r} xg_3(x)dx \quad \text{for } k \leq r$$

$$E(X) = K_4 \int_0^r xg_4(x)dx + K_5 \int_r^k xg_5(x)dx + K_6 \int_k^{m+r} xg_6(x)dx \quad \text{for } r < k \tag{8.46}$$

- The expected number of busy repair stands is:

$$E(\boldsymbol{B}) = k - \int_0^k (k-x)K_1 g_1(x)dx \quad \text{for } k \le r$$

$$E(\boldsymbol{B}) = k - \int_0^r (k-x)K_4 g_4(x)dx - \int_r^k (k-x)K_5 g_5(x)dx \quad \text{for } r < k$$

(8.47)

- The expected number of machines in a work state is:

$$E(Y) = m\sum_{n=1}^{r}(P_0 + P_n) + \sum_{n=r+1}^{m+r-1}(m+r-n)P_n$$

(8.48)

■ **Example 8.7**

There is a machinery system consisting of $m = 30$ machines directed to work and with r machines in a reserve. A maintenance bay is under construction and the number of repair stands k has not yet been finally decided. Machine parameters are known and they are as follow:

- the mean time of work is 18 h
- the standard deviation of work times is 15 h
- the mean time of repair is 4.5 h
- the standard deviation of repair times is 2.5 h.

An analysis of the system should be done by considering a different number of repair stands and a different number of spare units. The operating scheme is shown in Figure 8.16.

Notice at the very beginning of the analysis of the system that the steady-state availability A of machines in the system is 0.8.

It was presumed that the number of repair stands should be greater than the number of machines in reserve, $k > r$ (which is recommended). Therefore, formulas 8.39 were applied.

First, a number of machines in the reserve was assumed and the following system parameters were calculated:

- the expected number of machines in failure $E(X)$
- the expected number of busy repair stands $E(\boldsymbol{B})$
- the expected number of machines in a work state $E(Y)$.

Additionally, an estimation of the coefficient of flow intensity in the system was made by using formula:

$$\rho_s = \frac{m}{k}\xi$$

(8.49)

Recall, if $\rho_s > 1$ it is likely that a permanent queue of failed machines waiting for repair will be observed. If $\rho_s < 0.75$ the heavy traffic condition is not fulfilled making an estimation of the system parameters of low likelihood.

It was assumed at the beginning of the analysis that the reserve size would be $r = 5$ and k will vary from 6 to 9.

Figure 8.17 illustrates the outcomes of calculations of the expected values, and Figure 8.18 shows the course of the coefficient ρ of flow intensity to the repair shop.

The results of the calculations are interesting. In Figure 8.17 all parameters tend distinctly toward their limited values. The mean number of machines in a work state $E(Y)$ tends toward 28 which is obvious because $(m + r)A = 28$. Turning our attention to a machine's non-availability, we have $(m + r)(1 - A) = 7$ and this is the limit towards which both the mean number of machines in

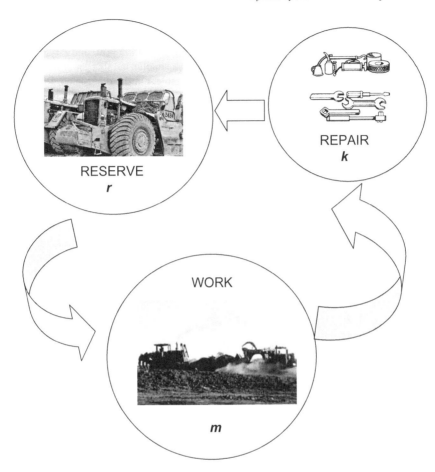

Figure 8.16. Operating scheme of a *G/G/k/r* system.

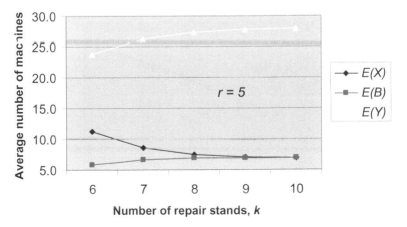

Figure 8.17. Relationship between the number of repair stands *k* and the mean number of machines in failure *E*(*X*), the mean number of busy repair stands *E*(*B*) and the mean number of machines in a work state *E*(*Y*) for *r* = 5.

failure $E(X)$ and the mean number of busy repair stands $E(B)$ are running. For a number of repair stands $k = 10$ the situation in the system is stabilised.

Looking at Figure 8.18 in turn, we can state that for the system $m = 30$ and $r = 5$ the number of repair stands k should not be less than 8 because for a smaller number an almost permanent queue of failed machines waiting for repair will be inevitable. On the other hand, for $k > 10$, the situation in the repair shop becomes comfortable—there is no heavy traffic situation. However, if so, the estimations obtained from the Sivazlian and Wang model will turn out to be of a lower likelihood.

Let us now consider a system of $m = 30$ machines directed to work and a repair shop with just $k = 10$ repair stands and in which the number of units in cold reserve varies. Presume that $r = 0, 1, ..., 9$ (because $r < k$). Figure 8.19 illustrates the outcomes of the calculations of the average number of machines in failure $E(X)$ and the mean number of busy repair stands $E(B)$ for the case being considered.

Information contained in Figure 8.19 is extraordinary. For small numbers of machines in reserve, the increase in both functions is identical and they run constantly along a straight line. For a larger number of spare units, the functions split; the average number of machines in failure increases in a slightly more intensive way than the average number of busy repair stands. At first glance it is difficult to find the reason for such a regularity. However, it is worth noticing that when the average

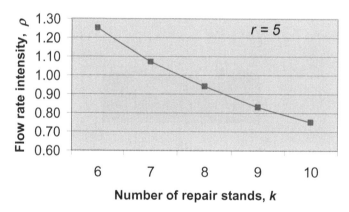

Figure 8.18.   Relationship between the number of repair stands k and the coefficient ρ of flow intensity to the repair shop.

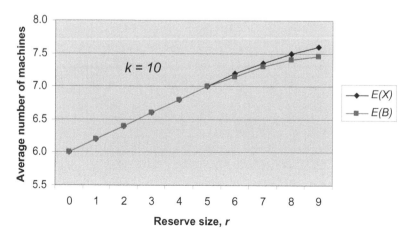

Figure 8.19.   Relationship between reserve size r and the mean number of machines in failure $E(X)$, the mean number of busy repair stands $E(B)$ for $k = 10$.

number of failed machines increases, the average queue of failed units waiting for repair standing in front of the repair shop also increases. Let us look at the conditional average time of a failed truck waiting for repair in relation to the reserve size r, calculated from the formula:

$$T_o = T_n \sum_{j=k+1}^{m+r} (j-k) \int_{j-0.5}^{j+0.5} K_6 g_6(x) dx \tag{8.50}$$

This relationship is shown in Figure 8.20.

It seems that just this increase in time caused the divergence of both functions.

Look now at Figure 8.21 which illustrates the relationship between the number of machines in reserve r and the average number of machines in a work state, $E(Y)$. Similar to the case of functions $E(X)$ and $E(B)$ versus r, the average number of units in a work state increases linearly for a small number of machines in reserve. When the number of spare units increases above 5 this function grows in a less intensive way and the speed of this increase is slowly going down. Obviously, this function tends to the limited value that is $m = 30$ machines.

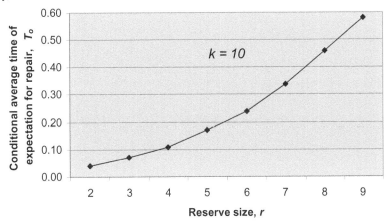

Figure 8.20.   Relationship between reserve size r and the conditional average time T_o of failed machine waiting for a free repair stand.

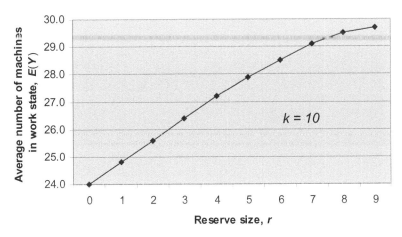

Figure 8.21.   Relationship between reserve size r and the average number of machines in a work state $E(Y)$ for $k = 10$.

To recap, the number of repair stands should be about 8 to 10 units. Obviously, the greater the number of these stands, the more comfortable the maintenance situation becomes. Additionally, a technical survey scheme for these machines should be taken into account before increasing this number. If the reserve size is considered, r should be about 5 machines. It is clear from considering the system parameters that when the number of spare units starts to grow from zero, system performance increases distinctly. For further increases in the size of the reserve the advantage becomes smaller.

All these investigations are conducted neglecting an extremely important aspect of system exploitation—profitability. Financial parameters and their relationships can change the results obtained dramatically. For this reason, the economic aspect of the system's operation must be considered by making use of the exploitation relationships obtained during the analysis. Thus, the analytical considerations should be conducted first. Their outcomes are appropriate tools with which to conduct an economic analysis.   ◄

CHAPTER 9

Combined systems

In our previous investigations all systems were homogeneous from a functional perspective. These systems were either continuous or cyclic ones. However, in investigating cyclic systems only those that are convenient to analyse by applying queue theory models were considered. Nevertheless, there are some systems in mining that are not homogeneous; one part is continuous and another is cyclic, or *vice versa*.[1] It cannot be excluded that some systems in the future will have a mixed structure—in a given system there might be cyclic and continuous subsystems or cyclic-and-cyclic subsystems; these cyclic elements will have significantly different properties. An example of such a future system could be a conventional cyclic blasting-and-loading system plus a system of hauling units of the skip type, which have their own propulsion and travel one-by-one on a rack path horizontally as well as vertically in shafts (Pepler 1989, Czaplicki et al. 1990). An example of a cyclic-and-cyclic system could be—again, perhaps in the future—a shovel-truck system and an inclined hoist of a TruckLift type working with a basic hauling system in a pit. This last system will be the point of analysis here.

We start our investigation with a shovel-truck system and in-pit crushing.

9.1   A SHOVEL-TRUCK SYSTEM AND IN-PIT CRUSHING

It is a well-known fact that the shovel-truck systems commonly applied in open-pit mining are very expensive, especially where operational cost is concerned. It has been assessed (Woodrow 1992, Bozorgebrahimi et al. 2003) that the cost of loading and hauling is approximately 60–65% of the total operating costs of open-pit mining, and the haulage cost alone is about 40–50% of total operating costs (Wang and Zhao 1997, Woodrow 1992). Some researchers are of the opinion that this latter figure is even as high as 60% (Fabian 1989). For these reasons every proposition to reduce these costs is most welcome and carefully analysed.

Nowadays there are two possibilities to modify a shovel-truck system in such a way as to reduce these huge expenses,[2] namely:

a. to apply in-pit crushing and conveying
b. to apply an inclined hoist of a truck-lift type.

---

[1] In some cases the inhomogeneity of a machinery system is caused by the different sizes of machines executing identical duties, but in mathematical modelling this increases the dispersion of determined random variables—see Chapter 10.

[2] A different possibility to cut costs is the application of a trolley assist system if it is profitable, but this is considered to be an improvement in the system, not a modification.

The first proposition has been known for years and has been applied successfully in several cases throughout the world. Perhaps the most popular system of this type is the one used at the Bingham Canyon Open Pit Mine, USA. In 1980 the management of this mine concluded that modernisation of the ore transport system was required in order to reduce costs. This was necessary to survive because of the depressed copper prices at that time. The old train transport was finally withdrawn and a truck system was introduced. To reduce expenses, a design concept was introduced into the in-pit crushing and conveying system in 1984–85. In 1988 the full system was in operation (Kammerer 1988).

The latter proposition (b) has not yet been employed in spite of the fact that this type of hoist was promoted intensively a few years ago in the major mining journals and on the internet. This system will be discussed later.

Let us now consider the problem of how to analyse a shovel-truck system with an in-pit crusher and a conveying subsystem delivering crushed mineral to a dressing plant. A different version of such a system can be a shovel-truck subsystem with an in-pit crusher plus a conveying subsystem delivering crushed overburden to a disposal subsystem (Kolonya et al. 2003). Let us denote a system of this type by S-T-C-C.

The problem can be formulated as follows. Let us presume that there is an open-pit mine in which a shovel-truck machinery system is in operation. Therefore, the number of working machines and their main exploitation parameters are known. The management of the mine has decided to change the hauling system by applying an in-pit crusher and a system of belt conveyors to deliver the crushed ore to a peripheral dressing plant unit. The problem is how to design this new component of the machinery system and to discover what kinds of changes will be seen in the 'old' system.

Before this problem is described in a mathematical form and analysis conducted, one point must be stated in advance. To evaluate a system of this kind, the procedure for the calculation of shovel-truck systems presented in the Czaplicki's monograph (2009) will be applied. At a certain stage of this procedure the accessibility of trucks is taken into account and the number of trucks needed for the system is enlarged by so-called 'surplus' units. These haulers fill the gap caused by the required truck fuelling, drivers' breaks, replacement of drivers, technical surveys, etc. All these 'enterprises' absorb the trucks, meaning that on average a certain number of haulers 'do not exist' for the execution of the transportation task. To mitigate this 'loss', surplus units are applied. These trucks are not considered in the further part of the calculation procedure apart from in the evaluation of repair shop requirements. We presume here—to make our considerations simpler and more communicative— that in the S-T-C-C system the number of stands in the repair shop is rationally selected and, for this reason, both surplus trucks and repair stands will be excluded from further analysis.

Let us formulate the problem to be investigated in a mathematical manner.

We have the system (see Figure 9.1):

$$\mathfrak{S} : < \mathfrak{H} = \mathfrak{H}_w \cup \mathfrak{H}_o; n = n_w + n_o, A_k; \mathfrak{W}: A_w; Q; b = 1,2, \ldots, m; r \mathfrak{Q}: k >$$

for which the exploitation parameters are:

$$\mathfrak{P}(\mathfrak{S}, \mathfrak{D}) : < B_k, Z', T_t, T_{dp}; \tau >$$

Let us read the notation.

The system $\mathfrak{S}$ of power shovels $\mathfrak{H}$ consists of n machines with the steady-state availability A_k, where n_w units load waste (set $\mathfrak{H}_w$) and n_o units load ore (set $\mathfrak{H}_o$). The system of dumpers $\mathfrak{W}$ of payload Q consists of m trucks fulfilling hauling duties, and r units are in the reserve.[3]

---

[3] The division of a truck fleet into two subsystems, say, m trucks directed to work and r units in reserve is important only for calculations. In practice both these numbers are random variables at any given moment of the system's operation. The only precise statement is that the sum of these variables is always constant and the same. An exception to this last rule is when new trucks arrive into the system or some trucks are finally withdrawn from further operation.

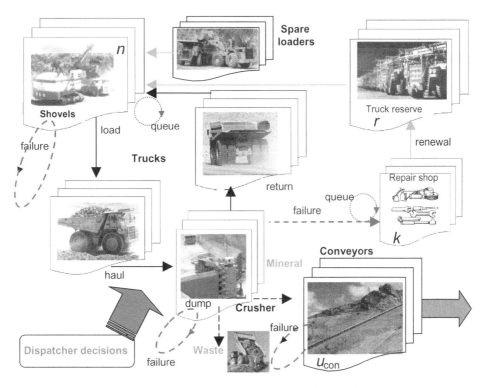

Figure 9.1.   Scheme of a shovel-truck system with in-pit crushing and conveying.

The steady-state availability of trucks is A_w. The system of repair stands $\mathfrak{D}$ consists of k repair units, and we assume that is it large enough so that the infrequent event of a queue of failed trucks waiting for repair can be ignored. Thus the problem of the proper selection of k is not considered here.

The second set of parameters, $\mathfrak{P}(\mathfrak{S},\mathfrak{D})$, describes the exploitation process $\mathfrak{P}$ of the machinery system that is a function of the properties of the equipment involved and the decisions $\mathfrak{D}$ made by a truck dispatcher. These parameters are the accessibility coefficient of shovels B_k, the mean loading time Z', the mean time of truck travel (haul—dump—return) T_t, the mean time of dumping T_{dp} and the mean loading time for spare loaders which is τ times longer than for shovels. During the system's operation—if no priority as to the type of material transported is implemented—it makes no difference as to which loader a given empty truck is directed and to which dumping point a given full hauler drives. If the new component is to become a part of the system, one part of the truck route will be strictly determined: all trucks loaded with broken ore will always be directed to the crusher.

We presume that the system was rationally selected (see, for example, Chapter 8 of the author's monograph 2009), that is the system's structural parameters $< m, r, k >$ were chosen in the proper way. Installation of the new proposed subsystem makes no immediate difference to the system of loading machines. The system of trucks—in turn—is divided into two subsystems, but in a *stochastic way*. It is not strictly ascribed that a given truck is only connected with the haulage of a given type of material. Sometimes the truck will transport waste, and sometimes ore. Nevertheless, we presume that the system of haulers is divided into two subsystems *numerically*—a certain number of trucks will on average serve the ore loading machines and a certain number on average will serve machines loading the waste. Because the whole truck system was designed for longer distances and has m units, we in fact have to recalculate this system from the beginning.

The statement 'longer distances' contains information that the original mean truck travel distance from the loading machine to the unloading point is now reduced because of the existence of the crusher; there is no need for the haulers to go up the pit to deliver the crushed ore to the dressing plant unit.

Thus, we presume that m_w trucks haul the waste on average and m_o units haul the ore to the in-pit crusher on average. We obviously expect that for the new system $m_w + m_o < m$. In the new system, the size of the reserve is $r_n = r_w + r_o$, and the forecast is that $r_n < r$. For this reason, the repair shop will not be as heavily loaded as before.

To design, analyse and calculate the modified system we need information on the following six parameters:

$$< T_{tc}, c, A_{cr}, B_{cr}, A_c, u_{con} >$$

where:

T_{tc} – the mean time of truck travel serving the crusher
c – the number of trucks that can unload simultaneously into the crusher
A_{cr} – the steady-state availability of the crusher
B_{cr} – the accessibility coefficient of the crusher
A_c – the steady-state availability of the conveyor
u_{con} – the number of conveyors used connecting an outlet of the crusher with the peripheral dressing plant unit.

Notice that the information on the value of c means that c truck payloads can be delivered in time T_{dp} to the crusher without any choke.[4] Thus, the relationship between the truck payload, the mean dump time and crusher capacity should be rationally matched. The procedure of analysis and calculation of the system can be performed according to the following scheme.[5]

1. Construction of the probability distribution of the number of ore shovels in a work state.
2. Calculation of the mean time of loading taking into account spare loaders—all for the ore system.
3. Recalculation of the whole system S-T-C-C considering the trucks.
4. Verification of the flow intensity rate in the system of loading machines.

A further part of the modelling, analysis and calculation scheme will concern the ore system exclusively—that is, the truck subsystem and the continuous system.

5. Construction of the probability distribution of the number of trucks in a work state.
6. Evaluation of the probability distribution of the number of trucks at loading machines.
7. Calculation of the unconditional probabilities of the ore truck system; using the assumption: no queue at the crusher.
8. Calculation of further unconditional parameters of the truck system.
9. Assessment of the system output.
10. Calculation of the steady-state availability of the continuous system.
11. Assessment of the potential crushing-hauling capacity of the continuous system.
12. Construction of the probability distribution of the number of trucks at the crusher.
13. Analysis of the basic measures of the system's efficiency.
14. Conclusions, recommendations and remarks.

---

[4] However, if during dumping into the crusher some large lumps halt the operation, the accessibility of the crusher is shortened. Such phenomena can be taken into account in the procedure by reducing the accessibility coefficient of the crusher.
[5] This procedure differs slightly from that proposed by Czaplicki (2008d). This one is more exact. The first trial of a calculation of a system of this type was described by Czaplicki (2004a).

Presuming that the procedure of modelling, analysis and calculation of shovel-truck systems is known (Czaplicki's monograph 2009) and, similarly, familiarity with the corresponding procedure for mine continuous systems (see Chapter 7 here), we are able to conduct analogous investigations for this combined system. However, keep in mind that some modifications have to be made. And, what is more, it was proved (Czaplicki 2008a) that each exploitation process of a shovel-truck system is a unique one due to dynamic truck dispatching. There is the possibility to construct a general procedure of analysis for these types of systems, but it has great complexity and its clear comprehensibility is doubtful. Usually, this is difficult to achieve in a correct application. For this reason it is better to consider each individual case separately. Nevertheless, the way of reasoning can be repeated when considering different case studies.

■ **Example 9.1**
Let us consider, for instance, the following machinery system:

$$\mathfrak{S} : < \mathfrak{H} = \mathfrak{H}_w \cup \mathfrak{H}_o : n_w = 5, n_o = 4, A_k = 0.870;$$

$$\mathfrak{W}: A_w = 0.762; Q = 120 \ t; b = 1, 2, ..., m = 142; r = 24 >$$

for which:

$$\mathfrak{P}(\mathfrak{S},\mathfrak{D}) : < B_k = 0.830, Z' = 1.6 \ min, T_t = 28.2 \ min, T_{dp} = 0.7 \ min; \tau = 2.6; B_w = 0.850 >$$

The supplementary system parameters are:

$$< T_{tc} = 8.2 \ min, c = 2, A_{cr} = 0.820, B_{cr} = 0.880, A_c = 0.998, u_{con} = 5 >$$

At the beginning of any analysis, a quick approximate evaluation of the system can be done:

a. 4 trucks can be loaded by shovels in 1.6 min, which means 2.5 trucks per min as far as the ore is concerned
b. 2 trucks can unload into the crusher in 0.7 min, which means 2.85 trucks per min.

1. We start our analysis with the system of loading machines which generate several discrete streams of mineral that must be taken by the system of haulers. The probability distribution of the number of shovels able to load[6] is described by the binomial distribution (see equation 6.1) because this is a system of machines operating in parallel in a reliability sense. The existence of spare loading machines must also be taken into account. We presume that if any shovel is down it is immediately replaced by a wheel loader. Additionally, we presume that the spare loader fleet is large enough so that if any machine of this kind is down it is replaced by the next wheel loader without delay. If so, the wheel loaders are totally reliable from the standpoint of the accomplishment of the loading and therefore all loading points are totally reliable. But the reliability of shovels has a direct influence on the mean time of loading. Notice the following regularity: the more reliable shovels, the shorter the mean time of loading and *vice versa*, which is obvious.

   Let us introduce the following notation: $P_{kd}^{(p)}$ will mean the probability that $d, d = 0, 1, ..., n$ shovels (k) are in a work state (p). In the case being analysed $n = n_o = 4$ and the probability distribution of the number of shovels in a work state is given by the following set of equations:

$$P_{kd=4}^{(p)} = (0.870)^4 = 0.573$$

$$P_{k3}^{(p)} = 4(0.870)^3(1 - 0.870) = 0.342$$

---

[6] It is stated *able to load*, which means the machine is in a work state and is also in state of availability for loading. However, it can happen that the machine does not load at a given moment in time because of the lack of a truck to fill.

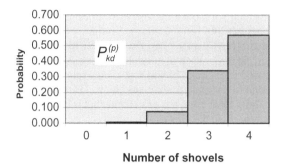

Figure 9.2. The probability distribution of the number of shovels in a work state.

$$P_{k2}^{(p)} = 6(0.870)^2(1-0.870)^2 = 0.077$$

$$P_{k1}^{(p)} = 4(0.870)(1-0.870)^3 = 0.008$$

$$P_{k0}^{(p)} = (1-0.870)^4 = 0.000$$

2. Now, we have to include the fact that when one shovel is down and a spare loader commences its duties the mean loading time for the system is changed, it increases. We have to apply the results of the reasoning contained in Chapter 12 of the author's 2009 monograph. Knowing the mean loading time for spare loaders, which is $\tau = 2.5$ times longer than for shovels, the mean time of loading for the system of n_o shovels if d shovels are down ($d \le n_o$) and all failed machines are always replaced by the same wheel loaders, can be calculated from the formula:

$$Z_n'(d,\tau) = Z_n' = Z'\left(1+(\tau-1)\frac{d}{n_o}\right) \tag{9.1}$$

For the system being considered:

$$Z_0' = Z' = 1.6 \quad Z_1' = 2.24 \quad Z_2' = 2.88 \quad Z_3' = 3.75 \quad Z_4' = 4.16 \text{ min}$$

This indicates that the mean truck work cycle varies and it is a function of the number of failed shovels and the coefficient τ. Moreover, different values of these average values have their own probability of occurrence following the binomial distribution. We can now evaluate the total mean time of loading for the whole system by applying the formula:

$$\bar{Z}_w = \sum_{d=0}^{4} Z_d' P_{kd}^{(p)} = 1.93 \text{ min} \tag{9.2}$$

3. We now recalculate the system of trucks.

• The waste subsystem—trucks are going up the pit.

  o The minimum mean number of trucks in a work state:

$$h = B_k n_w[(Z_w + T_{tc})/Z_w] = 64.8 \tag{9.3a}$$

Note that in this formula that the steady-state availability of a shovel is excluded because the system of loading machines is totally reliable (because there are spare units).

o The minimum number of trucks needed: $h/A_w \Rightarrow 86$

o Applying the Maryanovitch model (equation 8.23) and enlarging the result for trucks directed to the pit by 10% we have:

$$< 1.1 \text{ m} = 75; r = 18 >$$

o The mean number E_{wkd} of trucks in a work state for this system is 70.6.

- The ore subsystem—trucks drive to the crusher:

o The minimum mean number of trucks in a work state:

$$h = B_k n_o [(Z_w + T_t)/Z_w] = 17.4 \tag{9.3b}$$

o The minimum number of trucks needed: $h/A_w \Rightarrow 23$

o Applying the Maryanovitch model and enlarging the result for trucks directed to the pit by 10% we have:

$$< 1.1 \text{ m} = 22; r = 3 >$$

o The mean number E_{wkd} of trucks in a work state for this system is 19.

4. Calculating the flow intensity rate ρ in the system of loading machines.

Let us check at the beginning how intensively the loading machines operating in the old system are occupied. An estimation of this rate can be obtained from the formula:

$$\rho = \frac{E_{wkd} Z_w}{n B_k T_t} \tag{9.4}$$

which gives:

$$\rho = \frac{126.5 \times 1.93}{9 \times 0.83 \times 28.2} = 1.16$$

This result (ρ > 1) indicates that there will be a queue of trucks waiting for loading on average, but this is good as the loading machine should operate round-the-clock.

Now check what the situation is in this regard in the newly designed system.

For the system of trucks hauling the waste we have:

$$\rho_w = \frac{70.6 \times 1.93}{5 \times 0.83 \times 28.2} = 1.16$$

The situation is unchanged.

For the system of trucks hauling the ore:

$$\rho_o = \frac{17.4 \times 1.93}{4 \times 0.83 \times 8.2} = 1.23$$

This outcome indicates that the system of loading machines coping with the broken mineral should be busier—that is, working in a more intensive way—than its waste subsystem. However, taking into account that at the second service point (that is, the crusher) there is sometimes a queue of full haulers waiting for dumping, this higher intensity of occupation appears to be correct. (If the designer of the new system is penny-wise, he is going to recommend purchasing a few less trucks, but in some cases such an approach does not appear to be rational, which will be proved in a further part of the analysis of the system).

Now, we can formulate preliminary inferences.

1. After introduction and application of the crusher and conveyors, the number of trucks in the system is 48 units fewer than before.
2. Due to a smaller number of trucks in the system, the number of surplus units can be reduced appropriately.

The next part of the modelling, analysis and calculation scheme will concern the ore system exclusively, that is the truck subsystem and the continuous system.

5. Construct the probability distribution of the number of trucks (w) in a work state (p), $P_{wb}^{(p)}$, $b = 0, 1, ..., 22$ for the ore system by applying the Maryanovitch model (equation 8.23). The results of the calculations are given in Table 9.1 and Figure 9.3. The table and figure show all probabilities not less than 0.001.
6. Now evaluate the probability distribution of the number of trucks at loading machines for the ore system.

    In the calculation procedure of shovel-truck systems more information is required than just the mean times, regarding both loading and travel. The corresponding standard deviations are needed. There are two ways of estimating these parameters. The best solution is an estimation based on empirical data taken from the operation of the system being considered. The second solution, resulting in a much lower likelihood but not needing any data, is the presumption that the coefficient of variation remains approximately constant in relation to the corresponding average value. This means that the coefficient of variation is invariable. It can be presumed that the standard deviations are 0.3 to 0.5 of the mean. If the real value is a bit higher or lower than numbers in this range, the error generated is rather small and it can often be neglected.

Presume that the field investigation shows that the standard deviation of loading times is, say, 50% less than the corresponding average value, and the standard deviation of travel times equals 4 minutes.

Table 9.1.   The probability distribution of the number of trucks in a work state for the ore system <22, 3>.

b	12	13	14	15	16	17	18	19	20	21	22
$P_{wb}^{(p)}$	0.002	0.005	0.014	0.032	0.064	0.109	0.156	0.183	0.176	0.134	0.124

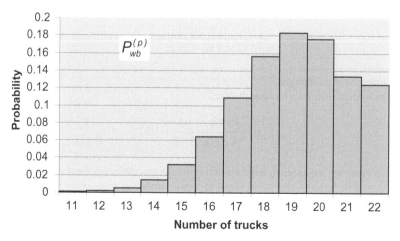

Figure 9.3.   The probability distribution of the number of trucks in a work state for the ore system $< m = 22, r = 3 >$.

Now we are in possession of all of the parameters required to commence an evaluation.

- The mean loading time Z'_d, $d = 0, 1, 2, 3, 4$
- The standard deviation of loading times $\sigma_z = 0.5 \, Z'_d \min$
- The mean travel time T_{tc}
- The standard deviation of travel times $\sigma_t = 4 \min$
- The probability distribution of the number of shovels in a work state $P^{(p)}_{kd}$
- The probability distribution of the number of trucks in a work state $P^{(p)}_{wj}$, $j = 0, 1,\ldots,22$.

Before a presentation of the calculation procedure for the ore shovel-truck subsystem one important point must be clearly stated: the calculations undertaken will consider succeeding cases for a variable number of shovels in failure, d. Therefore, the parameters and functions that will be presented at this point will be conditional ones—provided that there are a given number of shovels in a down state.

The procedure of calculation of the ore shovel-truck system is according to the following scheme (see Czaplicki's monograph 2009, Chapter 9.2).

1. Calculation of the squares of the coefficient of variations:

$$C_1 = \left(\frac{\sigma_j}{T_{tc}}\right)^2 \quad C_{2d} = \left(\frac{\sigma_z}{Z'_d}\right)^2 \tag{9.5}$$

2. Calculation of the coefficient of the relative intensity of loading:

$$\varpi_d = \frac{Z'_d}{T_{tc}} \tag{9.6}$$

3. Power exponents calculation:

$$\beta'_{3d} = \frac{2[1 + (C_{2d}/C_1)]}{\varpi_d C_1} \tag{9.7a}$$

$$\beta'_{5d} = \frac{2\varpi_d(C_1 + C_{2d})}{(C_{2d} - \varpi_d C_1)^2} \tag{9.7b}$$

4. Construction of the probability distribution of the number of trucks at ore shovels.

   a. Auxiliary functions:

   for $\quad 0 \le x \le 4$

$$f_{5d}(x,b) = \frac{1}{(b-x)\varpi_d C_1 + xC_{2d}} \left(\frac{(b-x)\varpi_d C_1 + xC_{2d}}{b\varpi_d C_1}\right)^{b\beta'_{5d}} \exp\left(\frac{2(\varpi_d + 1)x}{\varpi C_1 - C_{2d}}\right)$$

$$\text{if } \varpi_d C_1 - C_{2d} \neq 0 \tag{9.8a}$$

$$f_{5d}(x,b) = \frac{1}{b(\varpi_d C_1)} \exp\left(\frac{2x}{C_1} - \frac{(\varpi_d + 1)x^2}{b\varpi_d C_1}\right) \quad \text{if } \varpi_d C_1 - C_{2d} = 0 \tag{9.8b}$$

   and for[7] $4 \le x \le b$

---

[7] The number 4 is still in the limits and functions because there are 4 totally reliable loading points in the system.

$$f_{6d}(x,b) = \frac{1}{(b-x)\varpi_d C_1 + 4C_{2d}} \left(\frac{(b-x)\varpi_d C_1 + 4C_{2d}}{(b-4)\varpi_d C_1 + 4C_{2d}}\right)^{4\beta'_{3d}} \exp\left(\frac{2(x-4)}{C_1}\right) \quad (9.8c)$$

b. The coefficient taking into account function continuity:

$$\Psi_{bd} = \frac{f_{5d}(x,b)}{f_{6d}(x,b)} \quad (9.9)$$

c. The normalisation constant:

$$\kappa_{bd} = \left(\int_0^4 f_{5d}(x,b)\,dx + \Psi_{bd}\int_4^b f_{6d}(x,b)\,dx\right)^{-1} \quad (9.10)$$

d. The conditional probability that no truck will be at ore shovels:

$$P_{0d} = \sum_b^{22} P_{wb}^{(p)}\kappa_{bd}\int_0^{0.5} f_{5d}(x,b)\,dx \quad (9.11a)$$

e. Further conditional probabilities:

$$p_{1d} = \sum_b^{22} P_{wb}^{(p)}\kappa_{bd}\int_{0.5}^{1.5} f_{5d}(x,b)\,dx$$

$$p_{2d} = \sum_b^{22} P_{bj}^{(p)}\kappa_{bd}\int_{1.5}^{2.5} f_{5d}(x,b)\,dx$$

$$p_{3d} = \sum_b^{22} P_{wb}^{(p)}\kappa_{bd}\int_{2.5}^{3.5} f_{5d}(x,b)\,dx \quad (9.11b)$$

$$p_{4d} = \sum_b^{22} P_{wb}^{(p)}\kappa_{bd}\left(\int_{3.5}^4 f_{5d}(x,b)\,dx + \Psi_{bd}\int_4^{4.5} f_{6d}(x,b)\,dx\right)$$

$$p_{>4d} = \sum_b^{22} P_{wb}^{(p)}\kappa_{bd}\left(\Psi_{bd}\int_{4.5}^b f_{6d}(x,b)\,dx\right)$$

The last probability comprises all cases where the number of trucks at ore shovels is greater than the number of loading machines (>4).

5. The conditional average number of trucks at loading machines:

$$E_{wd} = \sum_b^{22} P_{wb}^{(p)}\kappa_{bd}\left(\int_0^4 xf_{5d}(x,b)\,dx + \Psi_{bd}\int_4^b xf_{6d}(x,b)\,dx\right) \quad (9.12)$$

6. An interesting system parameter is the ratio:

$$E_{wd}/4 \quad (9.13)$$

7. The conditional average time loss in a truck work cycle due to truck standstill waiting for loading:

$$\Delta_d = Z'_d P_{kd}^{(p)} p_{>4d} \quad (9.14)$$

Table 9.2. Parameters of the ore shovel-truck system $\mathfrak{S}_o$.

Conditional parameters					Conditional probabilities					
d	$P_{kd}^{(p)}$	Z'_{d-4}	E_{wd}	$E_{wd}/4$	p_{1d}	p_{2d}	p_{3d}	p_{4d}	$p_{>4d}$	Δ_d
shovels able to load		min	trucks	trucks/loading machine						min
4	0.573	1.60	3.2	0.8	0.023	0.217	0.412	0.237	0.111	0.1
3	0.342	2.24	5.2	1.3	0.002	0.036	0.156	0.216	0.590	0.5
2	0.077	2.88	7.7	1.9		0.005	0.031	0.068	0.896	0.3
1	0.008	3.75	9.7	2.4		0.001	0.006	0.017	0.976	0.1
0	0.000	4.16	11.1	2.8			0.001	0.005	0.994	0.0

Unconditional parameters						Unconditional probabilities				
T_C	E_{wlk}	E_{wk}	θ	W_{efk}	W_{efw}	p_0	p_1	p_2	p_3	p_4
min	trucks	trucks	trucks /loading machine	loaded trucks/h	truck work cycles/h					
11.13	3.4	15.2	9.9	105.7	102.4	0.000	0.014	0.137	0.292	0.557

E_{wd} - the conditional mean number of trucks at d shovels able to load

$E_{wd}/4$ - the conditional mean number of trucks per *l*loading machine

p_{bd} - the conditional probability that there are b trucks at d loading machines

$p_{>4d}$ - the conditional probability that there are more than 4 trucks at 4 loading machines

Δ_d - the conditional time loss parameter

T_C - the mean time of a truck work cycle including losses but without losses at the crusher

E_{wlk} - the mean number of loaded trucks

E_{wk} - the mean number of trucks at loading machines

θ - the mean number of trucks per loading machine

W_{efk} - the loading machine system's effective productivity

W_{efw} - the truck system's effective productivity

p_b - the probability that there are b trucks loaded, $b \leq 4$

The outcomes of the calculations of these probabilities and parameters[8] are shown in Table 9.2.

At the end of these calculations of conditional parameters some remarks should be made. When all shovels are down and only wheel loaders execute their duties, the mean loading time is longest (4.16 min) and on average there are about 11 haulers at loading machines (approximately 2.8 units per loader). We presume here that the truck dispatcher does not interfere in the machinery system. However, he may decide that there are too many trucks waiting idly for loading and for this reason he may withdraw some trucks from the pit.

8. Calculate the unconditional probabilities of the ore truck system.
The probability that no truck will be at ore shovels:

$$p_0 = \sum_d p_{0d} P_{kd}^{(p)} = 0.000 \qquad (9.15a)$$

The probability that only 1 truck will be loaded:

$$p_1 = \sum_d p_{1d} P_{kd}^{(p)} \qquad (9.15b)$$

The probability that 2 trucks will be loaded:

$$p_2 = \sum_d p_{2d} P_{kd}^{(p)} \qquad (9.15c)$$

---

[8] To be quite accurate probabilities are also system parameters but to make the results of the analysis more communicative probabilities are treated separately.

The probability that 3 trucks will be loaded:

$$p_3 = \sum_d P_{3d} P_{kd}^{(p)}$$

(9.15d)

The probability that 4 trucks will be loaded

$$p_4 = \sum_d P_{kd}^{(p)} (p_{4d} + p_{>4})$$

(9.15e)

All these probabilities are given in Table 9.2 and this probability distribution is shown in Figure 9.4.

Looking at Figure 9.4 some comments should be made. In only 56% of cases will all loading points be busy and in about 30% of cases only three loading points will be executing their duties. This does not seem to be very high. Especially since the analysis of the flow intensity rate suggested that the ore system was quite active because of the high number of haulers. What has caused this situation? The answer is simple: the stochastic nature of the exploitation process of the machinery system being analysed. In addition, one important observation, especially for designers of this type of system, should be given here:

The procedure for the selection of the number of trucks for a given system of shovels relies on '*just matching and only matching*' both systems; this does not guarantee a very high rate of utilisation of the loading machines.

Note that losses in time expressed by the values of parameter Δ are small. By the way, the time loss parameter depends on the standard deviations of the times of truck work cycle phases (times of loading and travel). In the system being investigated, the standard deviations are rather high. If their values are reduced to 30% of their corresponding means, for example, the masses of conditional and unconditional probabilities will be displaced resulting in a small reduction in time losses Δ and for this reason the ore truck system's effective output will be a little greater.

Notice additionally that the possible occurrence of a queue at the crusher and its influence on the system's performance has not yet been analysed but, unfortunately, it can make the exploitation situation worse.

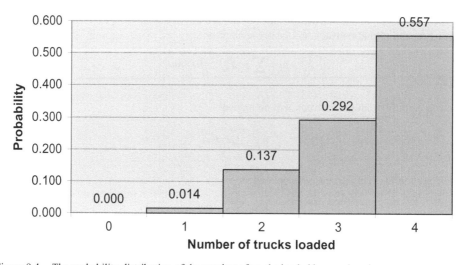

Figure 9.4.   The probability distribution of the number of trucks loaded by ore shovels.

9. Calculate further unconditional system parameters.
   The average number of loaded trucks at ore loading machines:

$$E_{wlk} = \sum_d dp_d = 3.4 \tag{9.16}$$

The average number of trucks per loading point:

$$\theta = \frac{1}{4} \sum_d P_{kd}^{(p)} E_{wd} = 1.1 \tag{9.17}$$

The mean time of a truck cycle including losses (however, without losses caused by the mean queue of trucks waiting for unloading at the crusher):

$$T_c = Z_w + T_{tc} + \Sigma\Delta_d = 11.13 \text{ min} \tag{9.18}$$

10. Now we make an assessment of the system's output.
    Several further parameters of the system can be defined that will give some additional information on the properties of the system but our point of interest is the calculation of the newly designed system with the crusher.
    Two measures of the system's output will be specified.
    The ore shovel system's effective output:

$$W_{efk} = \frac{60}{Z_w} E_{wlk} = 105.7 \text{ trucks/h} \tag{9.19}$$

This productivity measure includes the reliability of both shovels and trucks and losses due to the lack of a truck at a free shovel.
    The ore truck system's effective output:

$$W_{efw} = \frac{60}{T_c} E_{wkd} = 102.4 \text{ truck work cycles/h} \tag{9.20}$$

Thus, we can safely presume that the stream of ore that can be delivered to the crusher is:

$$S_{pcr} = W_{efw}Q \cong 12290 \text{ t/h} \tag{9.21}$$

11. Calculate the steady-state availability of the continuous system.
    Field investigations have shown that in many cases the crusher exploitation process of a work-repair type can be satisfactorily described by the Markov model. In some cases the distribution of repair times cannot be acceptably depicted by an exponential probability distribution and the process of the changes of the states is of the semi-Markov type (Chapter 7.6). However, keeping in mind the approximations given by Gneydenko et al. (1965) expressed by formulas 7.2 and 7.3 for series systems, we can still apply the principles of reduction for the series system.
    By knowing the relationship between the steady-state availability A and the repair rate κ (equation 5.25) we can calculate the steady-state availability A_{con} of the continuous system according to the second principle of reduction of the series system of a Markov type (equation 7.3):

$$A_{con} = (1 + u_{con}\kappa_c + \kappa_{cr})^{-1} = 0.813 \tag{9.22}$$

12. Now consider the potential crushing-hauling capacity of the continuous system.
    It is obvious that the crusher output and the transporting capacity of the conveyors should be identical. If all of these pieces of equipment are in operation, the stream of flowing mass must be uniform. Such conformability does not exist where their accessibility is considered. Belt conveyors can operate continuously for many days without any stoppage. A crusher must undergo some prophylactic actions. Therefore, the accessibility coefficient B_{cr} for a given crusher is given.
    If S_{cr} denotes the crusher and conveyors' output, the potential continuous system output S_{pc} can be calculated from the formula:

$$S_{pc} = S_{cr} A_{con} B_{cr} \qquad (9.23)$$

13. Construct the probability distribution of the number of trucks at the crusher.
    We analyse how busy the crusher will be by applying the proposed truck system for the given shovel system. The flow intensity rate for this sizer is:

$$\rho_{cr} = \frac{1}{2} \frac{T_{dp}}{T'_t} \frac{E_{wkd}}{A_{con} B_{cr}} = \frac{0.7 \times 19}{2 \times 10.43 \times 0.813 \times 0.880} = 0.89$$

where T'_t is the mean time of truck 'travel', calculated from the sum of the following average values: haul + load + return + time losses.

The outcome indicates that the crusher will be busy (heavy traffic situation) but not 'deadly' busy. This seems to be correct.

The probability distribution of the number of trucks at the crusher can be obtained by applying the Sivazlian and Wang model. Its main parameters as far as the crusher (as the service point) is concerned are:

- there are two 'service points' (two trucks can unload at the same time)[9]
- the mean time of truck 'travel' $T'_t = 8.2 - 0.7 + 1.93 + \Sigma\Delta = 10.43$ min
- the mean time of dumping $T_{dp} = 0.7$ min
- we presume that the corresponding standard deviation associated with the mean time of dumping is 20% of the average value.

After quick computations the parameters of the system can be found.

- The probability distribution of the number of trucks at the crusher:

$$q_0 = 0.027 \quad q_1 = 0.739$$
$$q_2 = 0.230 \quad q_{>2} = 0.003$$

- The average number of trucks at the crusher and the time loss coefficient:

$$\Xi = 1.2 \text{ trucks} \quad \Delta = 0.0 \text{ min}$$

14. Analyse the basic measures of the system's efficiency. The system consists of three subsystems:

- loading machines (shovels and spare units: wheel loaders)
- trucks
- crusher and conveyors.

---

[9] We reject the case that only one truck can unload.

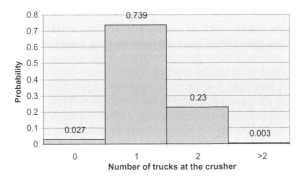

Figure 9.5. The probability distribution of the number of trucks at the crusher.

The effective output of the first two subsystems was assessed in point 9. Let us discuss the capacity of the third subsystem. By looking at the average number of trucks at the crusher, the stream of ore can be evaluated from the formulas:

$$W_{efcr} = \Xi \frac{60}{T_{dp}} = 102.9 \text{ trucks/h} \tag{9.24}$$

$$S_{ocr} = \Xi Q \frac{60}{T_{dp}} \cong 12340 \text{ t/h} \tag{9.25}$$

By comparing these three capacities, W_{efk}, W_{efw} and W_{efcr}, which are 105.7, 102.4, and 102.9 trucks/h, it can be presumed that the system capacity will be approximately 102.4 trucks/h (the lowest value), which gives 12290 t/h.

Now recall formula 9.23. By knowing the system capacity we can calculate the nominal capacity of the continuous system obtaining:

$$S_{cr} = \frac{S_{pcr}}{A_{con} B_{cr}} \cong 17200 \text{ t/h} \tag{9.26}$$

Remember:
The inaccessibility and unreliability of a given object removes time. To get the planned capacity, the nominal output of the object must be greater in order to accomplish a given task (hauling, extracting, dumping) in a shorter period time.

15. Conclusions, recommendations and remarks.

Looking at three outputs connected with the system, we may intuitively expect that all the figures should be approximately the same. In the case being investigated, the figures are close to each other, but in analyses of different systems of this kind dissimilarities may become more visible. The reasons for the generation of such differences originate from different sources. First, the procedure is a multistep one and at each step some rounding is inevitable. This can cumulate during calculations. Second, the nature of the method being applied also creates some differences. The model works but only under a heavy traffic situation. This condition is fulfilled with a different power for different service points in the system. Moreover, the character of the method is such that the outcomes obtained are approximate ones. In addition, there are two transitions: one, from a discrete space to a continuous one and—after transformations—there is a return from a continuous space to a discrete one. Each transition may generate some divergences. ◄

At end of this calculation procedure, we should undertake some financial analysis. The productivity of the new system must be compared with the productivity of the 'old' system. Savings associated with a reduction of truck fleet size must be compared with expenses connected with the construction and operation of the new subsystem and, obviously, time is important factor here—an assessment must be made about whether these expenses will be recovered through the planned savings before termination of the mine operation. Apart from the problem of the physical construction of new pieces of equipment in the mine, the problem of the crusher exploitation should be carefully analysed. As a rule, sizers need a lot of power, spare parts and maintenance. And additionally—and importantly—a sizer operates in series type of system. All these issues have to be considered precisely and in a penetrative way.

To conclude this section, a few additional explanations should be given because the case that has been analysed was a particular one. The reasoning was a bit simplified because the truck dispatcher does not interfere in the system, and this thereby changes the system's structure. It is worth remembering that there are two types of decisions made by a truck dispatcher:

a. decisions changing the system's characteristics persistently
b. current decisions reacting to a situation in the pit but not changing the system's characteristics.

To be quite truthful, each decision causes some changes in a shovel-truck system and its characteristics, but the majority of them make small changes, and only for a short period of time. From a calculations point of view, they are neglected. Evaluation of a system of this kind is based mainly on average values. Thus, there must be appropriately large data to get *good* estimations of these parameters; *good* in statistical terms. Their application also makes no sense if the operation of the changed system will not be appropriately long. This explains the term 'persistently'. Two types of decisions generate significant modification in the system's characteristics—changes in system's structure and changes in priority of type of rock hauled. The influence of these modifications on particular system characteristics is presented in Czaplicki's monograph (2009).

However, here is a hint about how to include possible truck dispatcher decisions concerning changes in the system's structure in the calculation procedure.

If the number of trucks in the pit is relatively large—that is, there are long queues of haulers waiting to be loaded—the truck dispatcher can take a decision to withdraw a certain number of vehicles from the pit. If so, the ratio of the number of trucks directed to work and the number of trucks in reserve changes. It is necessary to calculate the system with a new pair $< m, r >$. Modifications will be visible in the probability distribution of the number of trucks in a work state and in the probability distribution of the number of units at service points, both at shovels and at the crusher. Assessment of the system's efficiency will be different. Incidentally, if the number of haulers withdrawn is significant, the character of the reserve can change drastically making the system of trucks operating in the pit totally reliable because 'there will always be a machine to replace a failed one'.

9.2   A SHOVEL-TRUCK SYSTEM AND AN INCLINED HOIST
OF THE TRUCK LIFT TYPE

As was stated at the beginning of this chapter, there has been a continuous search for ways to reduce the huge costs connected with the application of shovel-truck systems. One proposal in this regard was the application of an inclined hoist of a new type,[10] called a TruckLift, which would allow the removal of a fully loaded truck from a pit. This proposal was made at the beginning of this century by a producer of mining equipment that has been well known for many years. This is

---

[10] The idea for this equipment was taken from the shipyard industry where huge slipways are used.

the most convincing information about the application of this type of hoist: '*The truck is hoisted to the surface in only 2 minutes while driving on the haul road incline takes some 20 minutes*' (in the case described, the pit had a depth of approximately 300 m and a 50° pit slope incline—www. siemag.de/new/en/schraeg.htm, 2004). It should be added here that during truck travel up a ramp a lot of fuel is consumed and emission of exhaust gasses is inevitable. A hoist, however, uses clean electric energy. In addition, during the truck's travel up the ramp its tyres are intensively worn and the surface of road is milled. Truck drivers work with great concentration. During hauler displacement by a hoist, a truck sits motionless, and the driver and engine have a rest. If use of this hoist lowers the intensity of traffic in the pit, it will lower the probability of a traffic accident and therefore increase safety. If there are many vehicles in a pit, they can disturb each other causing longer truck work cycles, thereby decreasing the system's output. The application of a hoist reduces the probability of the occurrence of this phenomenon. Furthermore, a reduction in the number of huge fuel tanks is also a great advantage. Some further merits can be enumerated here and generally the advantages of the application of this type of installation seem to be incontrovertible. However, the results of the advertising campaign created by the producer are somewhat surprising—until now no mine in the world has purchased and applied this type of hoist. The campaign was abandoned a few years ago. We will not analyse the arguments against the application here, noting only that the expected lifetime of a hoist of this type is estimated to be 40 to 50 years (perhaps longer) whereas the mean lifetime of an open pit mine does not exceed 20 years. There is no doubt that the application of the hoist makes economic sense if the whole mine enterprise is large enough to recoup the money spent on the purchase, delivery and installation in the pit. But the majority of mining projects are significantly shorter than the above-mentioned mean duration. For this reason, the range of possible purchasers is rather small. Also ignoring the mistakes made during the commercial campaign, one point seemed quite clear—there was no analytical description given by the producer stating what kind of changes in the machinery system parameters would take place following the application of this type of hoist. Another interesting topic in this regard is to consider how the significance of the hoist application will change with increasing mine development. An analysis of the hoist application is a subtle problem because when the depth of the mine increases the time of the platform travel carrying a hauler increases as well, which means that the transport capacity of the hoist decreases. Nonetheless, the average distance for a hauler to drive from the pit increases significantly and for this reason the advantage of the hoist application will also increase.

Now, with the benefit of the well-developed method of analysis and calculation procedure for shovel-truck systems given in Czaplicki's 2009 monograph, we are able to construct a method of analysis and estimation of the *benefit* of the application of a TruckLift type hoist. It is worth referring here to the first simplified procedure for an evaluation of the shovel-truck system with an inclined hoist. This procedure was described in a simplified way by the author in his 2004 textbook as well as in papers (2004c, 2008b).

A general idea of the system exploitation process that will be the point of our investigations is shown in Figure 9.6.

9.2.1   *Formulation of the problem and assumptions*

The application of the inclined hoist of the TruckLift type makes economic sense only when the whole enterprise is large, when great quantities of extracted rock will be removed from the pit, the final pit depth will be at least a few hundred metres and the machinery system involved will comprise several dozen or more machines. However, this means that the inclined hoist installed will take only a certain part of the stream of circulating trucks.

Let us write down the whole machinery system and the process of its exploitation in a mathematical form, providing additionally some explanations.

There is a given machinery system $\mathfrak{S}$ consisting of n ($n \geq 2$; $i = 1, 2, ..., n$) power shovels working in the pit. In order to accomplish the haulage task determined by the system of shovels m ($m \geq 2$; $b = 1, 2, ..., m$), trucks are directed to the pit and r trucks are in reserve. An inclined hoist is installed

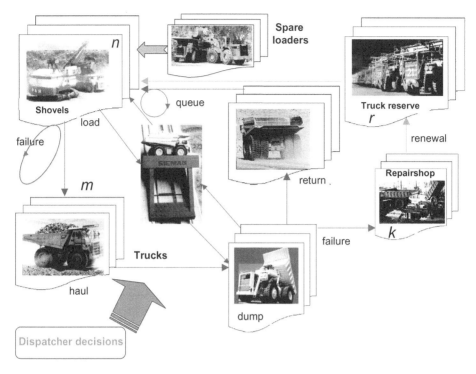

Figure 9.6.   Scheme of the flow of machines in the system.

in the mine. The basic system parameters are known: G_k is the shovel loading capability coefficient (this is the product of the shovel steady-state availability and the shovel accessibility coefficient), A_w is the steady-state availability of the truck, A_T is the steady-state availability of the hoist and U_T is the hoist's daily nominal utilisation (22.5/24 h).[11] The mine maintenance bay has k repair stands for trucks. We presume that structural system parameters $< m, r; k >$ were selected properly (the procedure for proper selection of these parameters is described in the Czaplicki 2009).

This system and its operational parameters determining the exploitation process $\mathfrak{P}$ can be described as:

$$\mathfrak{S} \cap \mathfrak{P} : < \mathfrak{H}: n \geq 2, i = 1,2, ..., n, G_k \, \mathfrak{W}: A_w, m \geq 2, b = 1,2, ..., m; r \, \mathfrak{J}: A_T, U_T; \mathfrak{O}: k >$$

where:

    $\mathfrak{H}$ – the set of power shovels
    $\mathfrak{W}$ – the set of trucks
    $\mathfrak{J}$ – the inclined hoist
    $\mathfrak{O}$ – the set of repair stands.

In order to accomplish an analysis and evaluation of the system several further parameters are needed. Let us list them.

---

[11] Notice the difference between object utilisation and nominal object utilisation, both expressed per unit of time (per day (24 h) in this case). A measure of the object utilisation can be the mean value, for instance, because it determines a random variable, whereas a nominal utilisation (deterministic value) describes how many hours per day the object can be used to execute its duties in order to accomplish its task. In fact, the nominal utilisation (a term used by the producer) determines the disposal time of the hoist.

The technical possibilities of the equipment being used and the pit configuration at the moment of our analysis are such that:

- the mean time of truck haulage up the ramp is T_o for all trucks
- the mean time of truck dump is T_w for all trucks
- the mean time of truck return by the ramp is T_r for all trucks
- the mean time of truck loading is T_z for all trucks and power shovels.

For a truck using the inclined hoist the appropriate times are:

- the mean time of loaded truck travel from the shovel to the hoist is T_{kT} for all trucks and shovels
- the mean time of loaded truck travel from the hoist to the dumping point is T_{Tw} for all trucks and dumps
- the mean time of empty truck travel from the dump to the hoist is T_{wT} for all trucks and dumps
- the mean time of empty truck travel from the hoist to the power shovel is T_{Tk}
- the mean time of truck travel by the hoist is T_T.

All times will be given in minutes if there is no other specification. For analysis of the system the doubled randomised Sivazlian and Wang model will be applied. Therefore, we need information on the appropriate standard deviations. For this reason, let us presume that:

- σ_t is the standard deviation of the times of truck travel not using the hoist
- σ_w is the standard deviation of the times of truck travel using the hoist
- σ_z is the standard deviation of the times of truck loading.

We now have to specify a few important assumptions needed before we can proceed with our considerations.

Ass. (1): The dumping points are properly arranged—that is, queues of trucks waiting for unloading are rare and negligible.

Ass. (2): The number of repair stands for trucks is properly selected—that is, the mean waiting time of a failed truck for repair is negligible.

Ass. (3): The repair stands are totally reliable.

Ass. (4): The truck dispatcher ensures that:

      a. the hoist is utilised at maximum
      b. any truck queue in the pit at the hoist loading point is short.

Let us briefly discuss the significance of these assumptions. Assumptions (1), (2) and (3) simplify the procedure greatly.

Suppose assumption (1) does not hold. If the exploitation situation in the pit is such that truck queues *can* occur at both loading and unloading points and queues at the dumps cannot be neglected, the calculation procedure is extended in comparison to the scheme given in the author's 2009 monograph. Consequently, the procedure must be a stepped-chain one. First, queues at dumps are identified and their effect on the elongation of the mean time of the truck work cycle is included in the calculation scheme. Next, the point of interest is the system of loading machines. Here a certain queue is advisable. Again, such a queue extends the mean truck time work cycle. This must be taken into account. Calculations must be repeated. Examples of this kind of procedure were presented in Czaplicki's 2004.

Suppose assumption (2) does not hold. If the number of repair stands is insufficient, a queue of failed trucks waiting for repair occurs. This decreases the mean number of trucks in a work state. In order to accomplish the haulage task, it is necessary to increase either the number of trucks or the number of repair stands. The first solution increases mine production to some degree; however, the mean number of trucks in failure increases. The best solution is enlargement of the maintenance bay. In some cases, application of trucks with better parameters should be considered. It is worth realising that even small changes in the reliability of haulers may lead to considerable changes in the system parameters. Many examples proving this statement are given in Czaplicki's monographs of 2006 and 2009.

If assumption (3) is rejected, an alternative pair of assumptions should be formulated from a theoretical point of view, stating that:

– the system of repair stands has a determined reliability
– the system of repair stands is available periodically.

These suppositions increase the number of calculations drastically. This opens a separate chapter of modelling and calculation. The assumption that repair stands can fail generates—from a mathematical point of view—a new randomisation of one of the parameters of the structural system. An example of this kind of modelling was shown in Czaplicki's monographs of 2006 and 2009. However, in mining practice the repair shop is open 24 hours a day as a rule.

Assumption (4) is obvious. The hoist is an expensive piece of equipment and, in a similar way to power shovels, it should be utilised extensively. Thus, the truck dispatching rule being used in the mine must be modified to assure maximum hoist utilisation. The second part of this assumption—the existence of a truck queue—should guarantee this high utilisation. However, this queue should not be too long, especially the one down in the pit. The time spent in the queue extends the mean time of the truck work cycle. Additionally, many trucks standing in the queue may block the pit transport road to a certain extent.

Two further assumptions should be formulated.

Ass. (5): The system has (or does not have) spare loading machines.
Ass. (6): The type of dispatching rule being used is known.

Consider assumption (5). The existence of spare units that can resume loading duties when one or more power shovels are down has vast repercussions. It not only means a simple increase in the system production (compared to a system without spares) but also significant changes in the system parameters, such as typically an increase in the mean loading time as was shown when investigating a shovel-truck system with crushing and belt conveyors. In a case where a hoist is installed in the system, possession of spare loading machines is highly advisable. The mean number of circulating trucks in a work state increases, as does the utilisation of the hoist. This will be proved in a further part of considerations.

Now consider assumption (6). Formulation of a dispatching rule for the machinery system has a colossal influence on the course of the operation of all of the machinery, the values of the measures of the system's efficiency, the utilisation of the equipment involved including maintenance devices, and the frequency and content of decisions concerning changes in the system's structure. During the exploitation process of a mineral deposit a situation sometimes occurs whereby it is necessary to speed up the removal of either ore or waste. This means that power shovels loading a certain type of broken rock which is being given priority are the main concern and empty haulers are directed—at first—to these machines. A problem of this kind of priority was discussed briefly in Czaplicki's 2009 monograph. This problem was also investigated more comprehensively recently (Czaplicki 2008e). Generally, the application of an inclined hoist should have great positive significance for the system operating under this type of priority. No publication on this topic has previously been issued, and it looks as though the hoist's producer missed this problem entirely during its advertising campaign.

In order to make our further investigations clear and communicative, and appropriately general, we will assume that:

• there are spare loading machines in the system
• there is no priority in the dispatching rule other than that associated with the hoist utilisation.

The calculation procedure for the system being considered can be as follows.

1. Determination of the probability distribution of the number of power shovels in a work state (binomial distribution), $P_{kd}^{(p)}$, $d = 0, 1, ..., n$.
2. Calculation of the mean time of loading as a function of the number of spare units working in the loading system and calculation of the total mean loading time for the system.

3. Calculation of the number of haulers in a work state that is needed to serve the system of loading machines. Determination of the probability distribution of the number of trucks in a work state, $P_{wb}^{(p)}$ $b = 0, 1, ..., m$.
4. Calculation of the flow intensity rate ρ in the system of loading machines.
5. Application of the Sivazlian and Wang model to get conditional system parameters for the following pairs $< d, (n - d) >$ (denoting the number of power shovels able to load and the number of loaders able to load replacing failed shovels); the values of some parameters will indicate when the system structure should be changed; when the system structure is changed the Maryanovitch model must be applied again to get the distribution $P_{wb}^{(p)}$ and the calculations must be repeated.
6. Evaluation of the probability distribution of the number of trucks at loading machines.
7. Determination of the unconditional system parameters and measures of the system's efficiency.

Calculation of a system with a hoist.

8. Calculation and analysis of the truck fractions $r_\tau(d)$ taken by the hoist.
9. Calculation of the mean time of truck travel.
10. Evaluation of the standard deviations corresponding with the mean values estimated in point 9.
11. Assessment of the system's conditional parameters at loading machines, with correction and repetition of calculations if necessary.
12. Assessment of the system's unconditional parameters at loading machines.
13. Evaluation of the system's productivities.
14. Conclusions, recommendations and remarks.

Let us formulate the essential parameters required to calculate the system.
The mean time of truck travel (haul—dump—return) not using the hoist is:

$$T_t = T_o + T_w + T_r \tag{9.27}$$

Let us now determine the mean time of truck travel using the hoist. A fully loaded truck starts to move away from a loading machine and the dispatcher instructs that the truck be directed to the loading station of the hoist. The hauler arriving at the station can face one of three possible situations:

a. the hoist platform is free and waiting for the load
b. there are no loaded truck/trucks waiting for the platform, but the platform is in motion with a fully loaded truck, or an empty truck is being lowered down into the pit
c. there is a fully loaded truck, perhaps two, at the hoist's loading station, and the platform is in motion with a fully loaded truck, or an empty truck is being lowered down into the pit.

Situation (a) is the most convenient for the truck; however, it is inconvenient from the standpoint of the recommended heavy utilisation of the hoist. In a good truck-dispatching regimen, this state should have a low probability of occurrence.

Situation (b) looks more probable although not convenient for high hoist utilisation. The truck waits a short time for the platform.

Situation (c) is most probable. The truck waits for the platform in a short queue.

Therefore, we can assume that there will be a certain mean time delay—the dumper waits for transportation. The value of this mean depends on the time T_r. Thus, we can introduce a coefficient increasing the mean time of truck travel. Let us denote it by ψ, $\psi > 1$, its value increases for dispatching of a lower quality (that is, a poor truck-dispatching regimen).

The mean time of truck travel using the hoist can be determined as follows:

$$T_{cT} = T_{kT} + \psi T_T + T_{Tw} + T_w + T_{wT} + \psi T_T + T_{Tk} \tag{9.28}$$

Let us now evaluate the respective service abilities of the loading and hoist systems.

The potential effective output of the hoist during its disposal time of 22.5 h/day, disregarding hoist standstill time because of a lack of trucks, is:

$$W_{pefT} = 60\, A_T/2\, T_T \text{ trucks/h} \tag{9.29}$$

The potential effective output of the loading system—neglecting the standstill of loading machines due to a lack of dumpers—for d power shovels able to execute their duties is:

$$W_{pefik} = 60\, n/Z'_n(d,\tau) \text{ trucks/h} \tag{9.30}$$

where $Z'_n(d,\tau)$ is the mean loading time.

Consider this ratio:

$$r(d,\tau) = r_\tau(d) = \frac{W_{pefT}}{W_{pefik}} = \frac{A_T}{n}\, \frac{Z'_n(d,\tau)}{2T_T} \tag{9.31}$$

Let us call this parameter a 'stochastic truck fraction of hoist'.

This quotient informs us which part of the stream of loaded trucks can be taken by the hoist assuming fault-free truck dispatching (in Czaplicki's 2009 monograph the concept of an *ideal dispatcher* was formulated). Notice that this is a discrete function that depends on two parameters d and τ. However, for a given machinery system, τ is fixed and for this reason this function depends exclusively on the number of power shovels in failure because this number is associated with the mean time of loading. Therefore, the notation $r_\tau(d)$ is introduced. If a different type of spare loader is used, this function changes because a different value of parameter τ must be included in calculations.

We should still keep in mind the maximum utilisation of the hoist.

Thus, the average time of truck travel for the whole population can be determined as:

$$\bar{T}_\tau(d) = [1 - r_\tau(d)]T_t + r_\tau(d)T_{cT} \text{ min} \tag{9.32}$$

To apply the Sivazlian and Wang model the standard deviation connected with this mean is required. Making use of the so-called variation identity (this identity determines the variance of the population divided into a certain number of groups, Sobczyk 1996 p. 46) and modifying it appropriately to meet our needs, we can write down the formula for the variance:

$$\sigma_\tau^2(d) = \sigma_c^2[1 - r_\tau(d)] + \sigma_2^2 r_\tau(d) + [T_t - \bar{T}_\tau(d)]^2[1 - r_\tau(d)] + [T_{cT} - \bar{T}_\tau(d)]^2 r_\tau(d) \text{ min}^2 \tag{9.33}$$

where σ_c^2 is the variance of times of truck travel without the hoist and σ_2^2 is the variance of times of truck travel using the hoist. Notice that $\sigma_2^2 < \sigma_c^2$ and this inequality strengthens when the mean time of truck travel increases.

Now, we are ready to analyse an example.

■ **Example 9.2**

Let us discuss the following system:

$$\mathfrak{S} : < \mathfrak{H}: i = 1, 2, ..., n = 9, A_k = 0.88\ \mathfrak{Q}: A_w = 0.77, b = 1, 2, ..., m;\ r\ \mathfrak{F}: A_T = 0.97\ \mathfrak{N}: k >$$

for which the exploitation parameters are as follows

$$\mathfrak{P}(\mathfrak{S},\mathfrak{D}) : < B_k = 0.82, Z' = 2.0 \text{ min}; \tau = 2.8; B_w = 0.86, T_t = 28.2 \text{ min}; U_T = 0.95 >.$$

Let us read these notations.

The system $\mathfrak{S}$ of power shovels $\mathfrak{H}$ consists of 9 machines with a steady-state availability of 0.88. The system of haulers $\mathfrak{Q}$ is not specified in number $< m, r >$ but the type has been selected in relation to the loading units and it has a steady-state availability of 0.77. The hoist $\mathfrak{J}$ has a steady-state availability of 0.97. The number of repair stands for trucks k has not yet been specified.

The second set of parameters, $\mathfrak{P}(\mathfrak{S},\mathfrak{D})$, describes the exploitation process $\mathfrak{P}$ of the machinery system that is a function of the properties of the equipment in the system and the decisions $\mathfrak{D}$ made by the truck dispatcher. These parameters have these values: the accessibility coefficient of shovels B_k is 0.82, the mean loading time for shovels Z' is 2.0 min, the mean loading time for spare loaders is 2.8 times longer than for shovels, the accessibility coefficient of trucks B_w is 0.86, the mean time of truck work travel T_l is 28.2 min and the nominal daily hoist utilisation is 0.95.

Let us commence the analysis and calculation of the system. At first, we consider the system without the hoist.

1. Analysis will begin from the system of loading machines. The term 'loading machines' is used rationally because spare loading pieces are in the system. We repeat the presumption used in Example 9.1 that the number of wheel loaders is large enough to make the loading system totally reliable from the standpoint of the accomplishment of the hauling task (there is always a loader to replace a unit in failure). Let notation $P_{kd}^{(p)}$ mean the probability that d, $d = 0, 1, ..., n = 9$ power shovels are in a work state. The following set of equations are important for further discussions about the system:[12]

$$P_{kd=9}^{(p)} = (0.880)^9 = 0.316$$

$$P_{k8}^{(p)} = 9(0.880)^8(1 - 0.880) = 0.388$$

$$P_{k7}^{(p)} = 36(0.880)^7(1-0.880)^2 = 0.212$$

$$P_{k6}^{(p)} = 84(0.880)^6(1-0.880)^3 = 0.067$$

$$P_{k5}^{(p)} = 126(0.880)^5(1-0.880)^4 = 0.014$$

$$P_{k4}^{(p)} = 126(0.880)^4(1-0.880)^5 = 0.002.$$

These probabilities are shown in Figure 9.7. The remaining probabilities are below 0.000.

2. Calculate the mean time of loading as a function of the number of spare units working in the loading system. We repeat formula 9.1 here:

$$Z_n'(d,\tau) = Z_n' = Z'\left(1+(\tau-1)\frac{d}{n}\right)$$

For the system being considered:

$$Z_0' = Z' = 2.0 \quad Z_1' = 2.4 \quad Z_2' = 2.8 \quad Z_3' = 3.2 \quad Z_4' = 3.6$$
$$Z_5' = 4.0 \quad Z_6' = 4.4 \quad Z_7' = 4.8 \quad Z_8' = 5.2 \quad Z_9' = 5.6 \text{ min}$$

Thus, the mean time of loading for the system of loading machines is:

$$\bar{Z}_w = \sum_{d=0}^{9} Z_d' P_{kd}^{(p)} = 2.4 \text{ min}$$

---

[12] Proportional coefficients in these equations can be taken from the Pascal's triangle for example.

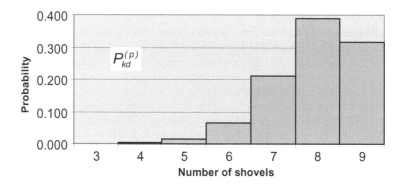

Figure 9.7.   The probability distribution of the number of shovels in a work state.

3. Calculate the number of haulers in a work state that is needed to serve the system of loading machines. Here we have:

$$h = B_k n[(Z_w + T_c)/Z_w] = 94.1 \text{ trucks}$$

The minimum number of trucks needed: $h/A_w \Rightarrow 123$

Applying the Maryanovitch model and enlarging the result on trucks directed to the pit by 10% we have:

$$< 1.1 \ m = 105; r = 28 >$$

and the average number of trucks in a work state is $E_{wkd} = 101.6$.

Conclusion: 133 trucks are required to operate in the mine.

4. Calculate the flow intensity rate ρ in the system of loading machines.

Applying the well-known formula we have:

$$\rho = \frac{E_{wkd}Z_w}{nB_kT_l} = \frac{101.6 \times 2.4}{9 \times 0.82 \times 28.2} = 1.17$$

Conclusion: the loading machines should be really busy.

5. Now determine the probability distribution of the number of trucks in a work state for the system <105, 28>. Applying the Maryanovitch model the distribution can be obtained and it is given in Table 9.3 and shown in Figure 9.8.

6. Evaluate the probability distribution of the number of trucks at loading machines. The procedure is a repetition of the reasoning used in Example 9.1. Two standard deviations are needed to conduct the calculations—the standard deviation of loading times and the standard deviation of travel times. Let us assume that the standard deviation for travel times is, say, 15 min and for loading times the following relationship holds: $\sigma_z = 0.3 \ Z'_n$.

Table 9.3.   The probability distribution of the number of trucks in a work state for the system $< 105, 28 >$.

b	89	90	91	92	93	94	95	96
$P_{wb}^{(p)}$	0.002	0.004	0.006	0.008	0.013	0.018	0.025	0.033
97	98	99	100	101	102	103	104	105
0.042	0.051	0.061	0.069	0.075	0.079	0.080	0.077	0.357

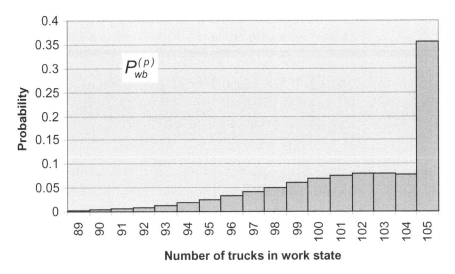

Figure 9.8. Probability distribution of the number of trucks in a work state for the system $< m = 105, r = 28 >$.

If so, we are in a possession of all the parameters necessary to carry out further investigations.

Formulas 9.5 to 9.7 remain unchanged. We need to make one change to formulas 9.8a to 9.10—the system analysed in Example 9.1 consisted of 4 ore loading machines; however, in the system being discussed here we have 9 loading points. Thus, the number 4 must be replaced by 9, and the number of equations has to be increased.

Conditional probabilities of the number of trucks at loading machines are obtained from this set of formulas:

$$p_{0d} = \sum_{j}^{105} P_{wj}^{(p)} \kappa_{bd} \int_{0}^{0.5} f_{5d}(x,b)dx$$

$$p_{1d} = \sum_{j}^{105} P_{wj}^{(p)} \kappa_{bd} \int_{0.5}^{1.5} f_{5d}(x,b)dx$$

$$p_{2d} = \sum_{j}^{105} P_{wj}^{(p)} \kappa_{bd} \int_{1.5}^{2.5} f_{5d}(x,b)dx$$

$$p_{3d} = \sum_{j}^{105} P_{wj}^{(p)} \kappa_{bd} \int_{2.5}^{3.5} f_{5d}(x,b)dx$$

$$p_{4d} = \sum_{j}^{105} P_{wj}^{(p)} \kappa_{bd} \int_{3.5}^{4.5} f_{5d}(x,b)dx$$

$$p_{5d} = \sum_{j}^{105} P_{wj}^{(p)} \kappa_{bd} \int_{4.5}^{5.5} f_{5d}(x,b)dx$$

$$p_{6d} = \sum_{j}^{105} P_{wj}^{(p)} \kappa_{bd} \int_{5.5}^{6.5} f_{5d}(x,b)dx$$

$$p_{7d} = \sum_{j}^{105} P_{wj}^{(p)} \kappa_{bd} \int_{6.5}^{7.5} f_{5d}(x,b)dx$$

$$p_{8d} = \sum_{j}^{105} P_{wj}^{(p)} \kappa_{bd} \int_{7.5}^{8.5} f_{5d}(x,b)dx$$

$$p_{9d} = \sum_{j}^{105} P_{wj}^{(p)} \kappa_{bd} \left(\int_{8.5}^{9} f_{5d}(x,b)dx + \Psi_{bd} \int_{9}^{9.5} f_{6d}(x,b)dx \right)$$

$$p_{>9d} = \sum_{j}^{105} P_{wj}^{(p)} \kappa_{bd} \left(\Psi_{bd} \int_{9.5}^{b} f_{6d}(x,b)dx \right)$$

This set is valid for a given number d.

Results of the evaluation of the conditional probabilities for a given set of parameters are given in Table 9.4.

The next conditional parameter to compute is the conditional average number of trucks at loading machines. This mean is obviously obtained from an equation similar to equation 9.12, namely:

$$E_{wd} = \sum_{b}^{105} P_{wb}^{(p)} \kappa_{bd} \left(\int_{0}^{9} x f_{5d}(x,b)dx + \Psi_{bd} \int_{9}^{b} x f_{6d}(x,b)dx \right)$$

Additionally, the ratio $E_{wd}/n = 9$ was calculated to get information about how many trucks there will be on average per loading machine. The outcomes of the computations are given in columns 6 and 7 of Table 9.4. Notice that when the number of successive power shovels in an up state decreases and the mean time of loading increases due to the use of wheel loaders as replacements for the power shovels, the number of trucks at loading machines also increases. This process is observed up to the moment when 4 power shovels are in failure and the mean time of loading jumps from 2.0 min to 3.6 min. This causes the number of haulers at loading machines to increase to 31.14. We presume that the truck dispatcher decided to withdraw 10 trucks from the pit, thereby changing the system's structure to $< m = 95, r = 38 >$. The calculation of the machinery system must now be repeated, apart from the system of loading machines which remains unchanged. Application of the Maryanovitch model gives a new probability distribution of the number of trucks in a work state and the average number of trucks in a work state, which is now $E_{wkd} = 94.9$ units. Withdrawal of these 10 haulers to the reserve means that the average number of trucks at loading machines is 24.38. Similarly, in the next case where 5 power shovels are in failure and the mean loading time increases to 4.0 min, it is presumed that the truck dispatcher again withdraws 10 trucks from circulation. After calculations, the average number of trucks at loading machines is 21.56. The last column of Table 9.4 reports the effects of the evaluation of the conditional time loss parameter, which provides information on how many minutes the mean time of truck cycle must be extended due to truck standstill while waiting for loading (cf. formula 9.14). Further cases have not been investigated because the likelihood of these events have a very low probability, below 0.001. All he outcomes of calculations are given in Table 9.4.

7. Now it is time to calculate the unconditional system parameters. Applying formulas similar to equations 9.15, with the modification that we have 9 loading machines here (previously were 4), the unconditional probabilities can be estimated. The results of these computations are also given in Table 9.4. Therefore, the probability distribution of the number of trucks loaded by the system of loading machines can be evaluated and this is shown in Figure 9.9.

The calculations of further system parameters can now be made.

The mean number of trucks at loading machines is determined by:

Table 9.4.   Parameters of system ℮ without hoist.

	Conditional parameters						Conditional probabilities								
m, r	E_{wkd}	d	$P_{kd}^{(p)}$	$Z'_n (d\,\tau)$	E_{wd}	$E_{wd}/9$	P_{3d}	P_{4d}	P_{5d}	P_{6d}	P_{7d}	P_{8d}	P_{9d}	$P_{>9}$	Δ_d
trucks	trucks	shovels able to load		min	trucks	trucks/loading machine									min
1	2	3	4	5	6	7	8	9	10	11	12	13	14	15	16
105, 28	101.6	9	0.316	2.0	6.74	0.75	0.001	0.019	0.112	0.288	0.334	0.184	0.050	0.011	0.0
		8	0.388	2.4	8.3	0.92		0.001	0.016	0.084	0.218	0.289	0.202	0.190	0.2
		7	0.212	2.8	12.76	1.42				0.007	0.033	0.078	0.103	0.777	0.5
		6	0.067	3.2	22.37	2.49						0.003	0.005	0.991	0.2
		5		3.6	31.14	**3.46**								1	
95, 38	94.9	5'	0.014	3.6	24.38	**2.71**								0.998	0.1
85, 48	85	4	0.002	4.0	21.55	2.39								1	0.0
														Σ	1.0

	Unconditional parameters					
T_C	E_w	E_{wlk}	E_{wk}	θ	W_{efk}	W_{efw}
min	trucks	trucks	trucks	trucks/loading machine	loaded trucks/h	truck work cycles/h
31.6	101.4	7.8	9.9	1.1	195.0	195.2

	Unconditional probabilities					
P_4	P_5	P_6	P_7	P_8	P_9	$P_{>9}$
0.006	0.042	0.125	0.197	0.187	0.122	0.321

E_{wd} – the conditional mean number of trucks at d shovels able to load
$E_{wd}/9$ – the conditional mean number of trucks at 9 machines able to load
P_{bd} – the conditional probability that there are b trucks at d loading machines
$p_{>9}$ – the conditional probability that there are more than 9 trucks at 9 loading machines
Δ_d – the conditional time loss parameter
T_c – the mean truck work cycle including losses
E_{wkd} – the conditional mean number of trucks in a work state
E_w – the mean number of trucks in a work state

E_{wlk} – the mean number of loaded trucks
E_{wk} – the mean number of trucks at loading machines
θ – the mean number of trucks per one loading machine
W_{efk} – the loading machine system's effective productivity
W_{efw} – the truck system's effective productivity
p_b – the probability that there are b trucks at 9 loading machines; $b \leq 9$
$p_{>9}$ – the probability that there are more than 9 trucks at 9 loading machines

Remark: Values of parameters in bold were the grounds to change the organisation of the system.

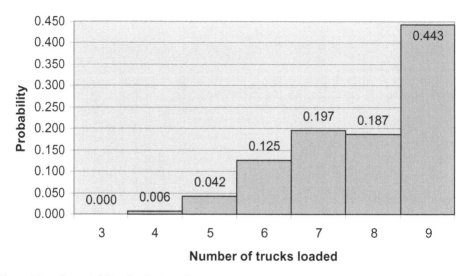

Figure 9.9.   The probability distribution of the number of loaded trucks.

$$E_{wk} = \sum_d P_{kd}^{(p)} E_{wd} \qquad\qquad (9.34)$$

and in the case being investigated equals 9.9.

Remember, two system parameters depend on decisions made by a truck dispatcher:

– the mean number of trucks in a work state
– the mean number of trucks at loading machines.

The average number of loaded trucks by the ore loading machines is:

$$E_{wlk} = \sum_d dp_d = 7.8$$

The average number of trucks per loading point is:

$$\theta = \frac{1}{9} \sum_d P_{kd}^{(p)} E_{wd} = 1.1$$

The mean time of a truck cycle including losses is:

$$T_c = Z_w + T_l + \Sigma\Delta_d = 31.6 \text{ min}$$

The mean number of trucks in a work state in the system is:

$$E_w = 101.6(0.316 + 0.388 + 0.212 + 0.067) + 94.9 \times 0.014 + 85 \times 0.002 = 101.4$$

Two productivities of the system will be considered:

• the loading machine system's effective productivity

$$W_{efk} = \frac{60}{Z_w} E_{wlk} = 195 \text{ trucks/h}$$

- the truck system's effective productivity

$$W_{efw} = \frac{60}{T_c} E_w = 195.2 \text{ truck work cycles/h}$$

It seems that we can presume that the productivity of the system should be approximately 195 trucks/h.

Let us now consider the case with the hoist installed in the system.

8. Let us start with the calculation and analysis of the truck fraction $r_\tau(d)$ taken by the hoist using formula 9.31. Let us present it as a percentage. The outcomes of the calculation are as follows:

$$r_\tau(d = n) = 5.4 \quad r_\tau(d = n - 1) = 6.5 \quad r_\tau(d = n - 2) = 7.5 \quad r_\tau(d = n - 3) = 8.6$$
$$r_\tau(d = n - 4) = 9.7 \quad r_\tau(d = n - 5) = 10.8 \quad r_\tau(d = n - 6) = 11.9$$
$$r_\tau(d = n - 7) = 12.9 \quad r_\tau(d = n - 8) = 14 \quad r_\tau(d = n - 9) = 15.1 \qquad \%$$

Looking at these figures at least three conclusions can be drawn.

- The hoist will take only approximately 5% to 15% of the stream of trucks circulating in the mine (this inference will be improved later).
- The more power shovels that are in failure, the greater the number of trucks that will travel by the hoist. The explanation is straightforward: the more power shovels in failure, the longer the mean time of loading; the greater the number of trucks at loading points, the fewer the number of travelling haulers.
- As the number of trucks travelling by the hoist increases, the mean time of truck travel decreases, but there will be increasing losses in the time of the truck's work cycle due to longer standstills caused by trucks waiting for loading.

9. Calculate the mean time of truck travel. Having estimations of the truck fraction taken by the hoist, an assessment of the mean time of truck travel can be made. By applying formula 9.32, all values of interest can be obtained. The results of the computations are given in column 8 of Table 9.5.

10. The next parameters to evaluate are the standard deviations corresponding with the average values. An estimation of these can be obtained by applying equation 9.33. Table 9.6 contains the outcomes of the calculations. Because the results are a little surprising—the standard deviation increases in spite of the fact that more trucks travel by means of the hoist—the outcomes of computations of particular components of equation 9.33 are shown in Table 9.6.

It is worth noting that only the first component in the standard deviation calculations decreases when number of trucks circulating decreases, which could be expected. Conversely, the remaining components increase with the growing importance of the hoist.

11. Now apply the Sivazlian and Wang model to obtain the parameters of the machinery system at loading machines.

We commence investigations by presuming that all power shovels are up and this means that we have the shortest mean time of loading. We use formulas 9.5 to 9.7 keeping in mind the new value of the mean as determined by formula 9.32 and given in column 8 of Table 9.5 for all cases being investigated. The corresponding standard deviations specified by formula 9.33 are shown in the second column of Table 9.6. These values obviously change when the hoist is used. Next, formulas 9.8 to 9.13 are applied and the results of the calculations are inserted into Table 9.5.

Table 9.5. Parameters of the system ℰ with hoist.

Conditional parameters / Conditional probabilities

m, r	E_{wkd}	d	$P_{kd}^{(p)}$	$Z'_n(d,\tau)$	$r_\tau(d)$	$T_\tau(d)$	E_{wd}	$E_{wd}/9$	P_{3d}	P_{4d}	P_{5d}	P_{6d}	P_{7d}	P_{8d}	P_{9d}	$P_{>9}$	s_9	Δ_d
trucks	trucks	shovels able to load		min	%	min	trucks	trucks/loading machine									trucks	min
1 – 2	3	4	5	6	7	8	9	10	11	12	13	14	15	16	17	18	19	20
105, 28	*101.6*	*9*	*0.316*	*2.0*	*5.4*	*27.2*	*6.98*	*0.78*	*0.001*	*0.081*	*0.249*	*0.343*	*0.224*	*0.072*	*0.019*	*2.04*		
cor. 105, 28	101.6	9	0.316	2.0	7.0	26.9	7.06	0.78	0.001	0.075	0.235	0.338	0.234	0.082	0.025	2.00	0.0	
		8	0.388	2.4	6.5	27.0	8.94	0.99		0.008	0.051	0.158	0.253	0.216	0.314	0.76	0.3	
		7	0.212	2.8	7.5	26.8	15.95	1.77		0.001	0.002	0.011	0.029	0.044	0.913	0.06	0.5	
85, 48	85	6	0.067	3.2	8.6	26.6	26.75	2.97					0.001	0.001	0.999	0.00	0.2	
		5	0.014	3.6	9.7	26.4	18.94	2.1						0.003	0.996	0.00	0.0	
		4	0.002	4.0	10.8	26.2	25.98	2.89							≅ 1	0.00	0.0	
																Σ	1.0	

Unconditional parameters

T_C	E_w	E_{wlk}	E_{wk}	θ	$\bar r_\tau(d)$	W_{efk}	W_{efw}
min	trucks	trucks	trucks	trucks/loading machine	%	loaded trucks/h	truck work cycles/h
30.3	101.2	8.1	11.2	1.2	7.1	202.5	200.8

Unconditional probabilities

P_4	P_5	P_6	P_7	P_8	P_9	$P_{>9}$
0.004	0.027	0.094	0.170	0.179	0.120	0.405

E_{wd} - the conditional mean number of trucks at d shovels able to load
$E_{wd}/9$ - the conditional mean number of trucks at 9 machines able to load
p_{bd} - the conditional probability that there are b trucks at d loading machines
$p_{>9}$ - the conditional probability that there are more than 9 trucks at 9 loading machines
Δ_d - the conditional time loss parameter
T_C - the mean truck work cycle including losses
E_{wkd} - the conditional mean number of trucks in work state
E_w - the mean number of trucks in work state

E_{wlk} - the mean number of loaded trucks
E_{wk} - the mean number of trucks at loading machines
θ - the mean number of trucks per one loading machine
W_{efk} - the loading machine system effective productivity
W_{efw} - the truck system effective productivity
p_b - the probability that there are b trucks at 9 loading machines; $b \leq 9$
$p_{>9}$ - the probability that there are more than 9 trucks at 9 loading machines
$r_\tau(d)$ - the stream of trucks taken by the hoist
$\bar r_\tau(d)$ - the mean stream of trucks taken by the hoist

Remarks: outcomes in grey (rather than solid black) are before correction (cor.) in calculations

Table 9.6. Results of the calculations of standard deviations $\sigma_\tau(d)$.

d	$\sigma_\tau(d)$	$\sigma_\tau^2[1 - r_\tau(d)]$	$\sigma_\tau^2 r_\tau(d)$	$[T_t - T_\tau(d)]^2[1 - r_\tau(d)]$	$[T_{cT} - T_\tau(d)]^2 r_\tau(d)$
p. shovels	min	min²	min²	min²	min²
n	14.24	185.44	0.49	0.91	15.98
$n-1$	14.28	183.33	0.58	1.3	18.74
$n-2$	14.32	181.21	0.68	1.74	21.36
$n-3$	14.35	179.1	0.78	2.25	23.85
$n-4$	14.38	176.99	0.87	2.81	26.2
$n-5$	14.41	174.88	0.97	3.43	28.42
$n-6$	14.44	172.76	1.07	4.1	30.51
$n-7$	14.46	170.65	1.16	4.82	32.48
$n-8$	14.48	168.54	1.26	5.59	34.32
0	14.5	166.43	1.36	6.4	36.04

Now, we are in possession of information on the losses connected with a lack of trucks at loading points.

The conditional average number of non-loaded haulers by loading machines is determined by a simple relationship:

$$s_d = \sum_{j=1}^{n} j p_{(n-j)d} \tag{9.35}$$

which, for the case being considered (when all power shovels are in a work state, gives 2.04. It is a high value and for this reason our previous calculations must be corrected and repeated.

The effective output of the loading system for d power shovels able to execute their duties can now be determined as:

$$W_{efd=nk} = \frac{n - s_n}{n} W_{pefik} \text{ trucks/h} \tag{9.36}$$

where W_{pefik} is given by formula 9.30.

Note that for $d = n$ we have:

$$W_{pefik} = 270 \text{ trucks/h}$$

whereas actually:

$$W_{efd=nk} = 208.8 \text{ trucks/h}$$

which means 23% are down.

Calculate the new truck fraction taken by the hoist. Now:

$$r_\tau(d = n) = \frac{W_{pefT}}{W_{efd=nk}} = 0.07$$

This means that the hoist will take 7% of the stream of travelling trucks, a slightly higher percentage than was previously assessed.

By duplicating the calculations, new outcomes are achieved and these are shown in Table 9.5. The last column of this table illustrates the mean time losses in a truck's work cycle due to the truck waiting for loading.

Having considered the case when all power shovels are in a work state, now it is time to consider the case where one power shovel is down. The line of reasoning remains the same. The results of the calculations are given in Table 9.5.

Notice that at the end of the calculations information on parameter s_d is obtained indicating that this loss is small. Consider now the ratio:

$$\frac{9 - s_8}{s_8} = 0.92$$

This result informs us that 92% of the service capabilities of the loading machines are now being utilised by hauling machines. This result is quite good and it can be presumed that there is no need to recalculate this case.

The next case to analyse is the situation where two power shovels are in failure. The results of the calculations are given in the next row of Table 9.5. It is easy to see that the loading points now are heavily occupied—in 96% of the cases all loading machines will be busy. This is the reason why there is no need to recalculate this case. The time loss is a little greater (0.5 min) than in the case with one power shovel in failure but still very small.

Now it is time to consider a situation where three power shovels are down. Because the truck dispatcher does not interfere and the mean time of loading increases (as more wheel loaders are in action), the number of haulers at loading points increases. With a probability of practically 1 all the loading machines are busy and the mean number of trucks at loading points is now about 3. The time loss is still small. However, it can be presumed that the truck dispatcher evaluates the exploitation situation in the pit as a bit unfavourable—too many trucks are waiting idly at loading points. Therefore, we may presume that when a new and worse situation occurs—four power shovels need repairs—he will decide to withdraw some haulers from the pit.

Thus, when considering the next case, it is assumed that 20 trucks are removed from circulation and the new structure of the system is $<m = 85, r = 48>$. This organisation of the system makes the mean number of haulers in a work state $\cong 85$ due to such a large size of the reserve (proof: the Maryanovitch model). Observe that if the number of trucks can be presumed as constant ($b = \text{const}$), formulas 9.8 to 9.12 are simplified and there is no need to make a summation over b. By applying these formulas, estimates of the basic conditional system parameters are obtained and the outcomes are given in the next row of Table 9.5.

The final case under investigation is a situation where only four power shovels are in a work state. Let us assume that the truck dispatcher does not make any changes in the structure of the system. Following a well-known way of reasoning, the conditional parameters of the system are evaluated and the results of the calculations are shown in the last row of first part of Table 9.5, which concerns conditional probabilities and parameters.

We are not going to analyse further cases because the probability of their occurrence is very small at below 0.001.

12. We are now able to assess the unconditional system parameters. Let us start from a calculation of unconditional probabilities. The set of formulas 9.15 (with a small adjustment in the number of loading points in operation) are useful here giving:

$$p_4 = 0.004 \quad p_5 = 0.027 \quad p_6 = 0.094$$
$$p_7 = 0.170 \quad p_8 = 0.179 \quad p_9 = 0.120 \quad p_{>9} = 0.405$$

Other probabilities are lower than 0.001.

The mean truck work cycle including losses at loading points is:

$$T_c = \sum_d P_{kd}^{(p)}[Z_n'(d,\tau) + \bar{T}_\tau(d)] + \sum \Delta_d \tag{9.37}$$

and $T_c = 30.3$ min.

The mean number of trucks at loading machines is determined by formula 9.34 and here we have:

$$E_{wk} = \sum_d P_{kd}^{(p)} E_{wd} = 11.2 \text{ trucks}$$

Note that this number depends on the decisions of the truck dispatcher.
The mean number of trucks in a work state is determined by the equation:

$$E_w = \sum_d P_{kd}^{(p)} E_{wkd} = 101.2 \text{ trucks}$$

The mean number of loaded trucks is:

$$E_{wlk} = \sum_d dp_d = 8.1 \text{ trucks}$$

Thus, the mean number of trucks per loading machine is:

$$E_{wk}/9 = 1.24 \text{ trucks/loading machine}$$

An interesting parameter is the mean stream of trucks flowing via the hoist, that is:

$$\bar{r}_\tau(d) = \sum_d P_{kd}^{(p)} r_\tau(d) = 7.1\%$$

which means that the hoist will take 7.1% of the total stream on average, but only in its disposal time.

13. Now calculate the output of the system.
    The effective output of the system of loading machines is:

$$W_{efk} = \frac{60}{Z_w} E_{wlk} = 202.5 \text{ loaded trucks/h}$$

This productivity measure includes the reliability of both shovels and trucks and losses due to the lack of a truck at a free shovel.
The truck system's effective output is:

$$W_{efw} = \frac{60}{T_c} E_{wkd} = 200.8 \text{ truck work cycles/h}$$

Looking at these two figures we can presume prudently that the effective system productivity should be approximately 200 trucks/h.

14. Conclusions, recommendations and remarks. These can be divided into particular conclusions and more general remarks.

    Particular conclusions are connected with changes in the values of the basic system parameters as a result of the application of the hoist. These are as follows.

- The mean time of the truck's work cycle was reduced by 4% from 31.6 min to 30.3 min:

  o the mean time of truck travel decreases to 26.9 min
  o the mean time spent in a queue for loading increases due to the greater number of trucks at loading machines.

- The mean number of loaded trucks increases by 4% from 7.8 to 8.1.

- The mean number of trucks at loading machines increases by 13% from 9.9 to 11.2; there is also information about the average truck queue length at loading points.
- The loading machines' effective output and the truck system's effective output increases by 2.6% from 195 to 200 trucks/h.

These are the more important general conclusions and that can be made.

- The effect of the application of the hoist in the system looks small at first glance—the mean time of a truck's work cycle decreases a little (by only 4%), however, this result applies on average to **all** (101) haulers 'continuously' in a work state in the pit.
- This decrease has many important and practical repercussions, including less fuel consumption, a smaller number of used tyres and other spare truck parts, a reduction in heavy traffic in the pit, higher safety, and fewer maintenance problems for mine roads.
- The utilisation of loading machines increases—there are more trucks at loading points (provided that the truck dispatcher does not withdraw some haulers from the pit).
- The productivity of the system increases, but there is the question as to whether this increase is sufficient to recoup the money spent to buy, install and operate the hoist; the time factor is extremely important here.   ◀

Perhaps, as a futuristic idea, one may assess the proposition of applying both an inclined hoist of the TruckLift type and an in-pit crusher with belt conveying for a large open pit enterprise. Today we have the mathematical tools to hand that would allow such a system to be analysed.

9.3   A STREAM OF EXTRACTED ROCK—SHAFT BIN—HOIST

An extremely important problem for every planned underground mine is the design of the hoist that will take the stream of extracted rock from a given underground production level. Here are two critical general problems:

- the designed hoist should be of such a capacity that it can take the total stream of broken rock delivered to the shaft housing the hoisting installation
- the designed hoist, together with the shaft bin, should be of such a capacity that allows for almost undisturbed cooperation with the horizontal haulage systems delivering streams of extracted rock to the shaft.

Let us look at the stream of rock delivered to the shaft. This is the sum of the single streams generated at working stopes or faces. Depending on the manner of rock extraction, two types of streams can be distinguished:

- a continuous stream in which a winning machine of a plough type, shearer or continuous miner are being used
- discrete-and-cyclic in which the winning process is done by blasting.

Notice that the names of these streams indicate what kinds of probabilistic properties are associated with the particular type.

Similarly, haulage of the generated stream can be:

- continuous (transport by belt conveyors, for instance)
- discrete-and-cyclic when rail haulage or tyre transportation is in use.

Let us consider both types of haulage more comprehensively.

Continuous Haulage

As has been proved (see, for example, Manula and Sanford 1967, Bucklen et al. 1968, Wianecki 1974, Pavlović 1988, Antoniak 1990), a stream of rock flowing through conveyor systems can be satisfactorily described by a Gaussian process. This type of modelling is still being used in spite of the fact that some issues associated with this type of modelling still exist.[13]

Let us look at the mathematical tools that will be used in the modelling. For obvious reasons, a normal distribution that can be applied for the imitation of the stream should be truncated double-sided and should have upper and lower limits. The left-hand boundary is obviously zero. The right-hand boundary is the point that determines the physical maximum of the stream that can be handled by the conveyor. In some publications investigating this issue no limits are given for the normal distribution of the stream of extracted rock. Such an approach can generate errors, and sometimes these can lead to incorrect decisions.

Thus, the probability density function of interest is:

$$f_z(z) = \frac{f(z)}{F(b) - F(0)} \quad \text{for } 0 \leq \mathbf{Z} \leq b = \mathbf{Z}_{max}$$
$$f_z(z) = 0 \quad \text{elsewhere} \tag{9.38}$$

where:

$$f(z) = \frac{1}{\sigma_z \sqrt{2\pi}} \exp\left[\frac{(z - \omega)^2}{2\sigma_z^2}\right] \quad \text{for } -\infty < z < \infty \tag{9.39}$$

and:

$$F(z) = \int_0^z f(v)\,dv \tag{9.40}$$

Parameters ω and σ_z have well-known meanings and they belong to the distribution before truncation. Notice in addition that the random variable $\mathbf{Z}$ has a physical meaning and denotes the stream of mass per time unit.

Let us define the two most important statistical parameters associated with the this density function. The following relationships hold (see, for example, Niewiadomska–Kozieł 1972).

- The expected value of random variable $\mathbf{Z}$ is:

$$E(\mathbf{Z}) = \omega + \sigma_z (I_1/I_0) \tag{9.41}$$

- The standard deviation of $\mathbf{Z}$ is:

$$\sigma(\mathbf{Z}) = \sigma_z \sqrt{\frac{I_2 - I_1}{I_0}} \tag{9.42}$$

---

[13] Due to the properties of the characteristic function of a normal process, this type of process can be comprehensively described by a two-dimensional distribution in which an important component is autocorrelation. The author of this book is not familiar with any publication showing results of statistical investigations concerning either the identification of a two-dimensional distribution or autocorrelation. Moreover, some publications (for example, Firganek 1973) apply different types of modelling functions with a memory of an autocorrelation type in their structure. And what is more, according to Zur (1979 p. 132) a stream of rock flowing through conveyors is a mixture of a normal process (the conveyor is loaded) and many zeros when there is no mass on the belt.

where:

$$I_0 = \Phi_N\left(\frac{b-\omega}{\sigma_z}\right) - \Phi_N\left(\frac{-\omega}{\sigma_z}\right)$$
(9.43)

$$I_1 = \phi_N\left(\frac{\omega}{\sigma_z}\right) - \phi_N\left(\frac{b-\omega}{\sigma_z}\right)$$
(9.44)

$$I_2 = I_0 + \frac{-\omega}{\sigma_z}\phi_N\left(\frac{\omega}{\sigma_z}\right) - \frac{b-\omega}{\sigma}\phi_N\left(\frac{b-\omega}{\sigma_z}\right)$$
(9.45)

and:

$\phi_N(x)$ is the standardised density function of random variable $N(0, 1)$
$\Phi_N(x)$ is the standardised distribution function of random variable $N(0, 1)$.

Notice that the total inconsistency mass when comparing the truncated and the non-truncated distributions is determined by the formula:

$$\Omega = \int_{-\infty}^{0} f(z)dz + \int_{b}^{\infty} f(z)dz$$

This mass is taken from the non-truncated distribution to get the truncated distribution, but the total mass is still closed to unity. In professional publications it has been stated (see, for example, Firkowicz 1970) that the application of a non-truncated distribution can satisfactorily describe empirical data if the inconsistency mass is relatively small, say, 5% or less of the total mass.

Let us denote by $J_i(t)$ a Gaussian process describing a stream of mineral flowing from the i-th working face. This process can be expressed as

$$\{J_i(t), 0 \le J_i(t) \le J_{i\,max}, \ t \ge 0; \ N(\bar{J}_i, \sigma_i)\}$$

If all of the streams from working faces are continuous (haulage by belt conveyors), then the total stream of the mass delivered to the shaft is also Gaussian. The average value of this total stream is the sum of all inflowing processes, and the variance is the sum of all variances. Thus:

$$\left\{J(t) = \sum_i J_i(t), \ t \ge 0; \ N_J\left(\sum_i \bar{J}_i \cdot \sqrt{\sum_i \sigma_i^2}\right)\right\}$$

Notice that when considering one working face, the limits of truncation can be important. However if the total stream is considered, the significance of truncation decreases. The mass of probability is shifted away from zero, indicating that the left-hand boundary can be neglected in the majority of cases. In addition, the more working faces generating streams, the less the probability that the total stream will be zero. Obviously, the maximum stream that can be taken by the main conveyor line must be appropriately greater.

Let us now consider as a time unit a 24 hour day. If so, we should analyse the following relationship:

$$Z = \int_t \sum_i J_i(t)dt, \quad t \ge 0$$

From a theoretical standpoint, finding the probability distribution of the random variable Z, which is a stochastic integer, from the executions of random processes is extremely difficult in spite of the fact that here we have well-known Gaussian processes and that, in addition, they are stationary ones. However, the situation is not hopeless. In several publications focusing on the properties of the total daily output from a given production level, called the *production function* (see, for example, Sevim and Qing Wang 1988a), it has been stated that this random variable can be satisfactorily described by the normal distribution. Let us presume that this random variable is determined by formula 9.38 together with its accompanying equations. Observe that this variable is a conditional one from an engineering standpoint—provided that the total stream generated by a certain device is being received by another piece of equipment. It is also important to add that the effects of the reliability of the means of haulage should be included in this variable.

Cyclic haulage

We will investigate classic rail transportation. Let us look at what happens at an unloading station where the mass from mine cars are dumped into a bin. Taking a long period of time into account, we can identify the process of unloading as a two-state one: unloading—lack of unloading (see Figure 9.10). This is a well-known process in reliability: it is a renewal process with a finite time of repair (see Chapter 5.1.2).

As mine investigations have shown these two random variables can be described by normal distributions. Based on this statement, it is easy to conclude that the sum of executions of random variable T_{un} is obviously normally distributed.

Summarising, it can be assumed that the stream of mineral delivered to a shaft daily is a random variable of the Gaussian type.

Now the problem is to consider how to model, analyse and calculate the remaining part of the system which includes a hoist and a shaft bin—and how to find the proper volume of this rock accumulator.

We will neglect here the procedure of the selection and design of the hoisting installation. This issue requires a separate and lengthy discussion, perhaps as comprehensive as this volume. Besides, this problem is far beyond the scope of this book. Nevertheless, generally speaking, in order to fulfil safety and technical requirements, we should search for a hoisting system—having the appropriate information on the production level capacity—that will *minimise the predicted cost associated with the hoisting of one tonne of rock per unit of time in the long term*. This differs slightly from the criterion formulated by Sevim and Qing Wang (1988a) which stated that the system being considered should *minimise the expected cost per tonne of ore hoisted*. At first glance this latter criterion looks clear and easily comprehensible. However, this statement is unusable for two reasons. First, when random phenomena are considered—and here we have such a case—the expression 'expected' is usually connected with the mean value (mathematical hope). Therefore, the term 'expected' should be replaced by, say, 'predicted' or 'forecasted'. Second, a time factor is missing. However, the formulas presented in the cited papers do include time: in

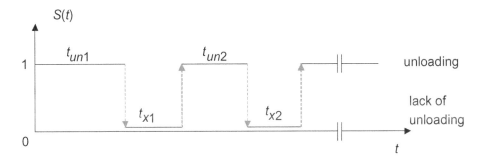

Figure 9.10. Process of mass flow at an unloading station delivering rock to a shaft bin.

Sevim and Qing Wang 1988a it is a year; in the second paper (Sevim and Qing Wang 1988b) it is a day. It is easy to see that bin volume sizing needs a day as the time unit. Let us now look at the first criterion. One element needs explanation: *cost associated with*. There are at least two reasons to include this expression. Initially, if the level production rate is greater than the hoisting capacity enlarged by the bin over a certain period, some working faces will be stopped causing losses. If rail haulage is applied at the level, or if there are a number of bunkers in the conveyor system, the probability of the occurrence of stoppage at a face is reduced. Nevertheless, possible production losses must be included in the criterion. Second, sizing and the cost of constructing a shaft bin is not directly connected with hoisting capacity. Bin size is first of all a function of the parameters of the hoist and the production rate. All these costs directly and indirectly connected with hoisting are named as 'associated costs' here.

Let us return to the main line of reasoning. A stream of broken rock being conveyed to a shaft must be taken by the hoist. This is a cyclically operated means of transport and, for this reason, a bin must be constructed before the entrance to this installation. Let us look at the transport possibilities of the hoist.

The theoretical hoist output (maximum) is determined by this formula:

$$W_{th}^{(h)} = \frac{Q_u}{T_c^{(h)}} \tag{9.46}$$

where:

Q_u is the skip payload
$T_c^{(h)}$ is the hoist time work cycle.

This productivity can be achieved by the hoist provided that there is no interruption during the hoist's accomplishment of the transportation task and that there is always a mass to wind.

The effective hoist output can be evaluated as:

$$W_{ef}^{(h)} = W_{th}^{(h)} E(X) A_h \tag{9.47}$$

where:

A_h is the hoist steady-state availability
$E(X)$ is the average value of the hoist's utilisation per day.

The effective hoist output is a deterministic value. This is the average productivity of the installation that can be expected to be achieved over the long term. However, it is a function that is dependent on the random variable X for which the expected value is a component of the right-hand side of equation 9.47. This is correct because the daily hoist output W is a random variable determined as:

$$W = W_{th}^{(h)} A_h X \tag{9.48}$$

The theoretical hoist output is constant. The steady-state availability of the hoist is also a deterministic value. Thus, the whole stochastic nature of the process of hoist's operation is associated with the daily utilisation of the hoist as expressed by the random variable X. Because these first two components of the right-hand side of formula 9.48 are deterministic we can join them together. This new hoist parameter is in fact the corrected hoist theoretical output; corrected by the hoist's steady-state availability. We presume now that from this moment all theoretical hoist productivities in this chapter will be corrected—that is, they will be the productivity that is a product of the hoist's theoretical output and the hoist's steady-state availability.

The value of X on a particular day is the ratio of the total time spent on the hoist accomplishing the transportation task (the numerator, a random variable) to the disposal time of the hoist (deterministic value). Note that the time during which hoisting is halted can be caused by different factors—sometimes it is a lack of rock, sometimes it is a failure of the device receiving the load from the skip, in some cases a measure-pocket is under repair, etc. Long ago it was stated (see, for example, Czaplicki 1977, and also Chapter 5.3) that a hoist's daily utilisation can be described by a beta distribution. Recalling formulas 5.31 we have:

$$X : Be(a, b) \quad a > 0 \; b > 0 \quad 0 \leq x \leq 1$$

$$f_X(x) = \frac{\Gamma(a+b)}{\Gamma(a)\Gamma(b)} u^{a-1}(1-u)^{b-1}$$

and the basic statistical parameters connected with this random variable are given by formula 5.32. We purposely repeat the mean here, which is:

$$E(X) = \frac{a}{a+b} \tag{9.49}$$

and an example plot of the function $f_X(x)$ (see Figure 9.11).

Keeping in mind relationship 9.48 and taking into account information on the random variable X, we can state that the probability density function of the random variable W can be determined as the modified beta distribution of the probability density function:

$$f_W(w) = \frac{\Gamma(a+b)}{\Gamma(a)\Gamma(b)} \frac{1}{W_{th}^{(h)}} \left(\frac{w}{W_{th}^{(h)}} \right)^{a-1} \left(\frac{W_{th}^{(h)} - w}{W_{th}^{(h)}} \right)^{b-1} \quad 0 \leq w \leq w_{th} \tag{9.50}$$

The problem that is of the greatest importance for the production design of a level should be the proper selection of the relationship between two density functions $f_Z(z)$ and $f_W(w)$. During the design process basic hoist parameters are selected: the skip payload Q_u and kinematic characteristics (travel speed, acceleration and deceleration) that determine the hoist time work cycle for a given depth of wind. In this way the maximum hoist output is determined. The variable X, in turn, can be determined during the hoist's operation, but the designer should have some idea of the mean value of X and—better—some idea of the corresponding standard deviation. These values can be based on mining practice—taken from the values of these parameters for comparable hoists operating in similar conditions. Usually, the designer can forecast the values of the mean and the standard deviation of the random variable X and can estimate the structural probability function parameters.

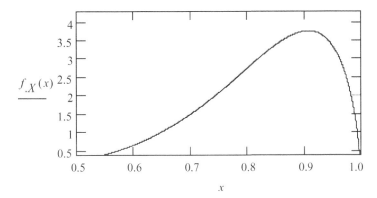

Figure 9.11. An example plot of the probability density function of hoist utilisation.

In practice, an experienced designer usually looks at the average forecasted production and at the depth of the wind which largely determines the hoist time work cycle, and he proposes the skip payload based on these two most important parameters. As a rule his knowledge in this regard is developed over many years' practice. Typically, he does not go deeper in his reasoning. Usually, his suggestion is quite a good one, but sometimes not well-aimed.

It is worth remembering that the density functions $f_Z(z)$ and $f_W(w)$ have different characters, especially where their dispersions are concerned (see Figure 9.12).

An idea of the system is shown in Figure 9.13.

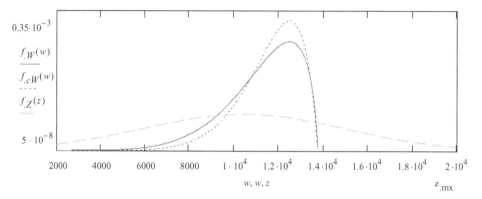

Figure 9.12.   The probability density functions of the daily production $f_Z(z)$, the daily transportation capacity of the hoist $f_W(w)$ and the corrected daily transportation capacity $f_{cW}(w)$.

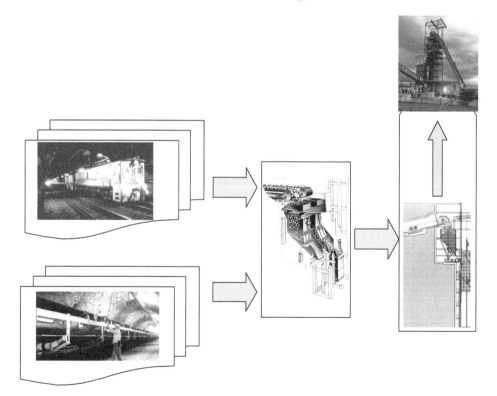

Figure 9.13.   A horizontal haulage—shaft bin—hoist system.

Let us now compare both probability properties that characterise the two types of haulage: horizontal and vertical. First, consider the following random variable:

$$U = Z - W \mid Z > W \quad 0 \le u \le z_{mx} \tag{9.51}$$

This is a conditional random variable characterising cases where production exceeds the transporting capacity of the hoist. To analyse this variable we need an auxiliary unconditional random variable:

$$\Xi = Z - W \quad -w_{th} \le \xi \le z_{mx} \tag{9.52}$$

The probability distribution of Ξ is obviously given by the formula:

$$F_{\Xi}(\xi) = \int_0^{w_{th}} \int_0^{\xi+w} f_Z(z) f_W(w) dz dw \tag{9.53}$$

Hence:

$$P(W > Z) = F_{\Xi}(0) \tag{9.54}$$

This measure multiplied by 100 tells us for what percentage of cases the transport capacity of the hoist will be greater than the stream of rock delivered to the shaft.

Now, we can define the probability density function of the conditional random variable U. We have:

$$f_U(u) = f_{\Xi}(u) \, / \, [1 - F_{\Xi}(0)] \tag{9.55}$$

Notice the expected value $E(U)$ is the conditional mean that gives the following information: if the delivered stream of rock is greater than the stream of rock hoisted (all per day), then the mean value of this excess equals $E(U)$.

Note as well a subtle problem. When analysing the problem of the economic evaluation of hoists, Sevim and Qing Wang (1988a, 1988b) stated that the production function $f_Z(z)$ sometimes has a long right-sided tail. They stated that in such a case the recommended hoist output can be very large. From a theoretical standpoint this reasoning is correct. However, from a practical perspective it does not work. If—based on the operational data—the function $f_Z(z)$ has a very high value of z_{mx} (a long right tail), this means that the haulage system has been over-dimensioned and such a case should be avoided in every instance because it means that the majority of transport drives of the production level being considered will be too large. In practice, frequently $z_{mx} \approx E_z + 2\sigma_z$. On the other hand, the density function $f_U(u)$ usually has a long right-sided tail (see Figure 9.14), has exponential character and indicates a large volume bin. This problem will be analysed later.

We are now able to evaluate the following average value:

$$\omega_0 = E(Z) - E(U) \, [1 - F_{\Xi}(0)] \tag{9.56}$$

which is the conditional mean of the hoist's output provided that the output of the flow of ore is greater than the haulage capacity of the hoist.

A parameter of great importance is the expected mass of rock hoisted out from the production level. This is determined by the formula:

$$L = E(Z) \, F_{\Xi}(0) + (1 - F_{\Xi}(0))[E(Z) - E(U)(1 - F_{\Xi}(0))] \tag{9.57}$$

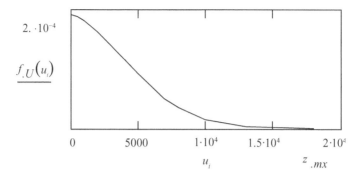

Figure 9.14. An example plot of the conditional probability density function of random variable *U*.

Observe that:

- the extracting-and-hauling system operating on the production level in relation to the existing mining and geological constraints *generates* the total stream of rock mined that should be delivered to the shaft; this stream is characterised by the density function $f_Z(z)$ with the mean $E(\mathbf{Z})$
- the hoist system is characterised by the transport capacity of function $f_W(w)$ with the mean $E(\mathbf{W})$
- constructing a system consisting of these two elements will result in the mean output of this system being *L*, where, alas, $L < E(\mathbf{Z})$ and $L < E(\mathbf{W})$, which should however have been expected.

Now assess what portion of the ore in percentage terms will be taken by the hoist on average. Construct a measure:

$$\chi_1 = 100 \, L/E(\mathbf{Z}) \tag{9.58}$$

We can also calculate the average rate of ore that will not be hoisted daily, that is:

$$\psi_1 = E(\mathbf{Z}) - L \tag{9.59}$$

However, it is enough to spend

$$T_{CL} = \frac{\psi_1}{Q_u} \frac{T_c^{(h)}}{60} \quad \min \tag{9.60}$$

to clear up the bin, on average. Some comments on this residual mass in the bin will made at the end of these considerations.

Let us consider the problem of bin volume *y*. The problem of shaft bin volume selection is rarely considered in professional publications. This was evident in Sevim and Qing Wang (1988b) and it would appear that this situation is constant. The reasons cited by Sevim and Qing Wang for not considering this problem were that hoisting systems are considered to be customised rather than standardised design products and the topic is so specialised that only a few manufacturing and consulting companies have developed the required expertise, which is proprietary. However, where bin sizing is concerned, one very important issue was missed by Sevim and Qing Wang (1987, 1988a, 1988b)—an analysis of the process of the inflow of the mass into this type of container, the temporary accumulation and reduction of the mass in the bin and the process of the outflow of the mass. This is a stochastic process of a very complicated nature with memory and lagging. For this reason, the analytical models applied here are more or less simplified. Sometimes this phenomenon is simply incomprehensible to some researchers, which makes their reasoning false (see, for example, Jaśkowski 1999). The nature of the process and the difficulties of identifying it adequately indicate that a good solution is the application of a simulation method to observe changes in the mass in a bin or to reproduce the operation of the whole haulage system. Simulation has proved its usefulness and is commonly applied in many countries (see, for example, Wianecki 1974, Zaikang 1985, Lineberry and Patsey 1987, Webster 1989). However, simulation has its own demerits (such as the lack of the possibility to estimate an error, and it requires, in some cases, a large number of probability distributions which can only be estimated with limited accuracy). For some specific drawbacks of simulation in connection with belt systems design see Thompson and Webster 1988, and Sevim and Qing Wang 1990a, 1990b.

A bin sizing problem can be investigated by presenting:

- an engineering approach, historically the oldest solution (see, for example, Peele 1945, Czechowicz 1952), based on engineering rules and hints gained from long experience of mining practice
- a simulation approach, historically the 'latest' arrival, based on an imitation of rock flow from production faces through the means of haulage to a bin, and sometimes to a further point of destination
- an analytical approach, based on mathematical models of various natures (see, for example, Czaplicki and Lutyński 1987, Sevim 1987, Sevim and Qing Wang 1988a, 1988b).

Generally, it looks as though when searching for a logical volume of a shaft ore accumulator at least four suppositions can be formulated.

a. Looking at the relationships 9.51 to 9.55, we can make the suggestion that y should be in relation to the conditional mean value $E(U)$, that is:

$$y = cE(U) \quad c \geq 1 \tag{9.61}$$

Keeping in mind that the probability function of the random variable U can be approximated by an exponential distribution and if $c = 1$ is assumed, such a bin volume allows the bin to only take approximately 63% of the cases when the stream of rock delivered exceeds the transportation capacity of the hoist (a property of exponential distribution).

Instead of relationship 9.61, the following equation can be taken into account:

$$y = E(U) + \sigma(U) \tag{9.62}$$

and such a bin volume permits the bin to take approximately 85% of the cases when the stream of rock delivered exceeds the transportation capacity of the hoist (again, a property of exponential distribution).

b. Bin volume should be in relation to the mean time of hoist repair T_n, bearing in mind the level's productivity, for example:

$$y \geq [E(Z) + \sigma(Z)] \, (T_n/T_d) \tag{9.63}$$

and this will allow approximately 83% of these cases (stream rates) to be taken (here a property of normal distribution is included). Generally, hoist failures halting the execution of the transportation task are not so frequent (less than one per month for modern hoisting installations). Note that this criterion differs from the two in supposition (a). These are based on everyday practice, whereas the criterion expressed in relationship 9.63 concerns relatively rare events.

c. A modification of supposition (b) can be considered. Presume that the stream of rock is average in value but the time of repair is extended to the mean value of repair plus the corresponding standard deviation. Presume a very inopportune case where the probability distribution of the times of repair is exponential and in such an instance the mean and the standard deviation are identical. If so, formula 9.63 is modified to:

$$y \geq (2T_n/T_d)E(Z) \tag{9.64}$$

d. Bin volume should accommodate the average rock mass that is delivered by the haulage system during n, $n \in \mathfrak{R}$, hours of its operation, that is:

$$y = n \, E(\mathbf{Z}) \qquad (9.65)$$

A few brief comments are necessary in relation to this last supposition. This stipulation is obligatory during design in many mining countries. The existing differences in this regard are connected with the value of n. In some cases n comprises a half shift, sometimes a whole shift. This principle is simple, and easy to remember, understand and apply. However, it ignores the relationship between the stream of rock delivered to the shaft and the hoisting capacity of the installation entirely. Thus, in some cases y determined equation 9.65 presuming n according to mining regulations in operation—which is a safe engineering solution—can be too large, and in some cases not enough.

Note that a few important questions can now be formulated.

- Which values should be preferred? Perhaps those values concerned with the majority of cases?
- What about those solutions that favour a *sparing* approach?
- What are reasons for deciding whether a bin problem is more or less important?

Let us start by answering the last question. We have an intuitive feeling that the more important the bin problem, the greater the tendency to prefer a solution accommodating more cases. On the other hand, one statement seems indisputable: the greater the hoist output in relation to the level of production, the less important will be bin problem (as determined by equation 9.61). This will be proved in Example 9.4. Therefore, the problem of the proper relationship between functions $f_Z(z)$ and $f_W(w)$ is the key issue here. Conversely, stipulation (b), which considers the hoist's reliability, does not appear to be affected by the problem of the relationship between these density functions. Thus, it seems that the following statements can be formulated:

- if a hoist's output is high in relation to the production function, the criterion expressed by equation 9.63 should be more important
- if a hoist's output is not so great in relation to the production function, the criteria expressed by equations 9.61 or 9.62 should decide the bin volume.

Now, a new problem arises. What kinds of measures should be applied to assess normatively the statement: *a hoist's output is high in relation to the production function*? We have two different probability distributions here. It appears at first glance that the expected values of both random variables should be compared. For this reason a measure $\beta = E(\mathbf{Z})/E(\mathbf{W})$ gives us information about the quantity of rock that will be taken by the hoist on average. In practice, the inequality $E(\mathbf{Z}) > E(\mathbf{W})$ holds. The lower the value of β, near 1, the smaller the hoisting capacity of the installation in relation to the level production. In such cases the bin volume should be large. Observe that the comparison of two probability distributions should not only contain a judgement of their mean values. From a theoretical standpoint a comparison of the corresponding standard deviations should also be done. However, it is obvious that $\sigma(\mathbf{Z}) \gg \sigma(\mathbf{W})$. Often, the coefficient of variation for the random variable $\mathbf{Z}$ is three times greater than that for the random variable $\mathbf{W}$. It is also worth noticing that while the designer has some influence on the mean values, he has almost no influence on the dispersions. Thus, a practical hint can be formulated: having no influence on the value of the dispersion, most attention should be paid to the dispersion of production. For a high value of the standard deviation of $\mathbf{Z}$ the bin volume should be large.

Let us now presume that the designer decides that the bin volume will be y tons and that such a bin will be deployed in the system. What kind of repercussions will this have for the system's parameters? Let us analyse three simple probabilistic parameters.

This formula:

$$100 \int_0^y f_U(u)\,du \qquad (9.66)$$

gives us information about how many cases (in percentage terms) will be taken by the hoist due to the application of the bin when production is greater than the hoist's transport capacity.

The distribution of the unconditional random variable Ξ in point y multiplied by 100, that is $100F_\Xi(y)$, gives us information on the percentage of cases in which the transport capacity of the hoist, 'enlarged' by the existence of the bin, is greater than the level's productivity.

The difference $F_\Xi(y) - F_\Xi(0)$ indicates in turn what increase in the transport capacity of the hoist is due to the application of the bin.

Let us now define a random variable R which concerns the bin in the system. We have:

$$R = W + y \quad y \le r \le y + w_{th} \quad f_R(r) = f_W(r - y) \tag{9.67}$$

We can repeat our line of reasoning from the beginning of this point and we have:

$$H = Z - R \mid Z > R \quad 0 \le h \le z_{mx} - y \quad y \le r \le y + w_{th}$$

$$H = Z - W - y \mid Z - W > y \quad H = \Xi - y \mid \Xi > y$$

$$f_H(h) = f_\Xi(h + y) / [1 - F_\Xi(y)] \tag{9.68}$$

and:

$$E(H) = \int_0^{z_{mx} - y} h f_H(h) dh \tag{9.69}$$

This expected value is a measure that informs us of the average mass of rock that cannot be hoisted in a case where the level's production is greater than the transport capacity of the hoist enlarged by the existence of the bin of volume y.

Similarly, the formula:

$$L_B = E_Z F_\Xi(y) + [1 - F_\Xi(y)] [E_Z - E_H(1 - F_\Xi(y))] \tag{9.70}$$

indicates the rock mass that will be transported by the hoist with a bin of volume y in the system.

Like before, we now assess what portion of the ore in percentage terms will be taken by the system hoist plus shaft bin, on average:

$$\chi_2 = 100 \, L_B/E(Z) \tag{9.71}$$

We can also calculate the average rate of ore that will not be hoisted daily, that is:

$$\psi_2 = E(Z) - L_B \tag{9.72}$$

This is the mean of the residual mass remaining after daily production. This is also an interesting system parameter indicating how good the proposed bin volume is. The value of this mean should be appropriately low. Remember that, generally, the magnitude of the residual mass in a bin is a random variable. If the time to wind up this mass is ascribed to this random variable, this time turns to be the discrete counterpart (the number of skips required to remove the residual mass is a natural number) for this mass with a stepped type distribution.

It would seem like all the critical problems of the operation of the stream of ore—bin—hoist system have been discussed, bearing in mind its stochastic nature. But this is not entirely true.

It is obvious that it would be impractical to design a hoist with such a large capacity that all the mass that can be delivered to shaft will always be taken by the hoist. There is no doubt that such an installation would be over-dimensioned. Hence, we can expect that the hoist will be designed

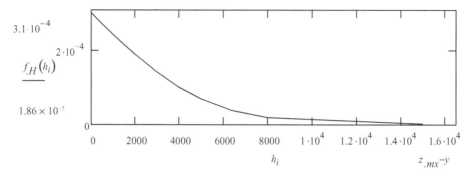

Figure 9.15. An example plot of the conditional probability density function of random variable **H**.

with a lower hoisting capacity and, for this reason, some losses are inevitable. But until now there has been no measure defined to describe this waste.

Look at the random variable **H**. It is a conditional variable describing the mass of rock that is the difference between the stream of rock delivered and the stream of rock hoisted provided that the level's production is greater than the hoisting capacity of the installation plus a bin of volume y. This variable has an upper limit that is $z_{mx} - y$. An example plot of the probability density function $f_H(h)$ is shown in Figure 9.15. This is a picture of the frequency of occurrence of the stream of rock that will not be hoisted, thus illustrating the losses. Observe that this function for the case being considered starts from the frequency 0.0003 and due to this fact we can be sure that such events are so rare that they can be neglected. This is also a hint for Sevim and Qing Wang that it is not necessary to design a hoist with a great capacity.

Now it is time to consider the effect of the fluctuation of the mass in a bin. There are two processes of interest that can be observed in a bin. There is a temporary accumulation of rock mass in the bin when the transport capacity of the hoist is not enough for the productivity of the stream of the mass, and *vice versa*. When the productivity of the stream is low, the accumulated rock mass will start to move out from the bin.

The probabilistic scheme that has been considered up to now was as follows. The total rock mass delivered by the haulage system of the level is z_1 on the first day and the total rock mass hoisted is w_1 on this day. We take into consideration and compare the following pairs: $\{z_i, w_i; i = 1, 2, \ldots\}$. Then the bin is included in the investigations.

Let us now presume a time unit $T_c^{(h)}$ —the hoist work cycle. In this time:

1. the hoist transported Q_u tonnes of rock
2. the haulage system delivered a certain amount ∇q of rock and if:

   a. $Q_u > \nabla q$, it means that there is less rock mass in the bin than before
   b. $Q_u = \nabla q$, it means the total mass in the bin remained unchanged
   c. $Q_u < \nabla q$, it means there is actually more rock in the bin than before.

The last point has limits. The bin has a finite volume and when it is full, point c does not hold. Likewise, when the hoist has taken the last tonnes of rock and the haulage system delivered significantly less than q tonnes of rock, the hoist's operation is stopped.

The problem now arises as to how to include these temporary fluctuations in mass in the calculation procedure.

If we approach this problem from a theoretical standpoint, the process of fluctuations of the mass is a stochastic process with lags, memory and limits. Proper determination and analysis of the problem is extremely difficult. An attempt at a description of this process and its examination in the field of time series analysis is futile. The only effective method of analysis of this kind of process is digital simulation, which has been used in mining for more than forty years. However, simulation has its own well-known demerits. And what is more, if the hoist—shaft bin system is designed in relation to the forecast level of production (all items non-existent at the design stage of a project), all the parameters are unknown and can only be predicted. We therefore have a prediction simulation characterised by a very large dispersion (uncertainty) as a rule. In this kind of situation, there is a high risk that during the design process random errors could accumulate giving doubtful or even incorrect outcomes. Any decisions based on these results would have a high degree of risk.

However, neglecting all of these problems, let us try to estimate the influence of these fluctuations of the mass in a bin.

We commence our considerations from the statement that due to these fluctuations the hoist's transport capacity will be greater, which is obvious. Values of the means $E(U)$ and $E(H)$ will be lower, whereas L and N will be greater. We can try to evaluate these changes taking into account the fact that the effect of fluctuations of the mass is equivalent to some extent to the increment in the degree of utilisation of the hoist. Hoist standstill time caused by a lack of rock to hoist will be reduced. Therefore, a modification of function $f_X(x)$ should be done. The mean value $E(X)$ should increase; a small reduction of the standard deviation $\sigma(X)$ should be made. We can presume that by designing a shaft bin of volume $y = E(Z) + \sigma(Z)$ we eliminate the majority of cases of hoist standstill because of a lack of rock (in some cases $E(Z) + 2\,\sigma(Z) > z_{mx}$).

For this reason the expected value of hoist utilisation can be estimated as:

$$E_c(X) = E(X) + \{y[1 - E(X)]/2[E(Z) + \sigma(Z)]\} \tag{9.73}$$

A small decrease in the corresponding variation may also be made, but this will significantly change the outcomes obtained. All calculations made up until now should be repeated.

Figure 9.16 illustrates the proposed procedure.

Additionally, it is worth having an idea about two regularities:

- if the shaft bin volume has been selected on the grounds of supposition (b), calculations should be repeated up to the moment when we have the mean hoist plus bin system output
- if the shaft bin volume has been selected on the grounds of supposition (a), the bin volume is usually a bit too large; the conditional mean value $E(U)$ will be a little lower in practice due to fluctuations of the mass—the bin volume should not be lower than the one resulting from reliability investigations.

■ Example 9.3

1. There is a given production function from a certain level of an underground mine based on daily performance. This function has these parameters: $\omega = 10500$ t/d, $\sigma = 4500$ t/d, $z_{mx} = 20000$ t/d. This function is shown in Figure 9.12. Because of truncation from both sides, approximately 3% of the total mass of probability was removed. Thus, the expected value dropped to $E(Z) \cong 10420$ t/d and the new standard deviation is also lower at $\sigma(Z) \cong 4130$ t/d. Note, the coefficient of variation $\sigma(Z)/E(Z) = 0.40$; before truncation it was 0.43.
2. The mean value of the utilisation was estimated at $E(X) = 0.82$ and the corresponding standard deviation was assessed at 0.12. Using these estimates to find the assessment of the structural density function parameters, we have $a = 7.58$ and $b = 1.67$. Notice that the maximum of the probability density function (beta) is $(a - 1)/(a + b - 2) = 0.91$. This function is shown in Figure 9.11.
3. Construct the probability density function of the hoist's daily output $f_W(w)$. During the design procedure of the hoist installation for a skip payload $Q_u = 22$ t, the skip time work cycle was assessed at $t_c = 92$ sec. Thus, the theoretical hoist output is $W_{th}^{(h)} \cong 13770$ t/d and the effective output $\cong 11300$ t/d. The density function of W is shown in Figure 9.12. The standard deviation of random variable W is approximately 1650 t/d and the coefficient of variation is 0.15.

First, we undertake calculations of the system without the effect of rock mass fluctuations in the bin.

4. The probability that the hoisting capacity exceeds the stream of mass delivered to the shaft $100P(W > Z) \cong 56\%$ cases. This means that the situation where the production goes above the

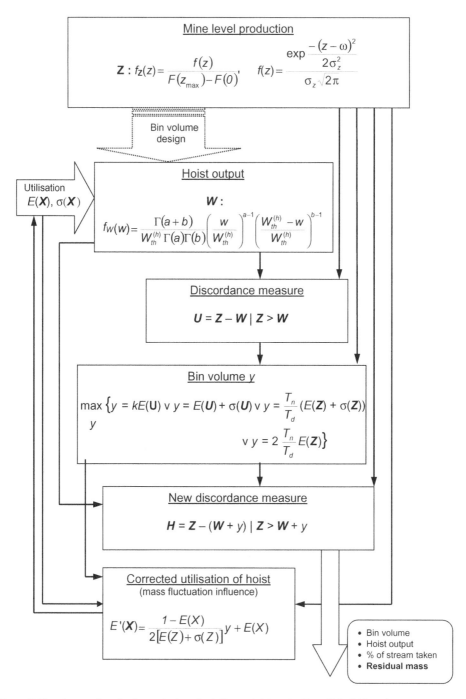

Figure 9.16. A procedure for the analysis, calculation and determination of shaft bin volume.

hoisting capacity will occur in 56% of the cases. The conditional mean value of this excess is $E(U) \cong 3600$ t/d. The plot of the probability density function $f_t(u)$ is shown in Figure 9.14. The conditional mean of the hoist output provided that the output of the flow of ore is greater than the haulage capacity of the hoist is $\omega_0 = 8840$ t/d. The average amount of rock hoisted is $L \cong 9730$ t/d. However, approximately only $\chi_1 = 93\%$ of the mean mass delivered to the shaft will be hoisted out. Converting this high value $E(U)$ into day-to-day practice, the outcome is that about $\psi_1 = 700$ t of rock will not be hoisted daily, on average. It will be necessary to spend $T_{CL} \cong 50$ minutes to wind up this mass. This is obviously the mean time, and it is quite high.

5. Now consider the recommended bin volume. Evaluate reliability.

- In equation 9.63 a case is considered where during the mean time of hoist repair there is such an accumulation of stream flow that the flow rate equals the mean value plus the corresponding standard deviation. The probability that the stream flow will be greater is only 0.17. This looks safe. If the next criterion is considered, due to the property of exponential distribution, the probability of the occurrence of a greater repair time than the one presumed is only 0.14. This also seems quite safe. Let us posit that the mean time of hoist repair is 2.3 h. These two criteria lead to quite different estimations, namely:

  o Formula 9.63: $y > 2100$ t
  o Formula 9.64: $y > 3000$ t

- However, it looks as if the suggested bin volume for this stage of reasoning is $y = 3600$ t which indicates the value of $E(U)$.

Repeat the calculations including the bin in the system.

6. First, an assessment should be made to get information in percentage terms of in how many cases will the mass delivered to the shaft be taken by the bin + hoist system. By calculating the value of distribution $F_g(y) = 0.82$ this means the increase is 26% compared to the hoist operating alone. We can also calculate in how many percentage of cases—when the mass delivered is greater than the hoisting capacity—the mass will be hoisted up because of the application of the bin. Using formula 9.65 we have 59%. The average amount of rock hoisted is $L_B \cong 10130$ t which means that $\chi_1 = 97\%$ of the rock mass delivered by the haulage system will be taken by the bin + hoist system. The average mass of rock that will not be taken by the hoist is now only 290 t/d. When compared with the previous corresponding value, 700 t/d, a positive change is clearly visible.

7. Evaluate the effect of rock mass fluctuations in the bin.

   7.1 The new mean of the utilisation factor can be estimated as 0.84 (from equation 9.73). We correct the corresponding standard deviation, say, to 0.10. For these statistical parameters the new density function parameters are 1.94 and 10.35. Both density functions—the function from the previous part of the analysis and the corrected one—are shown in Figure 9.17. The maximum of the new density function is at $x = 0.88$.

   7.2 Now, we repeat all the calculations starting from the new hoist transportation capacity determined by the new probability density function based on the results presented in point 7.1. The new assessment of the hoist's effective output is 11600 t/h. The corresponding standard deviation is reduced to 1380 t/d (previously 1650 t/d). All three density functions connected with productivities are given in Figure 9.18.

   The conditional mean value of mass that will not be hoisted up is 3470 t/d. Therefore, the new recommended shaft bin volume that can be suggested is 3500 t.

8. Performing calculations connected with the new bin volume again we obtain:

- $100F_g(y) = 83\%$, which means that the transporting capacity of the hoist plus the bin will be greater than the stream of ore delivered to the shaft
- the expected mass of rock hoisted out from the production level $L_B = 10370$ t/d is $\chi_2 = 99.5\%$ of the mass delivered from the working faces. The mean residual mass is only 55 t/d which can be taken by three skips.

The system looks well matched.

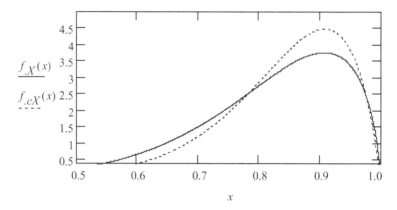

Figure 9.17.   The probability density function of hoist utilisation before and after correction.

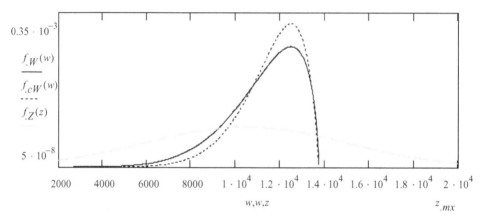

Figure 9.18.   The probability density functions of daily production $f_Z(z)$, the daily transportation capacity of the hoist $f_W(w)$ and the corrected daily transportation capacity of the hoist $f_{cW}(w)$.

The majority of these parameters are gathered together and they are shown in column 4 of Table 9.7. The procedure presented in this section and in Example 9.3 requires some comments.   ◀

- The mine production level is expressed by the mean and the corresponding standard deviation. It is the basic input information for both the design of hoisting installation and the selection of the shaft bin volume. Therefore, it is information of the utmost importance.
- The proposed method of analysis is not a precise one; it only provides approximate values for the majority of the parameters but it gives the possibility to include almost all the stochastic phenomena in the investigation being conducted.
- Many integrals used here cannot be expressed in an explicit form. Thus, the outcomes are produced by a computer using a specific software program. In some cases these outcomes have only a limited degree of accuracy. First of all, pay attention to the mass of probability. Sometimes the total mass of probability calculated by a computer may not equal 1 (usually it is a bit greater than 1). If so, it is necessary to implement a correction function. For this reason, some estimates of parameters are once again only approximate ones.
- The whole influence of rock mass fluctuation in a bin is indicated only by the modified probability density function $f_{cX}(x)$. This influence relies on the displacement of the mass of probability towards the right-hand boundary, which is 1. This mathematical tool is very useful but it is difficult actually to assess how good it is.

Table 9.7. Outcomes of calculations for three systems in relation to the basic system.

Parameter	Units	Formula	(a)		(b)		(c)	
1	2	3	4	5	6	7	8	9
z_{max}	t/d		20000	no change	20000	**18000**	20000	no change
$E(Z)$	t/d	9.41	10420	no change	10420	10150	10420	no change
$\sigma(Z)$	t/d	9.42	4130	no change	4130	3900	4130	no change
$E(X)$		9.48	0.82	no change	0.82	no change	0.82	**0.84**
Q_u	t		22	**26**	22	no change	22	no change
$T_c^{(h)}$	s		92	no change	92	no change	92	no change
$W_{ef}^{(h)}$	t/d	9.47	11300	13350	11300	no change	11300	11570
$100F_z(0)$	%	9.53	56	72	56	58	56	58
$E(U)$	t/d	9.55	3600	3100	3600	3880	3600	3520
ω_0	t/d	9.56	8840	9540	8840	8510	8840	8960
L	t/d	9.57	9730	10170	9730	9460	9730	9810
χ_1	%	9.58	93	98	93	93	93	94
ψ_1	t/d	9.59	700	250	700	690	700	610
T_{CL}	min	9.6	49	15	49	49	49	43
y	t	I SELECTION	3600	3000	3600	3900	3600	3500
$100F_z(y)$	%	9.53	82	88	82	83	82	83
L_B	t/d	9.69	10130	10310	10130	9880	10130	10170
χ_2	%	9.7	97	99	97	97	97	98
ψ_2	t/d	9.71	290	109	290	270	290	255
			CORRECTION		CORRECTION		CORRECTION	
$E_c(X)$		9.72	0.84	0.84	0.84	0.84	0.84	0.86
$W_{cef}^{(h)}$	t/d	9.47	11600	13660	11600	11640	11600	11840
$100F_z(0)$	%	9.53	59	74	59	61	59	61
$E(U)$	t/d	9.55	3470	2950	3470	3630	3470	3400
ω_0	t/d	9.56	9000	9660	9000	8700	9000	9100
L	t/d	9.57	9840	10230	9840	9570	9840	9900
y	t	II SELECTION	3500	3000	3500	3600	3500	3400
$100F_z(y)$	%	9.53	83	89	83	86	83	84
L_B	t/d	9.69	10370	10410	10370	10130	10370	10370
χ_2	%	9.7	99.5	99.9	99.5	99.8	99.5	99.5
ψ_2	t/d	9.71	55	15	55	25	55	50
T_{CL}	min	9.6	5	2	5	2	5	5

- The presented procedure relies on an in-depth analysis and evaluation of the stochastic properties and parameters that characterise the system under investigation: the stream of the mass delivered—shaft bin—hoist. It is necessary to include the economic aspects when looking at this problem.
- The statement that a certain amount of rock delivered to the shaft will not be hoisted because the level of production exceeds the transportation capacity of the hoist does not mean that this amount will not actually be hoisted at all. There are two possibilities for moving this rock out. One possibility relies on an extension of the disposal time (usually not recommended if this extension will be long). The second solution is to keep this mass in the bin and to wind it up at the beginning of the first shift the next day. Usually, at the very beginning of a shift production

is low and this residual mass of rock can be moved out without creating any disturbance to the level's production process.

■ Example 9.4

Having this helpful mathematical tool, a more in-depth analysis can be done. Let us presume that the system just considered is a basic one and we are interested in changes to the values of essential system parameters due to:

a. application of a skip of greater volume $Q_u = 26$ t
b. application of a main conveyor of smaller belt width, which means a reduction in the maximum stream rate delivered to the shaft $z_{max} = 18000$ t/d
c. greater skip utilisation $E(X) = 0.84$.

Because the method of obtaining an evaluation of particular system parameters has already been presented, it will not be repeated here. Table 9.7 contains the results of calculations. Some figures are quite interesting and they require comments.

Case (a)

- Skip volume has been enlarged by 18% and the effective hoist output increases identically.
- The transport capacity of the hoist increases significantly in relation to the stream of rock delivered: in 72% of the cases hoisting capacity is more than the stream of rock, which means an increase of almost 30% (72/56).
- The conditional average mass of rock that cannot be taken by the hoist because its hoisting capacity is below the requirements dictated by the level production decreases by 16% (3600 → 3100 t/d), which is near the percentage of the increase of skip volume.
- However, the real increase in the productivity of the whole system increases by only 4.5%, from 9730 to 10170 t/d.
- A significant decrease is observed in the residual mass—the reduction of the mean value is 64% from 700 to 250 t/d and this result is so striking that a supposition can be formulated that—if only this residual mass is concerned—we probably do not need any rock accumulator.
- This deduction means that the proposed bin volume should be considered by taking into account the reliability aspect and, for this reason, $y = 3000$ t was presumed.
- An effect of the construction of a bin of such a volume in the system causes only a small change in the values of the system parameters: for example, the system's productivity increases only from 10170 to 10310 t/d.

    An evaluation of rock mass fluctuation in the bin.

- The same increase in the hoist's utilisation was estimated; the mean value is now 0.84.
- The effective output of the hoist grows to 13660 t/d and the conditional average mass of rock that cannot be taken by the hoist because its hoisting capacity is below requirements dictated by the level's production remains almost unchanged at 2950 t/d.
- This result means that the bin volume was maintained: 3000 t.
- The system selected should give production of 10410 t/d, which is almost 100% of the mass delivered to the shaft because the mean residual mass was assessed as 15 t/d only.[14]

Case (b)

In this case we presume that the maximum of stream of rock was reduced by 10% due to some problems and for this reason $z_{max} = 18000$ t/d. The rest of the parameters remain the same.

- The average stream of rock delivered is reduced by 5%.

_____

[14] Incidentally these two means of residual mass are identical. Observe the numbers that are the components to calculate these average values are different.

- The skip hauling capacity now exceeds the stream of rock conveyed to the shaft to a greater extent, however this is a rather small increase from 56% to 58%.
- Nonetheless, the conditional mean of the mass of rock $E(U) \cong 3880$ t/d—an almost 8% increment—is caused by a reduction of z_{max}.
- Similarly, a troublesome change is noticed in the average amount of rock hoisted $L \cong 9460$ t/d (a 3% reduction). However, compared to the mean mass delivered to the shaft, approximately $\chi_1 = 93\%$ of it will be hoisted (no change). Converting this high value $E(U)$ into day-to-day practice, the outcome is that about $\psi_1 = 690$ t of rock will not be hoisted daily (compared with 700 t in basic case), on average.
- Looking at the conditional mean $E(U)$ the suggested bin volume is 3900 t (that is, 8% bigger)

The results when including the proposed bin in the system.

- Changes in the values of essential parameters are small but after correction of the mean utilisation $E_c(X)$ which was assessed identically (0.84) the effective hoist output was evaluated as greater than in the basic case—11640 t/d (compared with 11600 t/d).
- However, the mean conditional mass of rock $E(U)$ is now 3630 t/d and the average amount of rock hoisted $L \cong 9750$ t/d
- Because of $E(U)$ a final hint was formulated proposing a bin of 3600 t volume, which is approximately 3% less than the bin volume in the basic case. The hoist + bin system should take 86% of the stream delivered to the shaft giving a final output of 10130 t/d which is only 2% less compared to the system with greater z_{max}.
- The residual mean mass is almost negligible at 25 t/d.

Case (c)

We presume now that all the technical parameters are unchanged but—because of better organisation—the mean value of the utilisation factor was estimated at $E(X) = 0.84$ (only 2.4% up) giving an increase in the hoist's effective output by the same amount.

The main changes in the system parameters are as follows.

- At the first stage of investigations a bin volume of 3500 t was proposed and later this was reduced to 3400 t which means 3% reduction.
- The values of the remaining parameters remain unchanged or they only be such a remark:

Remark:

Variable T_{CL} is a discrete one and for this reason the time to remove the residual mass from the bin is a random variable that can be described by binomial or geometric distribution.

A few further examples that were examined concerning changes in the values of the utilisation gave similar results. This gave grounds to formulate the following suggestion: the shaft bin volume can be reduced if the utilisation factor of the hoist is higher. ◀

The procedure for analysing a stream of rock—shaft bin—hoist system can be a useful analytical tool to support and verify reasoning based on simulation.

CHAPTER 10

Special topic: homogeneity of a shovel-truck system

Homogeneity is normally understood as the quality of being alike or comparable in kind or nature. Homogeneity of a shovel-truck system is understood here as the homogeneity of machines—that is, their functions are the same, they execute identical duties and their basic parameters differ in a random way only.

Let us look at the problem of homogeneity more closely.

When a system of this kind is being considered, one can talk about homogeneity in the:

– volume of boxes of trucks
– volume of loading machines buckets
– time of loading
– time of unloading
– reliability indices of machines executing the same type of job
– weight, in the sense of the importance of functions executed by machines.

From the practical point of view the introduction of new machines, different from those already in use in a mine, generates inhomogeneity.

The application of new and different power shovels can change the system's characteristics. The system can have:

– a different volume of buckets
– different mean times of loading
– significantly different values of the reliability indices of loading machines.

Application of new and different trucks can cause the system to have:

– different payloads
– different mean times of loading

– different mean times of unloading
– significantly different values of the reliability indices of hauling machines.

Obviously, the system can change in both respects—by introducing different power shovels and different trucks. In this case, the inhomogeneity of the system is greater.

Looking at these lists one can distinguish:

– inhomogeneity of physical magnitudes: volumes
– inhomogeneity of parameters associated with the stochastic nature of the exploitation process of the system: means, reliability indices.

Usually, each larger system of this kind has spare loading machines. They are wheel loaders. If any power shovel is down a front-end loader takes over its duties by replacing the failed machine. We can say that at this moment a certain kind of inhomogeneity is introduced; wheel loaders usually have smaller buckets in comparison to power shovels and even when the bucket volume is the same, they have a longer mean loading time—they have to manoeuvre for loading. Moreover, a power shovel bucket empties by opening the flap of the bucket (an action taking a short time); it is usually necessary to tilt the bucket in order to empty a wheel loader bucket (an action taking a longer time).

If a truck dispatcher decides that a given type of extracted rock hauled from the pit—that is waste or ore—should have priority for removal for a certain period of time, inhomogeneity in the operation of the system is also introduced at this moment.

The consequences of the existence of reserve loading machines for the system are discussed in Czaplicki's 2009 monograph. It was assumed that the number of these spare units is high enough that it can be presumed these extra machines are totally reliable. This solution is based on the commonly known reliability principle: *if one can assume that there is almost always a spare unit to replace the failed one, the system of failing units can be treated as totally reliable.* Thus, it can be presumed that loading points are totally reliable; however, the reliability of the power shovels governs the value of the mean time of loading. It is necessary to note that a specific scheme is being examined: there is a system of homogeneous power shovels and there is a system of homogeneous spare loading machines (the wheel loaders). The shovels are of certain reliability, therefore they fail in a random way and if any power shovel is down, a wheel loader replaces the failed machine. Similarly, if any wheel loader is down, there will be a new unit of this type to resume its duties.

In Czaplicki's 2009 monograph one more type of homogeneity is examined as well—the difference in the importance of the type of rock being hauled. Its influence on the system's efficiency is presented in the monograph.

Before investigations were begun in connection with the operation of this type of machinery system, one of the assumptions was that all the loading machines have an identical payload of loading tools and all trucks have boxes of the same volume. However, this assumption sometimes does not hold true in practice.

It is a well-known fact that shovel-truck systems are used in open-pit mining, and indeed that this is a main area of their usage. However, there are some opencast mines that use this type of machinery system, such as the famous El Cerrejón coal mine in Columbia, where the largest machinery system of this kind is in operation. Nonetheless, this is an exception rather than the rule, because opencast mines usually apply a continuous type of machinery. At the very beginning of a mine's operation the principle holds that all the main loading machines should be the same and all haulers should be identical. But sometimes after a period of mine operation, a decision is made that due to developments in truck technology, there is an innovative offer on the market—a new truck with significantly better parameters than those used in the mine and all trucks will be successively replaced by the new better and bigger machines. In some cases, the mine decides to apply bigger machines in order to increase the mine's productivity (after an analysis that indicates that such an enterprise will be profitable).

This switch to better or bigger machines does not happen immediately. It has to be implemented successively. There are two types of factor that determine this course of action—economic and technical ones. It makes no sense to withdraw all of the trucks at the same time and replace them with new units. Usually, haulers in the worst condition are replaced, but some trucks that are still in a good technical state will remain in operation for a certain period of time. However, there is one more essential factor. These new trucks are bigger as a rule. If so, they need more room for driving. This means that the mine transport routes must be enlarged, and roads have to be widened. Usually it takes several months to adapt the hauling roads to meet the new requirements. If some routes are wider but some routes still have the old dimensions, some old trucks can be withdrawn and replaced by new units but some old haulers must remain in operation. Thus, the hauling machinery system is inhomogeneous.

As the number of the new bigger trucks increases, mine management must consider introducing bigger power shovels to boost production. New haulers have larger loading boxes and this means that new loading machines also need to have larger buckets. In such a way inhomogeneity in the loading systems commences. Obviously, this situation may be reversed.

Let us now analyse a system of this type in which one of the characteristic features is the inhomogeneity of machines.

It can be presumed that for a given type of loading machines, there is one probability distribution that describes the dispersion of a random variable—the truck loading time but for a given type of hauler. Therefore, there will be as many probability distributions as there are types of haulers used in the system. Following this line of reasoning, one can easily come to the conclusion that the total number of probability distributions of loading times is the product of the number of H types of loading machines and the number of J types of hauling machines. If we observe the exploitation process of the system for sufficient time, we obtain $J \times H$ groups of loading times. Field investigations have proved (see, for example, Kolonya et al. 2003, Wright 1998, Temeng 1988, Czaplicki 1989) that the probability distribution of loading times can be satisfactorily described by the normal distribution. For this reason, when considering the distribution of the random variable 'loading time' we have a mixture of normal probability distributions. To be more precise, the probability distribution of the random variable 'loading time' is a compound one.

Notice that if two random variables normally distributed are given and if one random variable occurs with the relative frequency w_1 and has a probability distribution function $f_1(x)$ and the second random variable occurs with the relative frequency w_2 and has a probability distribution function $f_2(x)$, then the compound random variable x of these two variables is also normally distributed its probability distribution function is determined by the equation:

$$f(x) = w_1 f_1(x) + w_2 f_2(x) \qquad (10.1)$$

Keeping this in mind a simple conclusion can be drawn: the probability distribution of the random variable 'loading time' can be treated as a Gaussian one.

For a system analysis and for calculation of the system, two basic statistical parameters are needed (Czaplicki 2009): the expected value and the corresponding standard deviation of the random variable being considered.

If $c = 1, 2, \ldots, H$ denotes the current indicator of the type of loading machine in operation and $i = 1, 2, \ldots, J$ denotes the current indicator of the type of hauler, the mean loading time in the system can be estimated from the expression:

$$Z = \sum_{i=1}^{J} \sum_{c=1}^{H} \bar{T}_{ci} w_{ci} \qquad (10.2)$$

where $\bar{T}_{ci}$ is the average time of loading of the i-th type of hauler by the c-th type of power shovel and w_{ci} is the relative frequency of the occurrence of this ci-th 'type' of loading.

As an estimator of the standard deviation of interest, the following formula can be used (*vide* Sobczyk 1996):

$$\hat{\sigma}_z = \sqrt{\frac{\sum_i \sum_c \hat{\sigma}_{ci}^2 n_{ci}}{JH} + \frac{\sum_i \sum_c (z_{ci} - \overline{z}) n_{ci}}{JH}}$$

(10.3)

where:

$\hat{\sigma}_{ci}^2$ is the estimate of variance within the *ci*-th group of loading times
n_{ci} is the number of loading times observed in the *ci*-th group
z_{ci} is the arithmetic mean (time of loading) of the *ci*-th group
$\overline{z}$ is the arithmetic mean of all of the observed times of loading.

The first component of this sum is the weighted arithmetic mean of the variance within groups, whereas the second component is the variance between groups. Looking at equation 10.3 a simple conclusion can be drawn—the more differentiated the population, the greater the value of its variance.

Knowing the principles of the analysis and calculation of the shovel-truck system (Czaplicki 2009), one important inference can be formulated at this stage of investigations:

The value of the mean loading time estimated from formula 10.1 has no important implications for the procedure of the analysis and calculation of an inhomogeneous system with the exception of the values of the parameters of the system; the value of the corresponding standard deviation of loading times is significantly greater than for a homogeneous system and this fact is important.

A similar investigation can be repeated taking into account a truck subsystem. Perceive that the point of interest here is a travel time for haul—dump—return. If we have different trucks, all these component stages of the truck work cycle can have significantly different times. Generally, different mean times of travel can be assigned for different types of haulers. Again, the standard deviations of travel times will be considerably greater for an inhomogeneous system than for a homogeneous one.

Therefore, it seems interesting to perform an analysis and evaluation of system efficiency by examining a large dispersion of random variables as components of the analysis.

- **Example 10.1**
Let us consider the following machinery system:

$$\mathfrak{S}_i : < \mathfrak{H}: n = 7, A_k = 0.84; \mathfrak{W}: A_w = 0.648, \delta = 0.077, \sigma_p = \delta^{-1}, \sigma_n = 0.35 \, \gamma^{-1} >$$

for which the exploitation parameters are:

$$\mathfrak{P}(\mathfrak{S},\mathfrak{D}) : < B_k = 0.82, Z' = 2.2 \text{ min}, \sigma_z = 0.8 \text{ min}, T_t = 18.5 \text{ min}, \sigma_j = 8.7 \text{ min}; \tau = 3 >$$

System $\mathfrak{S}_i$ consists of the subsystem $\mathfrak{H}$ of power shovels of $n = 7$ units. The steady-state availability of shovels $A_k = 0.84$ and the accessibility coefficient of shovels $B_k = 0.82$. System $\mathfrak{S}_i$ also consists of a subsystem $\mathfrak{W}$ of trucks with the steady-state availability $A_w = 0.648$. The intensity of failures of trucks is $\delta = 0.077$ h^{-1} with the standard deviation equalling the mean, which indicates the exponential distribution of work times. The intensity of truck repairs γ can be calculated from the steady-state availability A_w if the intensity δ is known. The standard deviation of repair times is 0.35 of the mean. The adjusted mean loading time is $Z' = 2.2$ min, with the standard deviation $\sigma_z = 0.8$ min. The mean truck travel time is $T_t = 18.5$ min, with the standard deviation $\sigma_j = 8.7$ min.

This system was analysed in Czaplicki 2009, Chapter 13; however the point of interest of those investigations was to show the influence of the existence of spare loading units for the system[1].

---

[1] Mistakenly, in the case analysed in that monograph the steady-state availability of loading points were presumed to equal A_k. It should have been 1 because of the existence of spare units.

Here we extend these considerations by taking into account the structural inhomogeneity of the system. We presume now that system $\mathfrak{S}_I$ is the first in our considerations. We will analyse a $\mathfrak{S}_{II}$ system for which the standard deviation σ_z will be 50% greater than for the $\mathfrak{S}_I$ system and the standard deviation σ_j will be 50% greater than for the $\mathfrak{S}_I$ system. We will also analyse a third system $\mathfrak{S}_{III}$ which has 100% greater standard deviations of both σ_z and σ_j compared to the $\mathfrak{S}_I$ system. All other system parameters remain unchanged.

Let us start the analysis. The beginning of the analysis of the $\mathfrak{S}_I$ system starts from a determination of the structural system parameters, including the number of trucks that should be directed to accomplish the transportation task determined by the loading machines, the number of trucks in the reserve and the number of stands in the repair shop (large enough that the queue of failed trucks waiting for repair can be neglected).

Let us recall the formula for the determination of the number h of trucks needed to accomplish the transportation task given by the loading machines. Looking at formula 9.3a, we have:

$$h = nG_k \frac{T_t + Z'}{Z'} = nG_k(1 + \varpi^{-1})$$

where:

G_k is the power shovel loading capacity coefficient
T_t is the mean truck travel time
Z' is the mean time of truck loading
ϖ is the coefficient of the relative intensity of loading.

Because it was presumed that spare loading units are easily available and that there will always be a wheel loader to replace a failed power shovel, the availability of loading points can be assumed to be 1. Thus $G_k = B_k$.

If the shovel-truck system is inhomogeneous, the formula must be modified to:

$$h = (1 + \varpi^{-1})\sum_{d=1}^{n} B_{dk} \tag{10.4}$$

The only change is that instead of multiplying the number of power shovels by the mean power shovel loading capacity coefficient, characteristic for all power shovels in the mine, we make a summation of all shovel loading capacity coefficients. In such a way the inhomogeneity of the exploitation shovel parameters is taken into account.

Let us recall that the number h is the expected number of trucks in a work state. Because h determines the number of ideal, totally reliable trucks, we have to convert our considerations into real mining practice.

If $\overline{A}_w$ denotes the mean steady-state availability of trucks in the whole system, the expression:

$$V = h / \overline{A}_w \tag{10.5}$$

determines the total number of trucks needed neglecting two items:

• division of trucks into units directed to accomplish the determined hauling task and a reserve
• truck accessibility.

A further principal part of the procedure comprises an analysis and calculation of the conditional system parameters, then the unconditional ones and, finally, the system output. Because the method of reasoning was presented in Chapter 9.1 and 9.2, it will not be repeated here.

It seems more interesting to present some system parameters that illustrate changes in the system performance due to changes in the dispersions of the selected random variables. Table 10.1 shows these changes. It contains cases where the number d of power shovels in a work state varies from 7 to 0. The second column shows the corresponding probabilities $P_{kd}^{(p)}$. Column three has the conditional average number E_{wd} of trucks at loading machines for a given number of power shovels in work. Obviously if d shovels execute their duties, this means that

Table 10.1.   Parameters of three systems with the different dispersion of times of loading and truck travel.

System $\mathfrak{S}_I$

d Shovels able to load	Conditional parameters										Δ_d Min
	$P_{kd}^{(p)}$	E_{wd} Trucks	p_{2d}	p_{3d}	p_{4d}	p_{5d}	p_{6d}	p_{7d}	$p_{>7}$		
7	0.249	5.2	0.002	0.036	0.200	0.378	0.277	0.087	0.021		0.01
6	0.383	7.3		0.002	0.028	0.129	0.249	0.223	0.369		0.40
5	0.252	12.1			0.002	0.015	0.049	0.072	0.863		0.75
4*	0.092	12.2					0.006	0.018	0.975		0.37
3*	0.020	7.7					0.180	0.265	0.505		0.05
2	0.003	9.9					0.029	0.079	0.888		0.01
1	0.000	12.3					0.001	0.007	0.992		0.02
0	0.000	14.4							1		0.02
				Unconditional parameters						Σ	1.63
The mean number of trucks loaded:	E_{wlk} 6.30			p_3 0.010	p_4 0.061	p_5 0.147	p_6 0.182	p_7 0.599			

System $\mathfrak{S}_{II}$

d Shovels able to load	Conditional parameters											Δ_d Min
	$P_{kd}^{(p)}$	E_{wd} Trucks	p_{1d}	p_{2d}	p_{3d}	p_{4d}	p_{5d}	p_{6d}	p_{7d}	$p_{>7}$		
7	0.249	5.4	0.002	0.019	0.086	0.199	0.265	0.218	0.118	0.093		0.05
6	0.383	7.8		0.003	0.016	0.059	0.126	0.174	0.161	0.462		0.50
5	0.252	12.4			0.002	0.009	0.027	0.050	0.064	0.848		0.74
4*	0.092	12.4				0.002	0.009	0.023	0.039	0.927		0.35
3*	0.020	8.2			0.005	0.026	0.079	0.147	0.174	0.568		0.05
2	0.003	10.2				0.006	0.023	0.055	0.088	0.827		0.01
1	0.000	12.4					0.004	0.013	0.027	0.954		0.02
0	0.000	14.4						0.002	0.006	0.992		0.02
					Unconditional parameters						Σ	1.74
The mean number of trucks loaded:	E_{wlk} 6.25			p_2 0.006	p_3 0.028	p_4 0.075	p_5 0.124	p_6 0.139	p_7 0.628			

System $\mathfrak{S}_{III}$

d Shovels able to load	Conditional parameters												Δ_d Min
	$P_{kd}^{(p)}$	E_{wd} Trucks	p_{0d}	p_{1d}	p_{2d}	p_{3d}	p_{4d}	p_{5d}	p_{6d}	p_{7d}	$p_{>7}$		
7	0.249	5.6	0.002	0.013	0.046	0.107	0.171	0.197	0.171	0.116	0.178		0.10
6	0.383	8.4		0.003	0.011	0.034	0.070	0.109	0.130	0.122	0.522		0.56
5	0.252	12.8			0.002	0.007	0.018	0.034	0.049	0.058	0.831		0.73
4*	0.092	12.8				0.003	0.009	0.020	0.034	0.046	0.886		0.33
3*	0.020	8.6			0.005	0.018	0.045	0.084	0.118	0.128	0.600		0.06
2	0.003	10.6				0.006	0.018	0.038	0.062	0.081	0.793		0.01
1	0.000	12.6					0.006	0.014	0.026	0.040	0.912		0.01
0	0.000	14.4							0.009	0.016	0.968		0.02

(Continued)

Table 10.1. Continued.

	Unconditional parameters							Σ	1.82
The mean number E_{wlk} of trucks loaded: 6.17	p_1	p_2	p_3	p_4	p_5	p_6	p_7		
	0.004	0.016	0.042	0.076	0.103	0.110	0.647		

E_{wd} – the conditional mean number of trucks at d shovels able to load
p_{bd} – the conditional probability that there are b trucks at d loading machines
$p_{>7}$ – the conditional probability that there are more than 9 trucks at 9 loading machines
p_b – the probability that there are b trucks loaded; $b \leq 7$
Δ_d – the conditional time loss parameter
E_{wlk} – the mean number of loaded trucks
* – the system structure is changed

$n–d$ wheel loaders support the loading action. In the next columns conditional probabilities are presented, and the last column provides information about losses Δ in a truck work cycle due to the queues of haulers at loading points. In the lower sections of the three sub tables of Table 10.1 unconditional probabilities p_b that give information on the probability that b trucks are loaded are presented. The outcome of the calculation of the average number of trucks being loaded in a given system is shown on the left-hand side of the lower sections.

It is necessary to note that two changes in the organisation of the system were made analogically as in the case that was considered in Czaplicki's 2009 monograph. The first system change was when four power shovels were able to load and the second was when three power shovels were able to load. This fact is indicated by the symbol*. In the first case (three power shovels down) the truck dispatcher decided to withdraw 10 from circulation and send them to the reserve; in the second case, he withdrew the same number again.

Let us look at the figures in Table 10.1 and make a few comments. It is easy to detect that when dispersions grow:

- the general conditional mean number of trucks at loading machines $P_{kd}^{(p)} E_{wd}$ increases
- the time loss in the mean time of truck work cycle increases
- the mean number of trucks loaded decreases, as does the productivity of the system.

Obviously the values are not important; the directions of the changes are crucial. It worth adding that interesting changes can be observed in the distribution of the mass of probability of the number of trucks loaded by the system of loading machines. When dispersions grow, the probability that seven haulers are filled by loading machines increases but the remaining probabilities decrease.

At the end of the investigations of a shovel-truck system's inhomogeneity, it is worth recognising that due to the fact that the random variable 'loading time' can be satisfactorily described by a normal distribution we can treat the whole machinery system as *homogeneous*. The estimate of the parameters of interest—that is, the mean and the corresponding standard deviation—should be directly based on a sample taken from an exploitation process of the system, and one that is ideally—obviously the best solution—a large sample.

References

Adachi, K., Kodama, M. and Ohashi M. 1979. K-out-of-n: G system with simultaneous failure and three repair policies. Microelectronics and Reliability, 19, 4: 351–361.

Adamkiewicz, W., 1982. Introduction to rational utilization of technical objects. WkiŁ, Warsaw. (In Polish).

Agnarsson, G. and Greenlaw, R. 2006. Graph Theory: Modelling, Applications and Algorithms. Prentice Hall.

Aiken, G.E. 1966. Bucket wheel excavators: how to choose the right one for the job. Mining Engineering, 76–81.

Albert, L., Cameron, A.M. and Gullick, J.W. 1975. Hoisting plants of International Nickel. AIME, Hoisting Conference, 29–52.

Aldous, D. and Shepp, L. 1987. The least variable phase distribution is Erlang. Commun. Statistical-Stochastic Models, 3, 3: 467–473.

Anderson, T. and Randell, B. (ed.) 1979. Computing systems reliability. Cambridge Univ. Press.

Antoniak, J. 1990. Transport means and their systems in underground mines. Śląsk. Katowice. (In Polish).

Antoniak, J., Czaplicki, J.M. and Lutyński, A. 1973–75. Reliability analysis of belt haulage systems of high output. Research work. Mining Mechanization Institute, Silesian University of Technology. Gliwice. (unpublished).

Antoniak, J., Brodziński, S. and Czaplicki, J.M. 1976–80. Reliability investigations of hosting installations. Research work. Mining Mechanization Institute, Silesian University of Technology. Gliwice. (unpublished).

Atis Telecom Glossary, 2000.

Aurignac, R., Engel, F. and Ferry, D. 1968. Etude du roulage general d'une exploitation miniere par simulation sur ordinateur. Revue de l'Industrie Minerale, Octobre.

Bailey, J.H. and Mikhail, W.F. 1963. Sequential testing of electronic systems. Proc. Ann. Symposium on Reliability, 391–400.

Barlow, R.E. and Proschan, F. 1975. Statistical theory of reliability and life testing. Probability models. Holt, Rinehart & Winston. New York.

Barret, V. and Lewis T. 1994. Outliers in statistical data. John Wiley.

Battek, J. 1965. Repair and work times of belt conveyors with reversion. Węgiel Brunatny, 2: 123–128. (In Polish).

Battek, J., Dyrka, K., Janczewski, K. and Rychlikowski, E. 1969. Mathematical model of a system KTZ (shovel-conveyor belts-stacker) of opencast mine. Węgiel Brunatny, 2: 137–145. (in Polish).

Baxter, L.A. 1985. Some notes on availability theory. Microelectronics & Reliability. R-25, 5: 921–926.

Beck, W. and Briem, U. 1995. Correlation between the estimated and actual service life of running ropes. Wire, 45, 6: 333–335.

Beichelt, F. and Fischer, K. 1979. On a basic equation of reliability theory. Microelectronics and Reliability, Vol. 19, No 4: 367–370.

Belak, S. 2004. Terotechnology. Oceans '04. MTS. IEEE Techno-Ocean'04. Vol. 1, Issue 9–12, Nov.: 201–204.

Belak, S. and Čičin-Šain, D. 2005. Evolution of the terotechnology concept. Journal of Maritime Studies 1. Vol. 19: 79–87. (In Croatian).

Będkowski, L. and Dąbrowski, T. 2006, Exploitation foundation. Part II. Exploitation reliability foundations. WAT, Warsaw. (In Polish).

Bhaudury, B. and Basu, S.K. 2003. Terotechnology: Reliability engineering and maintenance management. Asian Books.

Bielka, R.P. 1960. Availability—a system function. IRE Trans. Reliability and Quality Control, Vol. RQC-9, September: 38–42.

Bise, C.J. 2003. Mining Engineering analysis. SME.

Bishele, I.W., Gojzman, E.I. and Saratowski E.G. 1964. Имитрущая матиматическая модель работы рудничного транспорта. ИГД Скочинского, Москва.

Bobrowski, D. 1985. Models and mathematical methods in reliability. Examples and exercises. WNT, Warsaw. (In Polish).

Bojarski W.W. 1967. Introduction to reliability evaluation of technical objects. PWN. Warsaw. (In Polish).

Bousfiha, A., Delaporte, B. and Limnios, N. 1996. Evaluation numérique de la fiabilité des systèmes semi-markoviens. Journal Européen des Systémes Automatisés, 30, 4: 557–571. (In French).

Bousfiha, A. and Limnios, N. 1997. Ph-distribution method for reliability evaluation of semi-Markov systems. Proceedings of ESREL-97. Lisbon, June: 2149–2154.

Bozorgebrahimi, E., Hall R.A. and Blackwell G.H. 2003. Sizing equipment for open pit mining—a review of critical parameters. Mining Technology. Trans. Inst. Min. Metall. A, Vol. 112: A171–A179.

Brender, D.M. 1968a. The prediction and measurement of system availability: a Bayesian treatment. IEEE Trans. on Reliability, R-17, 3: 127–137.

Brender, D.M. 1968b. The Bayesian assessment of system availability: advanced application and techniques. IEEE Trans. on Reliability, R-17, 3: 137–146.

Breusch, T.S. 1979. Testing for Autocorrelation in Dynamic Linear Models. Australian Economic Papers, 17: 334–355.

Briem, U. 2000. Correlation between predicted and actual service life of ropes in mining installations. Mine Hoisting 2000. South African Institute of Mining and Metallurgy: 147–151.

Броди, С.М., Погосян, И.А. 1973. Вложенные Стохастические процессы в теории массового обслуживания. Кийев.

BS3811: 1993. Glossary of terms used in Terotechnology.

Bucklen, E.P., Suboleski, S.C., Preklaz, L.J. and Lucas, J.R. 1968. Computer applications in underground mining systems. Vol. 4, BELTSIM Program, Research Development Report No 37, Virginia Polytechnic Institute.

Бусленко, И.П., Калашников, В.В., Коваленко, И.Н. 1973. Лекции по теории сложнух систем. Изд. Советское радио. Москва.

Castillo, E. 1988. Extreme value theory in engineering. Academic Press, Inc. New York.

Çinclar, E. 1969. On semi-Markov processes on arbitrary spaces. Proc. Cambridge Philos. Soc., 66: 381–392.

Cooper, R.B. 1972. Introduction to queuing theory.

Corder, G.W. and Foreman, D.I. 2009. Nonparametric statistics for non-statisticians: a step-by-step approach. Wiley.

Cox, D.R. 1955. A use of complex probabilities in the theory of stochastic processes. Proc. Cambridge Philos. Soc. 51: 313–319.

Cox, D.R. and Smith, W.L. 1961. Queues. Methuen.

Cramér, H. 1999. Mathematical methods of statistics. Princeton University Press.

Crawford, J.T. 1979. Shovel and haulage truck evaluation. In: Open pit mine planning and design. SME.

Czaplicki, J.M. 1974. On a cyclic component in the process of changes of states. Exploitation Problems of Machines, 4: 545–555. (In Polish).

Czaplicki, J.M. 1975. Significant non-uniformity in the process of changes of states of an element or a system during work cycle. Exploitation Problems of Machines, 4: 525–531. (In Polish).

Czaplicki, J.M. and Lutyński, A. 1976. Reliability analysis of hosting installations. Papers of Silesian University of Technology, Mining, 69. Gliwice.

Czaplicki, J.M. 1977. Analysis of utilization of a disposal time In operation of mine winders. Conference: Mine winder modeling. Papers of Silesian University of Technology, Mining, 80: 95–101. (In Polish).

Czaplicki, J.M. 1978. Correlation analysis between selected exploitation properties of mine hoists. Papers of Silesian University of Technology, Mining, 92: 86–96.

Czaplicki, J.M. 1980. Automatic modelling of stream of mineral flowing through continuous transport systems. VI International Mining Autom. Conference, Katowice, Vol. II. May 12–17: 164–171. (In Polish).

Czaplicki, J.M. and Brodziński, S. 1980. Reliability testing of Polish winders. Colliery Guardian, January: 38–39.

Czaplicki, J.M. 1981. Exploitation process of mine winders. Conference on Exploitation of Mine Machinery. Wrocław: 273–281. (In Polish).

Czaplicki J. and Lutyński A. 1982. Vertical transport. Reliability problems. Textbook of Silesian University of Technology. No 1052. Gliwice. (In Polish).

Czaplicki, J.M. 1985. A certain model of reliability diagnostic of renewal objects. Microelectronics & Reliability, 4: 695–698.

Czaplicki, J. and Lutyński, A. 1987. Horizontal transport. Reliability problems. Silesian University of Technology. Textbook No 1330, Gliwice. (In Polish).

Czaplicki, J.M. 1989. Notes from reliability investigation of the Nchanga Open Pit 1986–1988, Copper belt. Mining Engineering Dept. of University of Zambia, Lusaka (unpublished).

Czaplicki, J.M. and Temeng, V.A. 1989. Reliability investigation of a shovel-truck system at the Nchanga open-pit. International Symposium on Off-Highway Haulage in Surface Mines. Edmonton, 15–17 May: 105–110.

Czaplicki, J.M., Lutyński, A., Carbogno, A., Palarski, J. and Zmysłowski, T. 1990. Theoretical basics for the Integrated Transport System for deep mines. Mining Mechanization Institute of Silesian University of Technology. Research work GR-726/RG-2/90.

Czaplicki, J.M. 1994. Critical remarks on analytical modelling of a belt conveyor mineral stream. Papers of Mining Institute of Wrocław University of Technology. Proc. of Conference on Basic Problems of Mine Transportation. Szklarska Poręba, 5–8 October: 95–107. (In Polish).

Czaplicki, J.M. 1999. Random component autocorrelation in process of accumulation of cracks of wires of hoisting ropes. Conference on Mining 2000. Papers of Silesian University of Technology: 71–77.

Czaplicki, J.M. 2000. Investigation of random component autocorrelation in the process of accumulation of wire breaks in hoisting ropes. III Conference on Energy-economical and Reliable Mine Machines. Papers of Silesian University of Technology, 246: 71–77.

Czaplicki, J.M. 2004a. Method of shovel-truck-crusher-conveyor system calculation. 5th International Symposium on Mining Science and Technology, Xuzhou, Jiangsu, P.R., China, 20–22 October: 787–792.

Czaplicki, J.M. 2004b. Elements of theory and practice of cyclic systems in mining and earthmoving. Silesian University of Technology. Textbook. Gliwice. (In Polish).

Czaplicki, J.M. 2004c. Method of „shovel-truck-inclined hoist" system calculation. Górnictwo Odkrywkowe (Surface Mining), 7–8: 71–75.

Czaplicki, J.M. 2005. On the system: "stream of mineral—shaft bin—hoist" in relation to the classical design of hoist installation. Proceedings of the International Conference "Shaft Transport 2005". Zakopane, 21–23 September: 17–31. (In Polish).

Czaplicki, J.M. 2006a. Reliability and properties of exploitation process of suspended diesel loco LPS-90 PIOMA. Mechanization and Automation of Mining, 3/422: 21–26. (In Polish).

Czaplicki, J.M. 2006b. On outliers in reliability investigations of mine equipment. Papers of Silesian Univ. of Tech. Mining, 274: 167–178. (In Polish).

Czaplicki, J.M. 2006c. Determination of function of accumulation of rope cracks of wires of hoisting installations by means of modified autoregression function. Xth International Conference on Quality, Reliability and Safety of Ropes and Rope Transport Means. Cracow 25–26 September: 69–84. (In Polish).

Czaplicki, J.M. 2006. Modelling of exploitation process of shovel-truck system. Papers of Silesian University of Technology, 1740, Gliwice. (In Polish).

Czaplicki, J.M. 2007. Model of hoist head rope worn in fatigue manner and its reliability in case of wear process of I kind. Mining Review, 4: 44–49.

Czaplicki, J.M. 2008a. On a possibility of comparison assessment of the shovel-truck system with spare loading machines and without spares. Papers of Silesian Tech. Univ. 1, Vol. III: 5–16.

Czaplicki, J.M. 2008b. A calculation method for a shovel-truck system with an inclined hoist of TruckLift type. International Journal of Mineral Resources Engineering, 1: 1–13.

Czaplicki, J.M. 2008c. Terotechnology versus Exploitation Theory—some remarks. Scientific Problems of Machines Operation and Maintenance. 2(154), Vol. 43: 45–58.

Czaplicki, J.M. 2008d. The analysis and calculation procedure for shovel-truck systems with a crusher and conveyors. XXI World Mining Congress, Cracow September 7–11. 24, 4/3: 117–124.

Czaplicki, J.M. 2008e. The shovel-truck system with priority for removal of a given type of rock extracted. Górnictwo Odkrywkowe, 1: 62–66.

Czaplicki, J.M. 2009. Shovel-truck systems. Modelling, analysis and calculation. Francis & Taylor.

Czaplicki, J.M. 2009a. Is a speed of accumulation of rope wire cracks a measure of rope weakening? Mining Review. 10. (In Polish).

Czaplicki, J.M. 2009b. On a memory of rope during its utilization—investigations based on mine data. Fifth International Conference 'Shaft Transportation 2009'. Szczyrk. 23–23 September. (In Polish).

Czechowicz, T. 1952. Synchronization of main transport with operation of main shaft. Ph.D. dissertation, AGH. (In Polish).

Czekała, M. 2004. Outlier groups—multidimensional test. In: 'Postępy ekonometrii'. Barczak (Ed.): 15–21. (In Polish).

Darnell, H. 1979. Guest Editorial. Terotechnica, Vol. 1, No. 1.

Das, P. 1971. Point-wise availability of an electronic system with reduced efficiency class of components. Microelectronics and Reliability, 10, 2: 61–66.

Dhillon, B.S. 2008. Mining equipment reliability, maintainability and safety. Springer.

Ditlevse, O. and Sobczyk, K. 1986. Random fatigue crack growth with retardation. Engineering Fracture Mechanics, 24, 6: 861–878.

Dłubała, M. 2009. Exploitation investigation of wear processes of hoist head ropes. M.Sc. thesis. Mining Mechanization Institute of Silesian University of Technology. Gliwice, Poland.

Dovich, R.A. 1990. Reliability statistics. ASQ Quality Press. Milwaukee, Wisconsin.

Downarowicz, O. 1997. Exploitation system. Management of technical resources. Gdańsk-Radom. (In Polish).

Draper, N.R. and Smith, H. 1998. Applied regression analysis. Wiley-Interscience.

Durbin, J. 1953. A note on regression when there is extraneous information about one of the coefficients. Journal of the American Statistical Association, 48, December: 799–808.

Durbin, J. and Watson, G.S. 1950. Testing for serial correlation in least squares regression I. Biometrica, 37, December: 409–428.

Durbin, J. and Watson, G.S. 1950. Testing for serial correlation in least squares regression II. Biometrica, 38, June: 159–178.

Eichler, P. 1968. Die Kostenberechnung der gleisgebundenen Hauptstrecken—Forderung auf digitalen Recheautomaten als Hilfsmittel der Planung und Überwachung. Schlegel und Eisen, 6.

Elsayed, E.A. 1996, Reliability engineering. Addison-Wesley.

Evans, D.W. 1974. Terotechnology—How it can work? I. Mech. E. Conference. University of Durham, UK. Sept. 1974.

Everitt, B.S. 2002. Cambridge dictionary of statistics. CUP.

Fabian, J. 1989. Cyclic mining systems in Czechoslovakian surface mines. Proceedings of the International Symposium on Off-Highway Haulage in Surface Mines, Edmonton 15–17, May: 205–209.

Fashandi, A. and Umberg, T. 2003. Equipment failure definition: a prerequisite for reliability test and validation. Electronics Manufacturing Technology Symposium. 16–18 July: 357–358.

Feller, W. 1957. An introduction to probability theory and its applications. J. Wiley and Sons, Inc.

Feyrer, K. 1994. Drahtseile, Bemessung, Betrieb Sicherheit. Berlin. Heidelberg. New York. Springer Verlag. ISBN 3-540-57861-7. (In German).

Finnistone, H.M. 1973. Terotechnology and management in BSC. Journal of the Iron and Steel Institute, July: 481–485.

Firganek, B. 1973. Optimal control of the production and transport processes in coal mines. IV International Mining Automation Conference.

Firkowicz, S. 1970. Statistical quality control. WNT, Warsaw. (In Polish).

Fisher, R.A. 1929. Test of significance in harmonic analysis. Proc. Roy. Soc. A 125: 54–59.

Franasik, K. and Żur, T. 1983. Mechanization of underground ore mines. Śląsk. Katowice. (In Polish).

Freeman, J. and Modarres, R. 2002. Properties of the power-normal distribution.

Fremlin, D.H. 2000. Measure theory. Torres Fremlin.

Giza, T. 2008. Reliability analysis of suspended loco with diesel engine. International Conference 'Modern, reliable and safe mechanized systems for mining' KOMTECH 2008, 17–19 November: (In Polish): 217–226.

Gładysz, S. 1964. Process of damages for opencast machinery systems. Węgiel Brunatny, 1: 62–74. (In Polish).

Герцбах, И.Б., Кордонский, Х.Б. 1966. Модели отказов. Изд. Советское Радио. Москва.

Гнеденко, Б.В. 1964. О дублировании с восставлением. АН СССР. Техническая кибернетика. 4.

Гнеденко, Б.В., Беляев, Ю.К. and Соловьев, А.Д. 1965. Математические методы в теории надёжности. Изд. Наука, Москва. (Translation: Gnyedenko, B.V., Belayev, Yu. K. and Solovyev, A.D. 1969. Mathematical methods of reliability theory. Academic Press. New York).

Гнеденко, Б.В. and Коваленко, И.Н. 1966. Введение в теорию массового обслуживания. Изд. Наука. Москва.

Гнеденко, Б.В. 1969. Резервирование с восставлением и суммирование случайново числа слагаемых.Colloquium on Reliability Theory. Supplement to preprint volume, pp. 1–9.

Godfrey, L.G. 1978. Testing Against General Autoregressive and Moving Average Error Models when the Regressors Include Lagged Dependent Variables, Econometrica, 46: 1293–1302.

Godfrey, L.G. 1988. Specification test in econometrics. Cambridge.

Gould, R.J. 1988. Graph theory. Benjamin—Cummings Pub. Co.

Goldberger, A.S. 1966. Econometric theory. J. Wiley & Sons.

Golosinski, T.S. and Boehm F.G. 1987. Continuous surface mining. Proceedings of an International Symposium. Edmonton, September 29–October 1. 1986.

Grabski, F. and Jaźwiński, J. 2001. Bayesian methods In reliability and diagnostics. WKŁ, Warsaw. (In Polish).

Granger, C.W. and Hatanaka H. 1969. Analyse spectrale des séries temporelles en économie. Dunod. Paris. (in French).

Granger, C.W.J. and Zhuanxin, D. 1996. Varieties of long memory models. Journal of Econometrics, 73, 1: 61–77.

Granger, C.W. and Rosenblatt, M. 1957. Statistical analysis of stationary time series. J. Wiley.

Gray, H.L. and Lewis, T.O. 1967. A confidence interval for the availability ratio. Technometrics, Vol. 9, August: 465–471.

Gross, D. and Harris, C.M. 1974. Fundamentals of Queuing Theory. John Wiley.

Gumbel, E.J. 1958. Statistics of extremes. Columbia University Press.

Halachmi, B. and Franta, W.R. 1978. A diffusion approximation to the multi-server queue. Management Science, Vol. 24, No. 5: 552–559.

Hamilton, J.D. 1994. Time series analysis. Princeton University Press.

Hannan, E.J. 1960. Time series analysis. Methuen.

Harvey, P.R. 1964. Analysis of production capabilities. APCOM, Colorado School of Mines Quarterly.

Харламов, Б. 2008. Непрерывныйе полумарковские процессы. Петербург.

Hewgill, J.C. and Parkes, D. 1979. Terotechnology—philosophy and concept. Terotechnica, Vol. 1, No 1: 12–19.

Hogg, R.V. and Craig, A.T. 1995. Introduction to mathematical statistics. 5th ed. Macmillan.

Hollander, M. and Wolfe, D.A. 1973. Nonparametric statistical methods. Wiley.

Hosford, J.E. 1960. Measures of dependability. Operation Research, Vol. 8, No 1: 204–206.

http://www.udc.edu/docs/dc_water_resources/technical_reports/report_n_190.pdf.

Hustrulid, w.A. and Crawford, J.T. 1979. Open pit mine planning and design. AIME.

Ibrahim, M.Y. and Brack, C. 2004. New concept and implementation of inter-continental flexible training of terotechnology and life cycle cost. IEEE Intern. Conf. on Industrial Technology. Vol. 1. Dec.: 224–229.

Jaszczuk, M., Markowicz, J. and Szweda, S. 2004. Analysis of the influence of time of utilization of powered support Fazos-22/44-Oz on number of failures. Proc. of Conference on Effectiveness of Operation of Powered Support: Overhauled or New? Ustroń 1–2 April. Papers of Central Mining Institute. 47: 75–81. (In Polish).

Jaśkowski, A. 1999. Two methods of bin size calculation for underground mine. Przegląd Górniczy, 9: pp. 13–24. (In Polish).

Johnson, R. 1984. Elementary statistics. 4th ed. PWS-Kent Publ. Co.

Jóźwiak, J. and Podgórski, J. 1997. Basic statistics. 5th ed. PWE, Warsaw. (In Polish).

Juckett, J.E. 1969. A coal mine belt system design simulation. 3rd Conference on the Application of Simulation. Los Angeles.

Kammerer, B.A. 1988. In-pit crushing and conveying system at Bingham Canyon Mine. International Journal of Surface Mining, Vol. 2, No. 3: 143–147.

Kaźmierczak, J. 2000. Exploitation of technical systems. Silesian University of Technology, Gliwice. (In Polish).

Keller, g., Warrack, B. and Bartel, H. 1987. Statistics for management and economics. A systematic approach. Wadsworth Publ. Co.

Kiliński, A. 1976. Industrial accomplishment processes. Principles of theory. WNT, Warsaw. (In Polish).

Kodama, M., Fukuta, J. and Takamatsu, S. 1971. Mission availability a 1-unit system with allowed down-time. IEEE Transactions on Reliability, 268–270.

Kodama, M. and Sawa, I. 1986. Reliability and maintainability of a multicomponent series-parallel system with simultaneous failure under preemptive repeat repair disciplines. Microelectronics and Reliability, 1: 163–181.

Kolonya, B., Stanic, R. and Hamovic, J. 2003. Simulation of in-pit crushing systems using AutoMod. 19th Mining Congress, 1–5 November, New Delhi: 517–531.

Kolmogorov, A.N. 1933. Grundbegriffe der Wahrscheinlichkeitsrechnung. Berlin: Julius Springer. (In German). (Translation: Kolmogorov, A.N. 1956. Foundations of the theory of probability. New York: Chelsea).

Konieczny, J. 1971. Introduction to theory of exploitation of devices. WNT. Warsaw. (In Polish).

Konieczny, J. 1973. Exploitation study for doctor's degree. Exploitation Problems of Machines, 1 (13), Vol. 8: 119–126. (In Polish).

Konieczny, J., Olearczuk, E. and Żelazowski, W. 1969. Elements of exploitation science. WNT. Warsaw. (In Polish).

Kopocińska, I. 1968. Mathematical models of cyclic queues. Inwestycje i Budownictwo. 18, Nr 12: 22–25. (In Polish).

Kopociński, B. 1973. An outline of renewal and reliability theory. PWN, Warsaw. (In Polish).

Kopociński, B. and Czaplicki, J.M. 2007. Problem of reliability of a rope. Lecture given on Seminar of Institute of Mathematics of Wrocław University. 4 January. (In Polish).

Korbicz, J., Kościelny, J.M., Kowalczuk, Z. and Cholewa, W. (Eds). 2004. Fault diagnosis. Springer.

Королюк, В.С., Турбин, А.Ф. 1976. Полумарковские процессы и их приложения. Наук. Думка, Киев.

Коваленко, И.Н., Кузнецов, Н.Ю. and Шуренков, В.М. 1983. Случайнуе процессы. Справочник. Наукова Думка.

Kowalczyk, J. 1957. Fatigue indicator for hoist head ropes. Papers of Central Mining Institute, 187. (In Polish).

Kowalczyk, J. and Steininger, J. 1963. Steel ropes. Śląsk, Katowice. (In Polish).

Kowalczyk, J. and Hankus, J. 1965. Fatigue durability of hoist head ropes of 20–40 mm in diameter. Papers of Central Mining Institute. (In Polish).

Kowalczyk, J. and Hankus, J. 1966. Safety indices of hoist head rope. Papers of Central Mining Institute, 390. (In Polish).

Koźniewska, I. 1965. Renewal theory. PWE. Warsaw. (In Polish).

Kruskal, W.H. and Wallis, W.A. 1952. Use of ranks in one-criterion variance analysis. Journal of the American Statistical Association, 47 (260): 583–621.

Lévy, P. 1954. Processus semi-markoviens. Pro. Int. Cong. Math. Amsterdam: 416–426.

Lee, A.M. 1966. Applied queuing theory, Macmillan.

Limnios, N. and Oprişan, G. 2001. Semi-Markov processes and reliability. Statistics for Industry and Technology, Birkhäuser.

Lindley, D.V. and Scott, W.F. 1995. New Cambridge statistical tables. 2nd ed. Cambridge Univ. Press.

Lineberry, G.T. and Patsey, J.D. 1987. Use of a microcomputer in the design and selection of materials hoisting systems. Mining Engineering, N.Y., 39, February: 127–129.

Malada, A. 2006. Stochastic reliability modelling for complex systems. Ph.D. dissertation, Faculty of Engineering University of Pretoria.

Маликов, И.М., Половко, А.М., Романов, Н.А., Чукреев, П.А. 1960. Основы теории и расчета надёжности. Судпромгиз.

Maliński, M. 2004. Verification of statistical hypotheses supported by computer. Silesian Technical University, Gliwice. (In Polish).

Manula, C.B. and Sanford, R.L. 1967. Planning belt conveyor networks using computer simulation. Special Research Branch NBC.

Martz, H.F. 1971. On single cycle availability. IEEE Transactions on Reliability. R-20, May: 21–23.

Марьянович, Т.П. 1961. Надійність системы зі змішаним резервом. Дон. АН УРСР, 8.

Markowicz, J. 2003. Influence of mine-and-geological conditions on failures of hydraulic supporting section. 21 International Conference DIAGO 2003. Ostrava: 247–255. (In Polish).

Matusita, K. 1951. On the theory of statistical decision functions. Ann. Inst. Stat. Math., Vol. III: 17–35.

McNichols, R.J. and Messer, G.H. 1971. A cost-based availability allocation algorithm, R-20/3: 178–182.

Melchers, R.E. 1999. Structural reliability analysis and prediction. J. Wiley & Sons.

Migdalski, J. (ed.) 1982. Reliability handbook. Mathematical fundamentals. WEMA, Warsaw. (In Polish).

Mishra, R.C. and Pathak, K. 2002. Maintenance engineering and management. PHI Learning Pvt. Ltd.

Mitchel, G.H. and Lee, T.R. 1965. Simulation applied to underground transport problems in collieries. Operational Research Branch NCB.

Moore, D.S. and McCabe, G.P. 1999. Introduction to the Practice of Statistics. New York: W. H. Freeman.

MSN Dictionary 2007.

Mutmansky, J.M. and Mwasinga, P.P. 1988. An Analysis of SIMAN as a General-Purpose Simulation Language for Mining Systems. Pre-print No. 88–185. SME Annual Meeting. Phoenix, AZ. Jan. 25–28.

Nagelkerke, N. 1991. A note on a general definition of the coefficient of determination. Biometrica, 78, 3: 691–692.

Nagy, G. 1963. The reliability of repairable systems. Proc. Ann. Symposium on Reliability, 93–108.

Nakagawa, T. and Goel, A.L. 1971. A note on availability for a finite interval. IEEE Transactions on Reliability, 271–272.

Niewiadomska-Kozieł, G. 1972. Ordinary moments of the chosen random variables which have truncated distributions. Przegląd Statystyczny, 19, 1: 75–90. (In Polish).

O'Connor, P.D.T. 2005. Practical Reliability Engineering. J. Wiley & Sons.

Olearczuk, E. 1972. An outline of theory of utilization of devices. WNT. Warsaw. (In Polish).

Orwell, G. Animal farm.

Palm, C. 1947. The distribution of repairmen in servicing automatic machines. Industritidningen Norden, 17: 75–80, 90–94, 119–123. (In Swedish).

Papoulis, A. 1965. Probability, Random Variables and Stochastic Processes. McGraw-Hill, Inc.

Pavlović, V. 1988. Design of continuous mining system elements applying bucket wheel capacity as a random process. Mine Planning and Equipment Selection Symposium. Calgary, November 3–4: 313–316.

Pavlović, V. 1989. Continuous mining reliability. Design and operation of mechanized systems. Ellis Horwood, J. Wiley & Sons.

Pawłowski, Z. 1973. Econometric predictions. PWN, Warsaw. (In Polish).

Peele, R. ed. 1945. Mining Engineers' Handbook. 36. Ore bins. John Wiley & Sons, Inc. pp. 12.126–12.131.

Pepler, D. 1989. Foreign engineers—are they superior to their South African counterparts? SA Mining, Coal, Gold and Base Minerals, October 5–9.

Pyke, R. 1961a. Markov renewal processes: definitions and preliminary properties. Ann. Of Math. Statist. 32: 1231–1242.

Pyke, R. 1961b. Markov renewal processes with finitely many states. Ann. Of Math. Statist., 32: 1243–1259.

Pyke, R. and Schaufele, R. 1964. Limit theorems fir Markov renewal processes. Ann. Of Math. Statist., 35: 1746–1764.

Piasecki, S. 1973. The exploitation theory as a science of specific type of activity. Exploitation Problems of Machines, 1 (13), Vol. 8: 73–77. (In Polish).

Rappini, G.E. 1973. Some aspects of Terotechnology at Italsider. Journal of the Iron and Steel Institute, July: 389–405.

Rist, K. 1961. The solution of a transportation problem by use of a Monte Carlo technique. APCOM 1, Tucson, University of Arizona, March: D1–D18.

Rohlf, F.J. and Sokal, R.R. 1981. Statistical tables. 2nd ed. Freeman & Co.

Ross, S.M. 1995. Stochastic processes. J. Wiley & Sons.

Ryabinin, I. 1976. Reliability of engineering systems. Principles and Analysis. Mir Publishers. Moscow.

Sajkiewicz, J. 1973. Reliability problems of machinery systems solved by theory of graphs. Papers of Mining Institute of Wrocław University of Technology. 5.

Sajkiewicz, J. 1974. General principle of reduction of systems of machines with continuous transport. Exploitation problems of machines. 4 (20), Vol. 9: 501–508. (In Polish).

Sajkiewicz, J. 1975. Application of numerical symbols for determination of technically possible work states for machinery system. Exploitation problems of machines. 2 (22), Vol. 10: 219–236. (In Polish).

Sargent, F.R. 1990. Mining and quarry shovels. In: Surface mining. (Ed. Kennedy, B.A.) SME.

Schuster, A. 1898. On the investigation of hidden periodicities. Terr. Mag.

Schuster, A. 1900. The periodogram of magnetic declination. Trans. Cambridge Phil. Soc., 18.

Sense, J.J. 1968. Equipment scheduling including utilization and availability. In Surface mining. 1st ed. AIME: 659–670.

Sevim, H. 1987. A heuristic method in bin sizing. Mining Science and Technology. Vol. 5: pp. 34–44.

Sevim, H. and Qing Wang. 1988a. A microcomputer model for efficient design of hoisting systems. Proceedings of the International Conference on Hoisting of Men, Materials and Minerals. Toronto, Canada, June: 129–133.

Sevim, H. and Qing Wang. 1988b. Design and economic evaluation of hoisting systems. Trans. Inst. Min. Metall., 97, July: A129–133.

Sevim, H. and Qing Wang. 1990a. Is simulation an answer to belt conveyor network design? SME preprint 90–161. SME Annual Meeting. Salt Lake City, UT, Feb. 26-March 1.

Sevim, H. and Qing Wang. 1990b. Reply on Thompson and Webster remarks. SME.

Siegel, S. and Castellan N.J., Jr. 1988. Nonparametric statistics for the behavioural sciences. McGraw-Hill.

Sing, C. 1980. Equivalent rate approach to semi-Markov processes. IEEE Trans. Reliability. R-29, 3: 273–278.

Sivazlian, B.D. and Wang, K.H. 1988. Diffusion approximation to the G/G/R machine repair problem with warm standby spares. Naval Research Logistics Quarterly.

Sivazlian, B.D. and Wang, K.H. 1989. System characteristics and economic analysis of the G/G/R machine repair problem with warm standbys using diffusion approximation. Microelectronics & Reliability, 29, 5: 829–848.

Smith, W.L. 1955. Regenerative stochastic processes. Proc. Roy. Soc. London. Ser. A, 232: 6–31.

Smith, W.L. 1958. Renewal theory and its ramifications. Journal Roy. Soc. Ser. B, 20: 243–302.

Smith, W.L. 1959. On the cumulates of renewal processes. Biometrica, 46: 1–29.

Smith, D.J. 2007. Reliability, Maintainability and risk. Elsevier.

Smith, R. and Mobley R.K. 2007. Rules of thumb for maintenance and reliability engineers. Butterworth-Heinemann.

So., M.K.P. 2002. Bayesian analysis of long memory stochastic volatility models. The Indian Journal of Statistics. Vol. 64, *B, Pt. 1: 1–10.*

Sobczyk, M. 1996. Statistics. PWN. Warsaw: 46. (In Polish).

Sobczyk, K. and Spencer, B.F. 1992. Random fatigue: from data to theory. Academic Press, Inc.: 139–147.

Sturgul, J.R. 1989. Simulating Mining Conveyor Belt Systems. Pre-print, 2nd World Congress on Non-Metallic Minerals. Oct. Beijing. China.

Шор, Я.Б., Кузъмин, Ф.И. 1968. Таблицы для анализа и контроля надежности. Изд. Советское Радио, Москва.

Takács, L. 1954. Some investigations concerning recurrent stochastic processes of a certain type. Magyar Tud. Akad. Mat. Kutato Int. Kzl. 3: 115–128.

Takács, L. 1955. On a sojourn time problem in the theory of stochastic processes. Trans. Amer. Math. Soc. 93: 631–540.

Takács, L. 1962. Introduction to the theory of queues. Oxford University Press.

Takahashi, Y. 1981. Maintenance-orientated management via total participation TPM. A new task for plant managers in Japan. Terotechnica. 2. Vol. 2. May: 79–88.

Tamaki, A. 1981. Guest editorial (on Total Productive Maintenance). Terotechnica, 1. Vol.12, January: 3–4.

Tan, S. and Ramani, R.V. 1988. Continuous Materials Handling Simulator: An Application to Belt Networks in Mining Operation. Pre-print No 88–179. SME Annual Meeting. Phoenix AZ. January 25–28.

Teicholz, P. 1963. A simulation approach to the selection of construction equipment. Technical Report No. 26. Construction Institute. Stanford University. June.

Temeng, V.A. 1988. Probabilistic analysis of reliability and effectiveness of elementary shovel-truck system at Nchanga Open-pit. M.Sc. Dissertation, University of Zambia, Mining Engineering Dept.

Thesaurus.com 2008.

Thompson, M. 1966. Lower confidence limits and test of hypotheses for system availability. IEEE Trans. on Reliability, Vol. R-15, 1: 32–36.

Thompson, S.D and Webster, P.L. 1988. *Design model for underground haulage systems. Discussion.* Transactions, SME, Vol. 284, pp. 1818–1822.

Tijms, H.C. 2003. Algorithmic analysis of queues. Wiley.

Tytko, A. 2003. Exploitation of steel ropes. Śląsk, Katowice. (In Polish).

Tytko, A. and Nowacki, J. 2006. Head rope turning around its axis observed in hosting installations. Conference on Safety Operation of Transport Means. Ustroń, 8–10 November: 116–124. (In Polish).

Uzar, A. 2001. Failure analysis of powered supports applied in 'Knurów' and 'Marcel' coalmines. M.Sc. Thesis. Mining Mechanization Institute of Silesian Technical University. Gliwice, Poland. (In Polish).

Wald, A. 1971. Statistical decision functions. Chelsea Publ. Co.

Wang, X. and Zhao, L. 1997. A study of haulage technology and equipment in large open pit mines. Proceedings of the International Symposium on Mine Planning and Equipment Selection, Ostrava, Czech Republic, 3–6 September: 511–514.

Webster, P.L. 1989. Stochastic models for sizing belts and bins in bulk conveyors network. M.Sc. Thesis. University of Illinois at Urbana-Champaign, Urbana.

Wianecki, A. 1974. Investigation of underground continuous transport systems by a simulation method. Ph.D. Dissertation. Silesian University of Technology. Gliwice. (In Polish).

Wiegel, W. 1974. Functional maintenance: a system approach. Metals Technology. January: 6–12.

Wold, H. (ed.), 1964. Econometric model building. Amsterdam.

Wolf, R.W. 2007. Stochastic modelling and the theory of queues. Prentice-Hall.

Wolfram, S. 2002. A new kind of science. Wolfram Media.

Wolski, J. and Golosinski, T.S. 1986. Performance factors of the bucket wheel excavator systems. AIME Transactions Vol. 280: 1904–1911.

Woodrow, G.B. 1992. Benchmarks of Performance for Truck and Loader Fleets. 3rd Large Open Pit Mining Conference. Macay, 30.08–03.09: 119–125.

Wright, E.A. 1988. Truck dispatching on a personal computer. Proceedings of the International Symposium on Mine planning and Equipment Selection, Calgary, November 3–4: 429–433.

www.asknumbers.com/WhatisReliability.aspx

www.businessdictionary.com 2008.

www.mathwords.com/o/outlier.htm 2008.

www.siemag.de/new/en/schraeg.htm 2004.

www.weibull.com 2007.

Zaikang, L. 1985. Computer simulation and its application in the extraction, conveyance and hoisting system of coal mines. Proceedings of the second conference on the use of computers in the coal industry. SME: 214–222.

Zeliaś, A. 1996. Methods of outliers tracing in economic investigations. Wiadomości Statystyczne, 8. (In Polish).

Zieliński, R. 1972. Statistical tables. PWN. (In Polish).

Ziemba, S. 1973. From the Editorial Board. Exploitation Problems of Machines. 1 (13), Vol. 8: 5–9. (In Polish).

Żur, T. 1979. Belt conveyors in mining. Śląsk, Katowice. (In Polish).

Subject index

A
Autocorrelation v, xi, 13, 36, 39–42, 60, 62, 96, 99, 102, 103, 113, 243

Availability xi, 4, 61–63, 68, 69, 71, 72, 85–87, 90, 91, 120, 131, 145, 146, 148, 157, 162, 164, 173–175, 188, 190, 192–199, 204, 210–214, 221, 226, 231, 246, 266, 267

C
Correlation xi, xii, 23–26, 36, 40, 42, 47–53, 91, 92, 98, 102, 159

D
Diagnostics, statistical v, 10, 12, 13, 53, 106

Distribution,
beta xii, 35, 87, 90, 193–195, 197, 247, 255
binomial 32, 33, 120, 173, 213, 214, 228
Erlang 18, 20, 21, 82, 89, 150, 170, 177, 178, 180, 183
exponential xi, xii, 18–20, 59, 65, 76, 81, 87, 89, 96, 97, 99, 109, 111, 112, 113, 121, 126, 130, 149–151, 162, 166, 170, 172, 175, 177, 178, 180, 183–185, 187–189, 221, 249, 251, 257, 266
gamma xii, 21, 59, 61, 89, 193
geometric 150, 193, 261
normal xii, 17, 21, 22, 24, 30, 41, 43, 49, 65, 73, 75–77, 83, 87, 108, 110, 135, 136, 148, 185, 188, 199, 243, 245, 251, 265, 269
Weibull 59, 61, 152, 153, 160

E
Embedded Markov chain 154, 164

Exploitation graph 12, 57, 58, 63–66, 68, 71, 74, 78, 127–129, 133, 138, 140–142, 144, 145, 151, 152, 155, 162, 163

Exploitation process v, xii, 1, 7–9, 12, 17, 23, 31, 39, 45, 47, 50, 55–60, 62, 63, 65, 69, 73, 74, 81, 83–85, 87–89, 91, 119, 121–123, 125, 126, 127, 128, 133, 138, 140, 152, 158, 159, 169, 188, 211, 213, 220, 221, 225, 226, 228, 231, 264, 265, 269

Exploitation repertoire xii, 12, 57, 65, 69, 72–74, 127, 128, 131, 132, 140, 141, 144, 151, 178, 180

Exploitation theory – goals 7

F
Failure, definition 93

H
Hazard function xii, 108, 109, 110, 115, 116, 159, 163

Heavy traffic situation vi, 199, 203, 204, 206, 222, 223, 242

Homogeneity v, vi, 13, 21, 22, 42, 45, 119, 123, 159, 193, 263, 264

L
Life of technical object/item – phases 8

Linear regression, assumptions 98

M
Machine, definition 6

Markov chain 150, 151, 154, 155, 160, 164

Median 15, 16, 26, 197

P
Pair of elements 9, 142–144, 148

Probability, definition 14

Process,
birth-death 121, 172
Gaussian 35, 102, 135, 243–245
intermittent 73, 74, 89, 158
Markov, definition 126
memoryless 82, 100, 126, 150,
regeneration 73, 74, 84, 158, 159, 188
semi-Markov, definition 150
with memory 42, 60, 96, 99, 100, 103–105, 243, 250, 254

R
Random variable, definition 15

Reliability function 72, 108–110, 114–116, 159, 162, 163